D0788446

THE ILLUSTRATED DIRECTORY OF

AMERICAN CARS

THE ILLUSTRATED DIRECTORY OF

AMERICAN CARS

CHARTWELL
BOOKS

This edition published in 2016 by

CHARTWELL BOOKS
an imprint of Book Sales
a division of Quarto Publishing Group USA Inc.
142 West 36th Street, 4th Floor
New York, New York 10018
USA

ISBN-13: 978-0-7858-3395-6

Printed in China

Contents

Introduction

Without a doubt, the automobile is one of the things that has made America great. As well as being a backbone of the economy, and a driver of general prosperity, the car has also given Americans an unprecedented level of personal freedom, virtually unknown before its invention.

With the development of mass production, America gave the car to the world at large. Since the covered wagons rolled West, Americans have been on the move to all corners of their great land, seeking jobs, freedom and personal happiness. In other words, pursuing the American dream. The car has also been a fantastic motivator of social change. For decades now, Americans have been able to choose to live any number of different lifestyles. The right car exists to facilitate a life in the city or the country, for singles, couples and families, be it a slick sports number or a rugged SUV. But as well as its inherent practicality, the car has also become a completely indigenous part of the American way of life, infusing all kinds of popular culture with its universal presence.

From the early years, when cars were really only available to the upper echelons of society, the car was enshrined in contemporary culture. It became a mainstay of American movies, television shows, photographs, novels, plays and paintings. From John Dillinger's letter to Henry Ford, congratulating him on the efficacy of his getaway car, to Willy Lomax's bitter-sweet relationship with his Studebaker in "Death of a Salesman", and the Dodge/Mustang car chase in Bullit - the history of the car is completely intertwined with the folklore and popular culture of America.

The built environment also reflects America's close relationship with the auto. The first drive-in movies, restaurants and banks were all conceived and built in the US, and many cities are specifically designed around our desire to drive.

Throughout the twentieth and twenty-first centuries, the car has both mirrored the development of society, and has been a force accelerating the rate of change for good or ill. The fantastically luxurious models of the Roaring Twenties foretell the straightened offerings of the Depression years. The interrupted production of the forties tells its own grim story, as do the brilliantly innocent and optimistic offerings of the fifties. Cars of the sixties reflected their decade as a cross over period, somewhat anticipating the nervous times of the oil crisis in the seventies. The eighties saw performance fall victim to fuel economy and emissions controls, whilst the nineties and 2000s, 2010 onward reflected an upswing in the US economy and the positive mood of the nation.

1901 Packard Model C Runabout

James Ward Packard didn't like his Winton Automobile, and told Mr. Winton so. The reply was somewhat testy, but ran along the lines of 'if you don't like my car build one yourself'. So Packard did. He wisely recruited two of Winton's engineers, George Weiss and William A. Hatcher, to help build the first Packard. The car was completed in November 1899, at the Ohio Automobile Company's workshop in Warren, Ohio. The vehicle was designated as the Model A. Five cars were built that year.

The Model A and Model B followed in 1900. Both used a (for the time) conventional single-cylinder engine, producing seven horsepower fitted with two-speed planetary gears. The rear wheels were chain driven. However, two early innovations that singled out these cars were the 'H' configuration gear change slot and an automatic advance on the spark mechanism. Both of these ideas, which we now take for granted, were groundbreaking in 1900.

Packard's reputation grew quickly. William D. Rockefeller jettisoned his collection of Wintons in order to buy two new Packards at the New York Automobile show of November 1900. Henry B, Joy, a Detroit businessman, was so impressed when he bought his second Packard Model F in 1902, that he purchased quantities of stock in the company. His involvement was one of the factors that prompted the removal of the Packard factory to Detroit in October 1903. An early slogan adopted by the company was 'Ask the man who owns one' which clearly exemplified the fact that Packard had got it right where others had failed.

Our featured car, the 1901 Model C Runabout, was exhibited at the New York Automobile Show of 1900. By this time, improvements had increased the available power to 12 horsepower. This model was the first Packard to replace the spade handle lever with a steering wheel - another feature the modern motorist takes for granted. Back in 1901 somebody had to invent this leap forward. A bulb horn was added but the foot operated chime was also retained. A pair of handsome brass Dietz oil lamps replaced the original single solar lamp. Seating at the front of the car was bench style. At the back, there was an optional forward-facing rear seat.

Packard entered three cars in the 1901 New York to Buffalo auto race and received First Class certificates for their performance. These models reached the dizzy heights of 25 mph with standard gearing and up to 30mph with the higher gear option. Not bad for 1901!

Engine: Single-cylinder
Displacement: 183.8-cid
Horsepower: 12 at 850 rpm
Transmission: Two-speed planetary
Base Price: $1500
Number Produced: 81

1902 Oldsmobile Curved Dash

Ransom Eli Olds was already an old hand at the automobile business by 1902. He had tested his first car, a steamer, in 1887. His first gasoline car soon followed. His early production was mainly centered on engine manufacture. This early start earned Oldsmobile the title of the oldest surviving automobile company in America, until its recent demise.

Capitalized in 1899 by wealthy lumber merchant, Samuel L. Smith, the company was consolidated as the Olds Motor Works. Olds himself worked on the development of several different cars, including electric models, but it wasn't until the factory burned down on March 9th, 1901 (fortuitously, some believed) that Olds focused his attention on the surviving car, a gasoline runabout. This was the basis for the future development of the company. It was the Curved Dash Oldsmobile.

The car was powered by a single-cylinder engine, fitted with spur geared, twospeed transmission. It quickly became America's first volume-produced car.

In those days there were no trailers and if you wanted to show your latest car at the New York Automobile Show you would probably have to drive it there. Roy Chapin did exactly this with the Curved Dash in the Fall of 1901. This was a feat, which added greatly to the credibility of the car, and its sales. 425 cars were sold in 1901. Whilst rival manufacturers like Alexander Winton were happy to rely on a lower volume of cars at higher prices, Olds went for a price that people could afford, $650, and watched his sales rocket.

Our featured car clearly shows its attractive wooden construction –the familiar curved toboggan-like front end giving the car its name. A high gloss finish in black, with cherry trim, was achieved by hand applied layers of paint rubbed down and recoated many times until the correct depth of sheen was reached. The car was available as a two seat runabout with the option of a dos a dos (literally back to back, which meant a rear facing seat for additional passengers). The car was steered by a tiller and was right hand drive. The starting crank is a fixed item protruding from the right hand side of the engine and can be used from the driving seat. The white Tubeless tires are 28 x 2½ inch on wire wheels. The car has a Surrey Top. The only thing missing is the horse in front!

Engine: Single horizontal cylinder

Horsepower: 4½ at 600 rpm

Transmission: Two-speed planetary gears chain drive and semi floating rear axle

Fuel System: Floatless carburetor-gravity fed

Weight: 650lbs

Wheelbase: 66 inches

Base Price: $650

Below: Still very much craftsman built with a wooden body the 1902 Olds lingered in the Nineteenth Century.

1903 Model A Ford

Engine: Two-cylinder opposed

Displacement: 100.4-cid

Horsepower: 8

Transmission: Two-speed planetary

Fuel System: Schebler carburetor

Body style: Open top runabout with bolt-on tonneau

Number of seats: 2/4

Weight: 1250lbs

Wheelbase: 72 inches

Base Price: $850, plus $100 for the tonneau

Number Produced: 1700

Below: Finished in bright red with tasteful brass fittings, including a business-like curly bulb horn.

The most famous of all car manufacturers was forty years of age when his first real mass production car was born. Early backers were disenchanted by Ford's seeming fixation on racing projects. The famous 999, now in the Ford Museum in Dearborn, was the first car to circle the Grosse Pointe track in less than a minute in 1901. But its construction resulted in his backers defecting to Henry Leland, the founder of Cadillac. This may have been one of the factors that resulted in Ford finally buckling down to full-scale production in 1903. The Ford Motor Company had been born. Model A's began rolling off the line by July that year, and a total of 1700 cars were built in the subsequent 15 months.

The basic layout of the Model A was that of a horse-drawn buggy. The engine was under the seat and a detachable tonneau bolted to the rear platform (for an extra $100) was available for accommodating two extra passengers.

It had two cylinders horizontally opposed, which was an improvement over its rival, the Curved Dash Olds, as it provided 8 horsepower as opposed to the Olds' 4-1/2 horsepower. This gave a top speed of 30 mph, only achievable in the Olds with optional gearing.

The engine was cranked by a handle inserted in the side of the body below the seat. Within a year the car would evolve into the model AC, complete with an upgraded engine. The model C came next. Unlike some of his competitors Henry Ford was not standing still!

1906 Cadillac Model H coupe

Named for Le Sieur Antoine de la Mothe Cadillac, the French explorer who discovered Detroit in the early 18th century, the Cadillac Automobile Company was founded by Henry Leland in 1902. This was the beginning of a auspicious heritage that has survived to this day.

Company sales manager William E. Metzger took 2286 initial orders at the New York Automobile Show that year, which seemed to indicate that the company had a big future.

The car proved to be a success right from the start because, by offering precision engineering at an affordable price: $750. By 1906, the original single-cylinder engined cars had been joined by a new four-cylinder unit with its 30 horsepower giving a significant increase in power.

Models available in 1906 include the Model K, a single-cylinder powered car, which sported steering column mounted spark control for the first time, and a cam-powered oiler, positioned on the hub of the flywheel. Even in these very early days, car design was under constant review.

Straight-sided Dunlop rubber tires were standard equipment on Cadillacs by 1906. The Model M was very similar to the Model K, but was built on a two inch longer wheelbase.

The Model L was the company's first seven-seater limousine body. This was achieved by the addition of two rear-facing seats. This car had the four-cylinder engine, which was machined out to five inches per bore, and was fitted with a float-feed jet type carburetor with a centrifugal governor attached to the throttle butterfly. The L's dashboard was pressed steel in place of the more usual wood.

Finally we come to our featured model, the H Coupe. This was the first production two-door coupe offered by Cadillac. It was based on the Model L mechanical package, and like the Buick Model G, its styling was still heavily rooted in the era of the horse-drawn carriage.

Engine: Vertical inline four-cylinder with L-head

Displacement: 392.7-cid

Horsepower: 40

Fuel System: Float-feed jet type with governor.

Body Style: Two-door Coupe

Number of Seats: 2

Weight: 2500 lbs

Wheelbase: 102 inches

Base Price: $3000

Number produced: 509

Below: Aerodynamic design wasn't really a consideration when the H Coupe was built. Nevertheless, the result is an extremely pretty little car.

1909 Brush Runabout 2 seater

Engine: Single-cylinder	
Displacement: 62.5-cid	
Horsepower: 10	
Transmission: Two-speed planetary	
Base Price: $485	
Number produced: (1907-1913) 13,250	

The Brush Motor Car Company (1907-1909), later the Brush Runabout Company (1909-1913), was based in Detroit, Michigan. The company was founded by Alanson Partridge Brush, who designed a light car with a wooden chassis (actually, wooden rails and iron cross-members), friction drive transmission and "underslung" coil springs in tension instead of compression on both sides of each axle. Power was provided by a large single-cylinder water-cooled engine. Two gas-powered headlamps provided light, along with a gas-powered light in the rear. The frame, axles, and wheels were made of oak, hickory or maple, and were either left plain or painted to match the trim. The horn was located next to the engine cover, with a metal tube running to a squeeze bulb affixed near the driver. A small storage area was provided in the rear, with a drawer accessible under the rear of the seat.

Brush's engines ran counter-clockwise instead of the usual clockwise, an unusual feature in those days before the invention of the electric starter, his intention to make them safer for a right-handed person to crank-start by hand. With clockwise-running engines, many injuries were sustained, most often dislocated thumbs and broken forearms, if the hand crank kicked back on starting, if the person cranking it did not follow correct safety procedures, including fully retarding the manual spark advance, keeping the thumb alongside the fingers instead of around the crank, and pulling the crank upward in a half turn, never in a full circle or pushing down.

Below: A superb example of only 200 Brush automobiles thought to survive today. This is an early single-cylinder car. Note the hickory front axle.

Runabout advertising stated that the car would reach a speed of 35 mph. One ad claimed "40 mph if you want it". The two-cylinder ads claimed a speed of 50 mph. Few people today would want to ride on a Runabout going 40 or 50 mph even on modern roads. Most Brush Runabouts will attain a top sped of about 25 mph on level ground, faster downhill.

The Brush Runabout Company, along with Maxwell-Briscoe, Stoddard-Dayton, and others formed Benjamin Briscoe's United States Motor Company from 1910, an attempt at a General Motors style group, sadly ending when that company failed in 1913. The last year for sale of the Brush automobile was 1912. Runabouts, in general, fell out of vogue quickly, partly due to the lack of protection from the weather.

1909 Maxwell Model A

The Maxwell Company got started when backer Benjamin Briscoe diverted his cash away from Buick. He believed that the Buick Company would never amount to much. He now invested his money in company with Jonathan D. Maxwell an engineer who had worked at Olds. With further support from Banker J. P. Morgan, they launched the Maxwell-Briscoe Motor Company in 1904.

The company started manufacturing at a factory in Tarrytown, New York and this was where Maxwell's first model, designated the 'H', started production in 1904 when ten units were produced. Their design featured a twin-cylinder water-cooled engine, with a honeycomb radiator mounted up front, and with thermosyphon cooling. This was a system that Henry Ford also adopted on his Model T. The car also had two-speed gears, and in common with many other cars at this time, was equipped with right hand steering.

The company's early years were enhanced by a series of sales stunts organized by Sales Manager Cadwallader Washburn Kelsey. In the year we are featuring, Kelsey persuaded a party of young society women to drive across the country, from Hell's Gate in New York to the Golden Gate in San Francisco, in a Maxwell (of course). At the time, this raised unprecedented interest, as it was considered almost impossible for women to even attempt such a thing.

Largely as a result of this kind of publicity 20,000 Maxwells were sold in the following year. By 1909, a four-cylinder 30 horsepower engine had joined the range, but the two-cylinder engine (with a slightly up-rated power output) still accounted for most of the sales.

Maxwells were available in a number of different body options: a touring car and limousine, a tourabout, runabout and most notable the 'Doctor Maxwell' a car which was specially designed for physicians. This was a new slant on targeting your car to suit the customer. Our example is a rare delivery car, a utility vehicle converted from Maxwell's basic Model A runabout with an out-sourced aftermarket body. It uses the reliable two-cylinder ten horsepower unit, which seemed low on power even at the time. However, a team of Maxwell's had cleaned up in the Light Car race preceding the Grand Prize in 1908. The on-paper performance might not have looked so good, but it worked well in practice.

Maxwell were now third in the car sales league behind Ford and Buick, but in subsequent years the company suffered mixed fortunes, and later became Walter P. Chrysler's platform from which to launch his eponymous company in 1925.

Engine: Water-cooled Opposed two-cylinder

Horsepower: 10

Fuel System: Float-feed carburetor

Wheelbase: 82 inches

Base Price: $500

Number produced: 9,460

Right: Red paintwork and artillery-spoke wheels really show off the jaunty lines of this Delivery Car, based on the 1909 Model A Runabout.

1909 Ford Model T

Engine: L-head four-cylinder	
Displacement: 176.7-cid	
Horsepower: 22 at 1600 rpm	
Transmission: Planetary two-speed	
Fuel System: Kingston five-ball carburetor	
Body style: Open Top Tourer	
Number of seats: 5	
Weight: 1200lbs	
Wheelbase: 100 inches	
Base Price: $825	
Number Produced: 10,660	

At last, here was the car that Henry Ford meant to build all along! Its dual hallmarks were simplicity and reliability. Launched in the fall of 1908, the first Tin Lizzie two-person runabout (as the car became affectionately known) went on sale for just $825. This was an entirely new car compared to the previous Fords. It had a four-cylinder side valve power unit, with a cast iron block and removable cylinder head. Many of the ancillary engine and transmission parts were stamped, or drawn, rather than cast. This kept down both the weight and cost. Essentially, Ford had approached the building of his car as an engineer, not as a financier or marketing person. The revolutionary techniques displayed in the model were the result of two decades of engineering trial and error. The car's suspension was by way of two transverse leaf springs. Although this was basic, it gave long service and little trouble, contributing to the car's longevity and popularity. Childe Harold Wills, one of Fords leading metallurgists, had developed Vanadium steel. This was a lighter and more durable steel, which was used for the Model T chassis and engine components. This gave the car an enormous weight advantage over its competitors. Ford strongly believed in innovation, and many of his chosen comrades (like Edison) helped to keep him on the cutting edge of new technology.

The Model T wheels were 30 x 3 inch on the front and 30 x 3½ on the rear (giving extra grip to the driving wheels). This was an innovation, which is still in use in high-performance cars almost a century later.

The hood was plain and fabricated from sheet aluminum. A wide range of factory-fitted options was available including headlamps, tops, windshields, speedometers, Prest-o-lite gas tanks, footrests and autochimes. This made the car eminently flexible to the requirements of the individual customer. This was a lesson that other manufacturers had yet to learn.

Opposite page: Many Model T's found their way into rural locations as this picture depicts. The car's rugged simplicity, coupled with good mechanics, kept it running down on the farm for many years.

Right: The L-head four cylinder power unit was rugged and reliable contributing these qualities to the Ford Model T itself.

Ford
1909
MODEL T

1912 Buick Model 36 Roadster

Engine: Four-cylinder Inline	
Displacement: 201-cid	
Horsepower: 25.5	
Transmission: Three-speed sliding gear	
Fuel System: Marvel E carburetor	
Body Style: Two-door Open Top Roadster	
Wheelbase: 101.75 inches	
Base Price: $900	
Number Produced: 1,600	

Buick's success in intermediate-level automobile racing was an undoubted factor in their growing popularity. Pressure on cash had led to the company's bankers to force Billy Durant to step down as President of General Motors, which he had built up on the foundation of Buick. Durant made sure that the job of Buick President went to Charles W. Nash who occupied the Chair until Walter P. Chrysler took over in November 1912. The early days of the automobile industry saw a kind of musical chairs for the top positions of famous name companies.

Durant now linked up with Louis Chevrolet, and took off to pursue his other interests. But he left a legacy of belief in racing behind him. This acted as an important shop window for the company products, and contributed to Buick's strong reputation in the marketplace. By 1912, Buick was placed fourth in the top eight manufacturers. Needless to say, Ford was number one! A choice of four-cylinder engines was on offer for the year: a 165-cid of 22.5 horsepower, a 201-cid unit with 25.5 horsepower and the big daddy of 318-cid that developed 48 horsepower. All these power units featured spark plugs, canted into the head at forty-five degrees instead of the old horizontal position.

Other improvements included improved oil ways in the engine block that increased push rod lubrication, and beefed up lubrication points for suspension, steering and clutch controls, which allowed for a smoother ride and longer life. It could also be argued that these small improvements were

Right: Buick's advanced four-cylinder engine was the result of development due to intermediate racing experience. Even so they were in fourth place in the manufacturing stakes.

Below: An original chassis plate from 1912 adds to the car's originality.

the result of lessons learnt in competition. Our featured car is one of three roadsters on offer. It is finished in its original paint colors of blue and gray with blue/black hood, fenders and fuel tank. The wooden artillery wheels are picked out in white. By this time, the doors on all Buick Models were fully functional opening doors, unlike the false doors offered on the Ford Model T, or the cutaway coach style bodies of the early 1900s. The doors gave the car a more finished and modern look.

This car has the 201-cid engine, fitted with a three-speed gearbox, a powerful option for the times.

Above: 1600 of these handsome Buick Roadsters were sold in 1912 and the car looks at home outside this period building.

1912 Chevrolet Classic Six

Engine: T-head twin cam six-cylinder	
Displacement: 299-cid	
Horsepower: 40	
Fuel System: Stromberg carburetor	
Body Style: Four-door Roadster	
Number of seats: 5	
Wheelbase: 120 inches	
Base price: $2250	
Number produced: 2,999	

When Billy Durant was forced to step down as head of General Motors in 1910, he teamed up with Louis Chevrolet to form a powerful new force in the automotive world. It was his intention to regain control of GM at some future point. Chevrolet was already well known for his motor racing exploits. Durant considered that Chevrolet's European background and motor racing prowess made him the ideal person to produce a groundbreaking new model with European-styling.

By 1910, intense competition between the leading nations in Europe had produced a forward surge in design that was led by France and Germany. Durant hoped to tap into these technological advances to give his cars an edge in the American domestic market. He assembled a collection of factories to help him found the new company, including the Little Car factory in Flint, Michigan and the Republic Motor Company in New York City. A car had already appeared under the Little banner in 1912, but where was the Chevrolet? When the car finally appeared later in the year, it was a monster. The 120-inch wheelbase made the car one of the largest Chevrolets ever produced. This massive 299 cubic inch power plant was not overtaken by another Chevrolet engine until 1958! It was configured as a six-cylinder T-head twin overhead camshaft. Despite the engine's size and advanced technical design, its power output was a modest 40 horsepower, which rendered the car a sluggish performer.

The car's size and specification meant that it had to be marketed at well over

Below: None other than Louis Chevrolet at the wheel of the car that bears his name.

$2000, which limited sales. This over-specified model was the unfortunate result of giving an engineer free rein to produce the kind of car of his dreams rather than designing the car with the customer in mind.

Despite this, the car featured many excellent modifications such as dual ignition, self-starter, a gas gauge, speedometer with illuminated dial, electric lights, forged front axle, cone clutch with three-speed gears, semi-elliptic front springs and three-quarter rear platform springs. The engine was mounted at three points. All of this added up to a very well engineered car, which was just too expensive.

Above: The distinctive Silver Radiator cowl and nickel-plated electric headlights give the first Chevrolet a very modern appearance.

1912 Hudson Model 33

Engine: Four-cylinder inline

Displacement: 226-cid

Horsepower: 33

Transmission: Three-speed sliding gear

Fuel System: Stromberg carburetor

Body Style: Two-door Roadster.

Weight: 2631 lbs

Wheelbase: 114.5 inches

Base Price: $1600

Number Produced: 5708 [based on actual shipments to dealers]

Opposite page: This two-door Roadster looks very much at home outside a grand mansion. The distinctive triangular Hudson badge can be clearly seen. It was an inexpensive car with a lot of class. Note: Non-standard wire wheels and electric headlamps which are later additions.

Below: The clean lines of the Hudson four-cylinder power unit and the dash layout are testament to a quality product.

The men behind the Hudson Company were seasoned graduates of the 'Olds' school. Roy D Chapin made his name when he drove from Detroit to New York City for the National Automobile Show in 1901 in record time in an Olds Curved Dash and was joined at Hudson by fellow Olds veterans Howard E. Coffin, George W. Dunham and Roscoe B. Jackson. The latter was most important to the success of the enterprise, as his wife was the niece of the wealthy Joseph L. Hudson, the owner of Detroit's most successful department store. Hudson gave his name to the car company as well as a $90,000 investment.

The resultant car –the nippy little Model 20 Roadster-arrived in 1910 with a good turn of speed (50 mph) and an affordable price tag of just $900.With the Olds team's combined knowledge of the market the car had been developed to hit it dead right first time. In fact over 4000 cars were sold in the first trading year making Hudson the most successful auto company start up yet encountered. Joseph L must have been pleased indeed.

In 1912 the Model 20 was dropped, and a new upgraded car was introduced – the Model 33. A choice of seven different body styles was offered, including the sensational Mile-A-Minute Roadster, which guaranteed a heady 60 mph. A stripped down version of the normal roadster, this was the first factory Hot Rod with 32-inch special wheels, a 100mph speedo, and was supplied minus doors and windshield.

At the same time the company raised the issue of safety standards in the minds of the customers by mounting a campaign warning of the dangers of buying other makes which the folks at Hudson took it upon themselves to regard as 'unsafe motor car purchases'. You were quite safe if you stuck with them! Quite how much this was swallowed in 1912 is a matter for conjecture.

Prices of the Model 33 averaged out at $1600 except for the coupe and limousine versions, where specially made bodies increased the price.

HUDSON
LF 5591

1912 Oldsmobile Autocrat

- **Engine:** Four-cylinder inline
- **Displacement:** 471-cid
- **Horsepower:** 40
- **Fuel System:** Rayfield Model D carburetor
- **Wheelbase:** 126 inches
- **Base Price:** $3500
- **Number Produced:** 500 across the entire model range

Paradoxically a company that set out under its original leadership to become a volume-producing budget car manufacturer, was the maker of some of the most gargantuan monsters in the history of American cars by 1910. Under the influence of the Smith family, who ousted R.E. Olds from the company, production turned to large models with extremely low production runs.

By 1912 and now under the ownership of General Motors the trend had reached its most extreme conclusion in terms of large, opulent cars. Names like The Special, The Autocrat, The Defender, and The Limited gave an instant clue to the intention of this series of cars.

These were cars that appealed to buyers who valued exclusivity and elitism.

The cars were outrageous in every way, an affront to the man who bought the unostentatious Olds Curved Dash. The Autocrat, not even the worst of the bunch, sported a total length of just below 15 feet, making the Chevrolet Classic Six look like a compact. Its four-cylinder 471 cubic inch engine produced just

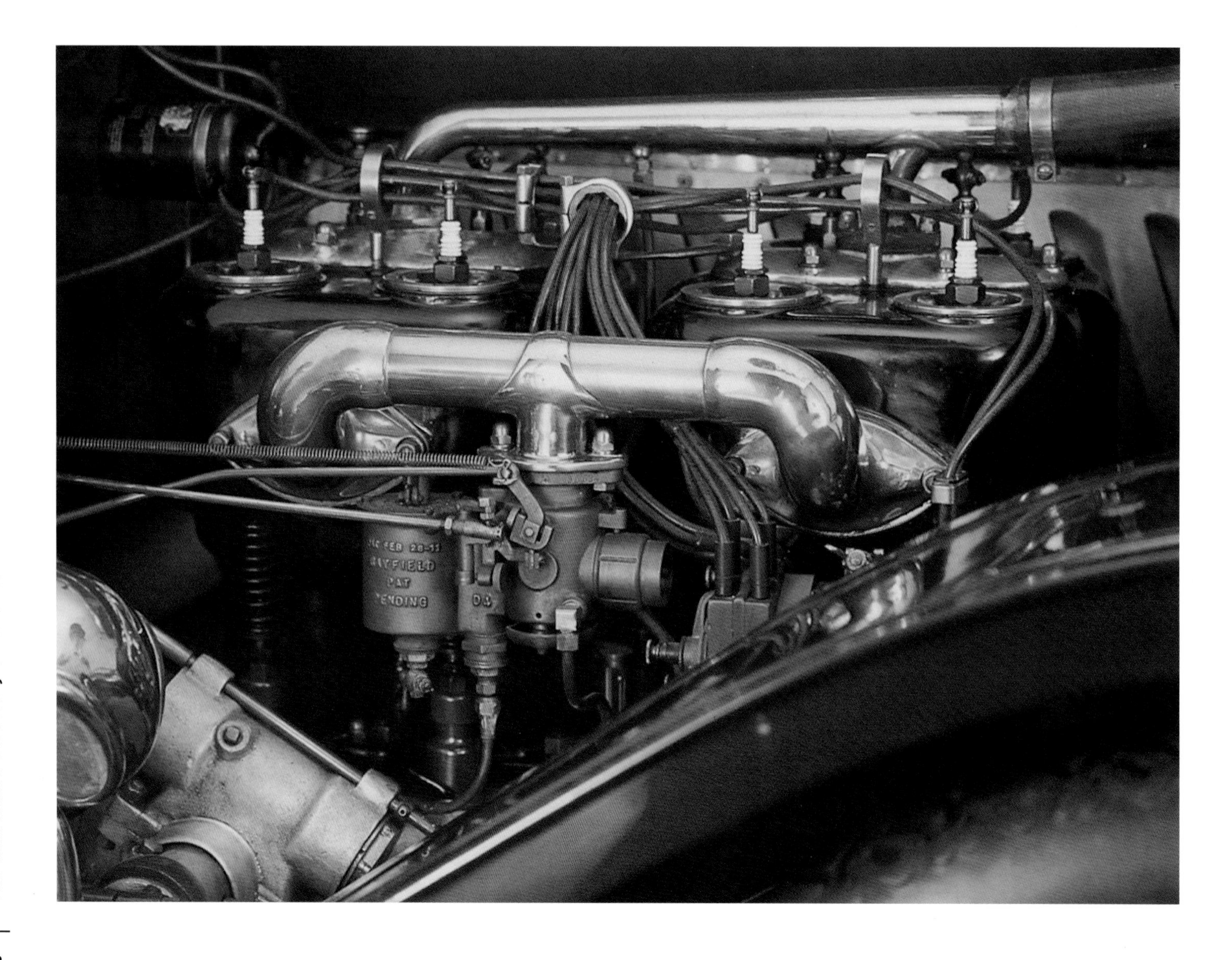

40 horsepower and one can only guess at its gas mileage. Its sister car, the Limited, equipped with the six-cylinder variant of the Autocrat's engine [same bore and stroke –just two more cylinders] inspired the famous action painting by William Hamden Foster - 'Setting the Pace'. The pace was 70 miles per hour even in 1911. The Limited's engine was 706 cubic inches! In years to come this trend was reversed, but this period left us with some of the most remarkable cars in our collection.

Above: The 1912 Autocrat sits confidently on its 38-inch Firestone tires.

Opposite page: The Autocrat was offered with two power units, a four-cylinder and a six. Seen here the four-cylinder 471 cubic inch engine produced just 40 horsepower. The six was identical in design but had two extra cylinders.

1914 Mercer Raceabout

'The Steinway of the Automobile World' was how the Mercer was described in 1914, and this title gives us a clue to the rather exclusive world that the car belonged to. Clearly, the model's very name tells us what the car was about. No comfort for starters. Everything was sacrificed on the altar of performance and you could certainly 'raceabout' in it. At least, you could if you had the necessary $2500. This was steep for a fairly basic car.

The Mercer Automobile Company, named for Mercer County, NJ, was established by the wealthy and entrepreneurial Roebling Family. This was the family, which gave us the Brooklyn Bridge so they were used to thinking big. Their first cars were manufactured in Trenton, New Jersey in 1909.

The Mercer started out with a four-cylinder L-Head Beaver engine, which was upgraded in 1911 to a T-head design, then all the rage, as President C.G. Roebling's son, Washington, pointed out. Washington Roebling took a keen interest in the company and provided much of its marketing thrust, knowing the kind of car that people in his 'set' would respond to. Stalwart engineer Finlay Robertson Porter sourced the better engine, and became a leading light in the company for the next three years. As well as the celebrated Raceabout, the company also offered a five-seater touring car, a Roadster and a Sport Touring model. 1914 production was 600 cars, of which around 90 were Raceabouts. The capable T-head four-cylinder engine was nominally rated at 34 horsepower. When this was coupled to a stripped-down lightweight body (with very little in the way of creature comforts) and power delivered smoothly and precisely by way of a four-speed selective transmission and oil immersed clutch, it made for an exhilarating driving experience.

But shadows fell across the future of the company. Company guru Washington A. Roebling went down with the Titanic, and from that moment in time, it seemed as though the company was drifting. Engineer Findlay Porter departed in 1914 and his replacement Eric H. Delling seemed obsessed with adding uncalled-for comforts to the car such as full-length windscreens, cumbersome faired in bodywork and a bench arrangement instead of the old bucket seats. This was all moving away from the essential spirit of the car.

Engine: Four-cylinder T-head.

Displacement: 300-cid

Horsepower: 34

Wheelbase: 108 inches

Base Price: $2600

Number Produced: 90

1914 Stutz Bearcat

Engine: Four-cylinder T-head	
Displacement: 389-cid	
Horsepower: 50	
Transmission: Four-speed rear axle mounted	
Fuel System: Schebler carburetor	
Body Style: Two-seater Sports	
Number of Seats: 2	
Wheelbase: 120 inches	
Base Price: $2000	
Number Produced: 647	

Ideal had an outstanding success at the inaugural Indy 500 race in 1911. Its first car, the 'A' series roadster, was driven to an 11th place finish by racer Gil Anderson. The Ideal Motor Car Company was now set up for full-scale production to capitalize on the reputation gained. The 'car that made good in a day', stemmed from the collective experience of Harry C. Stutz. Stutz was an automotive engineer, who started out in much the same way as Henry Ford, bolting agricultural scrap parts together in a home workshop in Dayton Ohio. He produced an engine by 1902 that somebody wanted to manufacture under license and he moved to Indianapolis that year to supervise its production. This didn't last long - like many of Stutz's jobs in this period. He moved around the industry working in tire, carburetor and engine plants gaining experience and giving him a clear industry overview, which came in very handy when he started to make his own cars.

The 'A' model was followed-up with a more comprehensive model range in 1912. This included the car that would make his name famous for all time. The Bearcat was a true hairy-chested, macho sportscar. Rumor had it that Stutz specially designed the clutch with extra strong springs to prevent females getting behind the wheel. There was a choice between four- and six-cylinder power units, both of these were outsourced T-head Wisconsin engines. Stutz didn't make its own engines until 1917. Power outputs were competitive with 50 and 70 horsepower respectively developed by the two engines.

Solid racing success was quick to follow. Of thirty events entered in 1912 the Bearcat triumphed in twenty-five. The 'White Squadron' as the Stutz racing team came to be known was the deadly rival of the Mercer Runabout boys. A whacky races scenario!

The steering wheel was on the right hand side and remained there until 1922. Artillery wheels with detachable rims were standard in 1914, but wires were available in the following year.

1915 Cadillac Type 51

Engine: 90-degree L-head V-8	
Displacement: 314.5-cid	
Horsepower: 70	
Transmission: Three-speed sliding gear selective	
Fuel System: Cadillac carburetor under CF Johnson patents.	
Body style: Two-door Sedan	
Number of seats: 5	
Wheelbase: 122 inches	
Base Price: $2800	

This model was the lucky recipient of the new V-8. The use of four-cylinder power units was now banished for the next 67 years establishing Cadillac firmly as a luxury car manufacturer that didn't dally with economy engines. Although not the first to come up with a V-8 option, this was certainly the first engine to go into serious production in the US. It displaced 314 cubic inches and produced a healthy seventy brake horse power. The engine remained in service until 1927.

The Type 51 also saw the change to left drive (although right drive was still available as an option). This was a gradual process, which slowly filtered through the car industry. Henry Ford's early decision to go left on the Model T must have had a big influence on those who followed on, but some die-hards like Stutz held out 'on the right' until 1922.

Type 51 also showed modern influences in its body styling. The hood was more rounded off with top panels blending smoothly into side panels, unlike the sharp-edged, box-like shapes of yesteryear. Closed bodies like the Landaulet coupe were sleeker with shallow profile glass in side windows and windshields and lower rooflines. The overall effect was that the closed car compartment now appeared as an integral part of the design - rather than a square cabin bolted onto a chassis. Type 51 offered no less than nine body styles. Two touring

cars with either 5 or 7 seats, a two-door salon with a passageway between the front and rear seats, a two-door roadster, the pretty two-door Landaulet coupe, the remarkable center-door 'sedan for five passengers', two limousines and an Imperial. The latter model had a glass partition between the seats for really grand people.

All closed cars had three-piece 'rain vision' windshields, a kind of early wraparound design. Open cars with the 'one man top' had side curtains, which opened with the doors.

A lot of thought had gone into the Type 51!

Above: The four-passenger Phaeton version of the type 51.

Opposite page: All of these models used a new L-head V8 engine, one of the first V8 engines ever mass-produced and a substantial differentiator for the marque.It remained in service until 1927.

1915 Chevrolet H series Roadster

Engine: OHV four-cylinder inline
Displacement: 171-cid
Horsepower: 24
Transmission: Three-speed selective sliding gears
Fuel System: Zenith two-jet carburetor
Body style: Two-door Roadster
Weight: 2100 lbs
Wheelbase: 106 inches
Base Price: $985
Number Produced: 13600 total for series

William C. Durant had learned some lessons from the launch of his first Chevrolet, the Classic Six. Designed by Louis Chevrolet, the car was overpriced and over engineered.

He made more money from the launch of the Little, a budget car assembled at one of the many Chevrolet plants. Durant now planned to consolidate his efforts back in Flint Michigan –his old stamping ground. He had Arthur Mason, a former Buick engineer who now ran the Mason Motor Company (an engine facility already owned by Chevrolet), design a down to earth four-cylinder unit. The side benefit to the arrangement was that it resulted in Louis Chevrolet storming off to make Frontenacs. Under Chevrolet's leadership this company ultimately went bust. Durant was probably relieved at his departure.

The new overhead valve unit that Mason designed produced a respectable 24 horsepower out of 171 cubic inches. But most importantly, the engine was economical to produce. A much more practical proposition than Louis Chevrolet's complicated 299 cubic inch twin cam six-cylinder monster!

Coupled with the H series compact body style with a wheelbase of 106 inches (against 120 inches for the Classic Six), Durant had got back into the real world of volume car production. He was able to price the new range at under $1000. Small improvements also found their way into the H series design. Concealed door hinges, larger tires, no windshield braces, electric starter and headlights all conspired to give the car a modern appearance. Three model options were offered: a two door roadster with a flat deck, named The Royal Mail, a four-door touring car called the Baby Grand (this car was notorious for a stunt where it climbed the State Capitol steps at Des Moines, Iowa), and a Roadster known as the Amesbury Special.

Right: The clean and uncomplicated Chevrolet H series dash. Note the knurled nuts holding the windshield in place.

Above: The very attractive Amesbury Special Roadster version of Chevrolet's 1915 'H' series. The car has the optional rear dickey seat with separate windshield and Houk Quick-detachable wire wheels.

1915 Dodge Brothers Model 30-35

Engine: Four-cylinder L-Head	
Displacement: 212.3-cid	
Horsepower: 35 at 2000 rpm	
Transmission: Three-speed	
Body Styles: Four-door Tourer, Two-door Roadster	
Weight: 2200 lbs	
Wheelbase: 110 inches	
Base Price: $785	
Number Produced: 45,000	

John and Horace Dodge established a small machine shop in Detroit in the early years of the century. They had already made a good living and a reputation for themselves by supplying the likes of R. E. Olds and Henry Ford. Their products were respected for quality and reliability. The brothers were smart enough to see the writing on the wall when large players like Ford sought to control costs by manufacturing their own components. The days of outsourced engines, and gearboxes in particular, were drawing to a close. Interestingly, the Dodge brothers were also Ford shareholders at a time when Henry Ford was planning to implement his dream of buying out non-family shareholders so that total control of FoMoCo would revert to the Fords.

When The Dodge Brothers announced that they would be introducing their own line of cars in 1914, the automobile industry was delighted. Their reputation gave them an enormous advantage.

The first Dodge offering was developed from their direct experience of competitive products. They knew exactly how their models could improve upon what was already available. Of course, they were particularly knowledgeable about the Model T, having supplied many of its components. Their Model 30-35 was sturdy in both body and design. It had an all-steel body made by Budd, and an early 12-volt electrical system, which included headlights and a self-starter as standard. An 'unburstable' L-Head engine of 212.3 cubic inches gave a reliable 35 horsepower. The three-speed gearbox was enclosed, and employed vanadium hardened gears. Oddly, its configuration was back to front. Leather seats were also standard on the car, as were a folding top,

Right: The origin of the Dodge Brothers badge has been the subject of much speculation. The simplest explanation is the design represents two interlocking Greek letter "deltas" or "D s" for the two Dodge brothers.

windshield, speedometer and demountable wheel rims. The Model T wasn't fitted with these options until well into the '20s.

Two model options were offered, a four-door tourer and a two-door roadster. Priced at $785, the models were an outstanding package, and the range immediately jumped into the number three sales slot for the year.

Above: The neat looking Dodge Brothers Model 30-35 sold 45,000 units in its first year.

1918 Cadillac Type 57

Engine: Ninety-degree L-head V-8	
Displacement: 314.5-cid	
Horsepower: 31.25	
Body Style: Four-door Open Tourer	
Number of Seats: 5	
Weight: 3925 lbs	
Wheelbase: 125 inches	
Base Price: $2590	
Number produced: 45146 (including war production)	

World War 1 was a busy time for Cadillac. After rigorous testing, they were awarded the

status of 'standard model' for war service. This special status involved the supply of staff cars, ambulances, hearses, and police patrol vehicles for the services. Much of the reason for this was the recognized strength, power and reliability of the V-8 engine, which the company had adopted as its standard powerplant three years earlier. Busy too, because the founding Leland family finally left the company. This was also portrayed as a war-related issue. The Lelands left the company as a protest, over the fact that William Durant, (head of parent company General Motors) had refused to build much needed Liberty Aero engines for the war effort. Whether the family feelings on this matter were governed by patriotism, or the desire to profit from war work is a moot point. But the ultimate result of their dissatisfaction was that the family left to found Lincoln, which immediately tooled up to produce Liberty engines. Once the Lelands had gone, Billy Durant relented and he also agreed to make the engines...

Type 57 was a larger, more comfortable car than earlier Cadillac models. Styling improvements meant that the hood flowed more smoothly into the windshield cowl, in a continuous straight line. Nine louvers adorned each side of the hood, raked back at a subtle six degrees. The windshield on open cars was slanted at a matching angle. This was a gradual move away from the upright boxiness of earlier cars. Other small improvements were also incorporated into the car, such as tilting headlight reflectors controlled by a lever.

The V-8 was also subject to modernization. It now had detachable heads and ultra-lightweight pistons together with a brand new transmission arrangement.

Right: The Cadillac L-head V8 engine was central to the car's reputation and success. Its reliability was proven in WW1 when it was used widely for ambulances and staff cars.

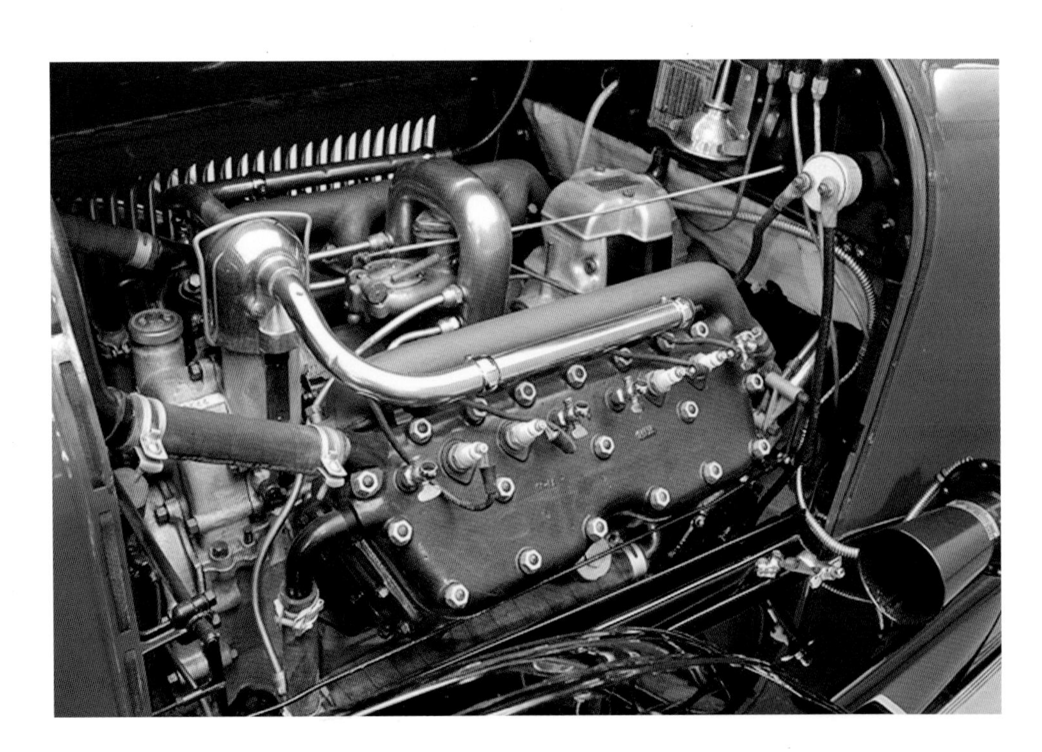

Priced at under $3000 this car represented excellent value for money.

This engine, first fitted to the type 51, was a 90-degree L-head with (by 1918) detachable heads and two cast iron four-cylinder blocks on an aluminum crankcase making a 314.5-cid unit which had 3 1-7/8 in main bearings, a 3-1/8 x 5-1/8 in. bore and stroke. With a 4.25:1 compression ratio, it produced 70 hp at 2400 rpm and 180 lbs.-ft. at 1800-2200 rpm. The engine had rockers with roller cam followers, a 1.5 gallon crankcase and a 5.25-gallon cooling system. Its Johnson float feed carburetor had auxiliary air control. The water jackets and combustion chambers were integral with the blocks.

Coolant circulation and temperature control relied on an impeller pump with a thermostat for each block. A single camshaft with eight cam lobes was used. A silent chain drove the camshaft and generator shaft. The generator and distributor were rear mounted since a two-cylinder power tire pump was up front.

Above: Our featured model is an Open tourer, clearly showing the modern styling and spacious interior, which made it the 'standard model' for service in World War 1.

1918 Chevrolet Series D

Engine: Ninety-degree overhead valve V-8	
Displacement: 288-cid	
Horsepower: 55	
Fuel System: Zenith Double jet carburetor	
Body style: Two-seater Roadster	
Weight: 3150 lbs	
Wheelbase: 120 inches	
Base Price: $1550	
Number Produced: 2781	

Billy Durant, by now a regular player in our drama, had managed to buy back General Motors in 1915 with the aid of Chevrolet. But while he was maneuvering it seems that he took his eye off the ball at Chevrolet. The result was that the Model D, an admirable new project, didn't get the full marketing thrust that it deserved.

The car, which came in two models: a five-passenger Touring and a four-passenger Roadster. The Roadster was offered in Chevrolet Green with pleated French leather upholstery and had lots of exposed mahogany, varnished to a fine finish. Materials alone lift it out of the under $2000 category that it actually occupied. At just $1400 for the launch model the car was astonishingly good value.

Its body style was very much in vogue with its upscale big brother, the Cadillac Type 57. Hood and cowl now flowed smoothly into each other without an obvious step and there were slanted louvers on the hood sides. At 120 inches, the wheelbase was only 5 inches down on the Cadillac. This meant that the Chevrolet offered roomy accommodation too. The open car was well equipped, with a one-man waterproof top and side curtains. The body was described in advertising copy of the time as a 'delight to the eye' and 'a series of curves that blend harmoniously'. Whatever the value of the purple prose

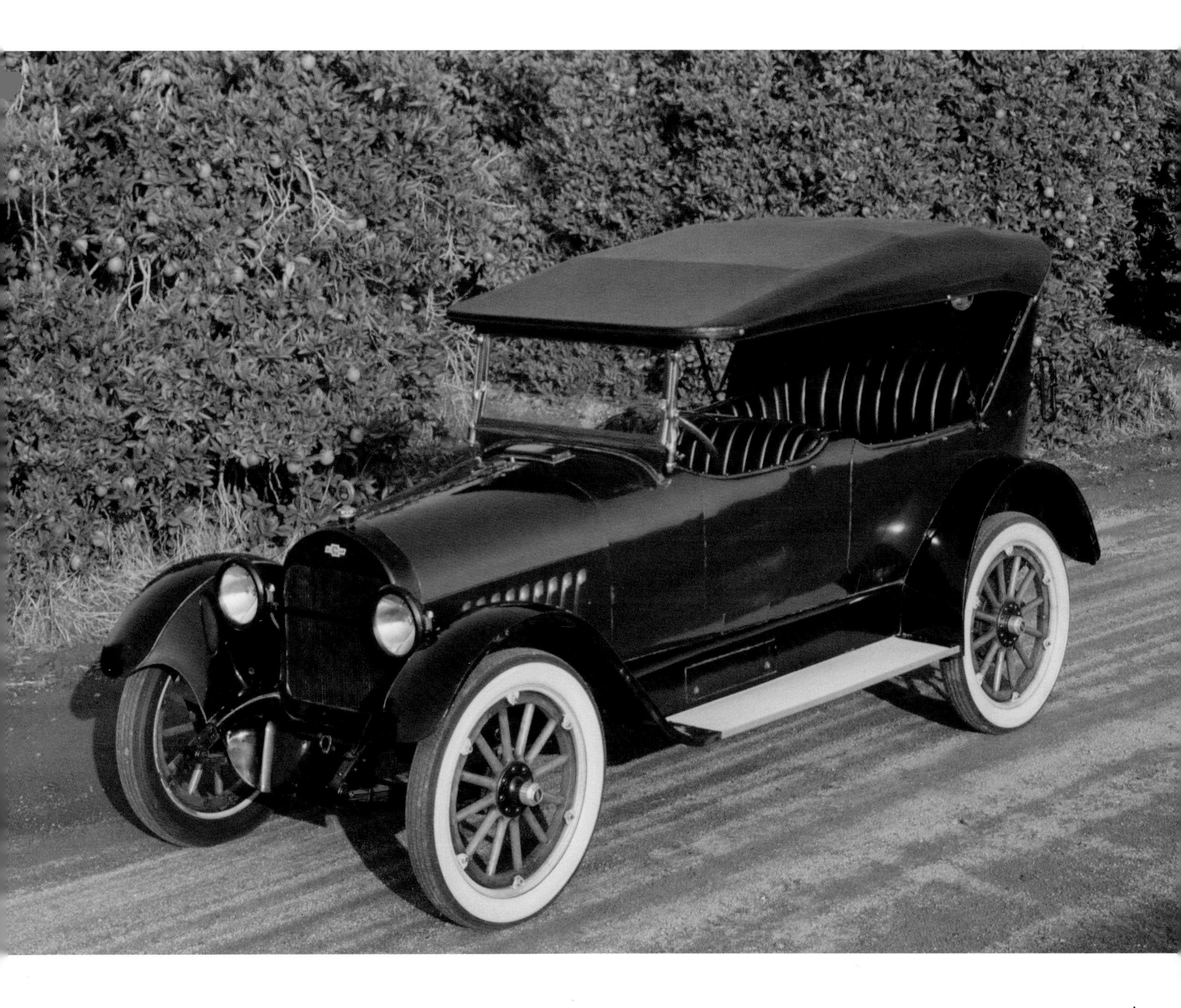

this was a good-looking set of wheels. Technologically too, the car was all there. Its floor pan was pressed steel and all its metal parts were now nickel-plated. The real kick came from the spanking new V-8 designed by Arthur Mason, the man who baled out the company with his sensible four-cylinder engine in 1913. It was an ambitious ninety-degree V-8 with overhead valves and detachable cross-flow cylinder heads, probably the most advanced engine of its kind around at the time. Its 288 cubic inches gave a guaranteed 55 brake horsepower.

By 1919 the model had disappeared. 'Why' is a complete mystery. Perhaps customers just couldn't get their heads round paying over $1000 for a Chevrolet, a lesson already learnt with the Classic Six. Personally, if I had been around at the time I'd have bought one. Chevrolet would not offer V-8 power again until the fifties.

Above: The D4 Roadster resplendent with leather, wood, 16 candle electric headlights and distinctive 'bowtie' radiator badge. It looks every inch a Chevrolet.

Opposite page: The advanced overhead valve 288 cid V8 designed by Arthur Mason produced a guaranteed 55 horsepower and was potentially one of the best engines around at the time. Chevrolet would not offer V8 power until the 1950s.

1922 Duesenberg Model A Roadster

Engine:	Single overhead camshaft straight eight-cylinder.
Displacement:	260-cid
Horsepower:	88 at 3000 rpm
Transmission:	Three-speed Manual
Body Style:	Two-door Roadster
Number of seats:	2
Wheelbase:	134 inches
Base Price:	$6500
Number Produced:	600

This manufacturer's roots can be traced back to March 1903 when one Fred Duesenberg, 'a bicycle builder from Iowa' filed for bankruptcy. Like all good businessmen, Fred bounced back, realizing that the demand for bicycles was about to be eclipsed by that for automobiles. To this end, he threw himself into the design of the Mason, which was named for Edward R Mason, a local attorney in Des Moines who put up the money. The 'Mason' featured a two-cylinder unit to begin, with but this was replaced by a four-cylinder unit designed by Dusenberg himself. At this time he enjoyed a good deal of success racing the cars including almost qualifying for the 1912 Indy 500.

By 1913 he had teamed up with his brother, August, to form the Duesenberg Motor Company in St Paul, Minnesota. This was the first of many different locations and incarnations of the company that bore their name. Strangely the brothers were never actually the owners. They preferred to remain as employees, while the owner/financiers did the worrying for them. This left them free to get on with the design, production and racing of the cars. This was what they really enjoyed.

By 1920 the Duesenberg Automobile and Motors Corporation was incorporated with funding of $1.5 million, and was about to launch its first car. The prototype, which had its debut at the Hotel Commodore, created a sensation. It had four-wheel brakes, which was a first. The car was powered by a 260-cid eight-cylinder overhead valve engine, which produced a respectable eighty-eight. In another quirky move Dusenberg decided that he would upgrade the engine to single overhead camshaft for the production model. The prototype used the Duesenberg patented horizontal valve and rocker arm set-up. This change delayed the flow of cars into the dealers until late 1921.

The new 'Model A' as it was designated, came in a variety of factory bodies

Right: Duesenberg's straight-eight was specced up at the last minute to run a single overhead cam shaft rather than overhead valves.

including a Tourer, Roadster, Sedan and Town car. All these models were all based on a 134-inch wheelbase chassis. The featured car is probably the nicest, a very attractive two-door Roadster with a host of chrome-plated details set off by the dark coachwork. White wall tires on spoked wire wheels and a white hood also add glamour. The rear deck is in an early boat style. The car heralds the classic styling of the 20s.

Above: Only 600 of the classic 'Model A' were built in five years. The result is that this Roadster model is now one of the cars most sought after by American collectors.

1923 Studebaker Light Six

Engine: L-Head inline six-cylinder	
Displacement: 207.1-cid	
Horsepower: 40 at 2000rpm	
Fuel System: Stromberg one-barrel carburetor	
Weight: 2730 lbs	
Wheelbase: 112 inches	
Base Price: $1395	

Opposite page: The next light six coupe in black awaits its owner.

Below: The L-head six –cylinder engine ran an aluminum head and was designed by skilled engineers Zeder,Skelton and Breer who defected to Chrysler soon afterward.

According to company boss Albert Russel Erskine, six-cylinder cars 'were what made Studebaker famous.' Indeed, the Light, the Special and the Big Six were legends in their own lifetime. But the Studebaker Company had been around a lot longer than that. The company was around during the Civil War, and supplied the North with horse-drawn vehicles. It continued to prosper. There were plenty of small wars around the globe in the late nineteenth century, where the company's product was in demand. Their entry into the gasoline automobile business was rather more tentative. John M. Studebaker was something of a die-hard. He thought gas cars were smelly, and that electric was the way forward. But he was finally persuaded to go into car production by his rather more go-ahead son in law, Fred Fish. Studebaker subsequently took over the General Car Company in 1911. The Studebaker Corporation was now formed with Fred Fish as President.

The company continued to manufacture worthy, medium priced cars - profitably. Fred Fish retired in 1915, and Albert Erskine took over the management of the company. World War1 had begun in Europe and there was a great demand for military vehicles. Studebaker got their share of government supply contracts, which put the company in a healthy financial position to face the changing times in the '20s. Studebaker Corporation had some very able engineers on board, including Fred M. Zeder, Owen R. Skelton and Carl Breer. These men were responsible for designing the six-cylinder powerplant that kept the company's sales buoyant in the '20s. Subsequently, this engineering team defected to Chrysler, taking their know-how with them.

The Light Six had a new all steel body for 1923, in common with many other cars of the time. Studebaker held out against four-wheel brakes, considering them dangerous, so the car retained the two-wheel brakes at the rear. Body options were the two-door roadster, four-door tourer, two-door coupe, and four-door sedan. Our featured car is the cute coupe. Standard wheels were the artillery wood spoke type and the windshield on open cars was raked back provocatively. The engine was an L-head Inline six with an aluminum head, of 207.1 cubic inches, giving a credible forty horsepower.

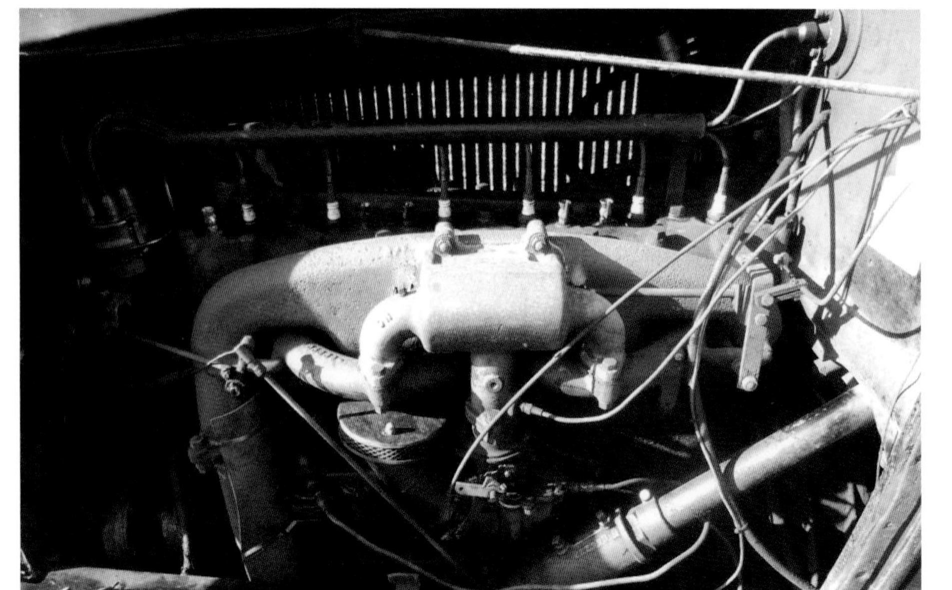

Like all straight sixes, the resultant smooth power of the engine mounted in a relatively light body must have made the car a delight to drive.

1504

1925 Cadillac V63 Brougham

Since the introduction of V-8 power to its lineup in 1915, Cadillac hadn't looked back. Valuable contracts in World War 1 had boosted the company's position as one of the most profitable divisions of the General Motors group. With the departure of the founding Leland family, however, there was a period in the late teens and early '20s when the top management was in some disarray. Richard L. Collins, Herbert H. Rice and Lawrence P. Fisher all occupied the 'number one' position at the company during this period, and only the latter survived longer than a couple of years. Luckily, the Chief Engineer's job remained stable with Ernest Seaholm in charge from 1923 up to World War II.

By the time of our model, the V-8 had detachable heads and a balanced crankshaft. This, together with other detailed improvements would carry the engine right through to 1936. The V63 was also was the first Cadillac to be fitted with four-wheel brakes. New too was the Custom concept –a special collection of non-standard body styles. Sedan, Imperial Sedan, limousine, Imperial Suburban, Town Brougham, and a brace of coupes were all available for either two or five passengers.

New technology in paint finishes meant that many more colors were available, and this launched cars forward into a new realm of brighter hues. Nickel-plated radiator shells and headlights also improved the look of most model ranges. The Custom series also pioneered the scrolled-style radiator detail, which was soon adopted by the entire Cadillac range.

The result of all of this was to produce a car that needed to be sold at around $4000. This high price had the potential to have had a negative effect on sales. But customers loved the whole 'Custom' concept, and sales actually rose steadily. Cadillac found buyers for as many as 20,000 units per year throughout the middle years of the '20s.

Engine: L-Head V-8

Displacement: 314.5-cid

Horsepower: 85

Fuel System: Float fed carburetor-Cadillac M/F to Johnson Patent

Weight: 4530 lbs

Wheelbase: 138 inches

Base Price: $4950

1925 Hudson Super Six

Hudson powered out of the postwar sales slump in fine style. Profitability was a given at Hudson. The company was properly financed by Joseph Hudson from the word go, and was properly run by a strong and stable management team led by Roy D. Chapin. This was unlike the fragmented management pattern at many other car manufacturers who hired and fired top people after a couple of years, or else they became disaffected and left to found their own companies. Hudson was either very wise or very fortunate, as they managed to keep a great team together over a long period. The family-feel of the company may have contributed to this. Roscoe B. Jackson was President for many years, alongside Roy Chapin, who had married into the Hudson Family. This happy situation meant that shareholders were also well looked after. The regularly paid dividends resulted in increasing confidence in the company. This feel good factor enabled the company to weather periods of depression better than most. They also cut prices to boost sales whenever necessary.

Hudson now had a good foothold in the domestic market, thanks to its marketing strategy. The Hudson marque took care of mid-section sales whilst the Essex line, launched in 1919, provided budget-end sales. The combined sales of the two divisions put the company at third place in the year's sales charts, behind Ford and Chevrolet. The Essex range was not a cheapened form of Hudson but a completely unique range, with its own separate design and power options. Essex also had a reputation for sportiness. Hudson, on the other hand, relied on a more solid, well-heeled image.

The car that sustained Hudson sales in the '20s was the Super Six. It was based on the company's L-Head inline 289 cubic inch six-cylinder power unit, which gave an impressive 76 horsepower. The car offered fantastic value, priced at just over $1000. Hudson continuously developed the standard power plant, and by 1925 it sported aluminum pistons and chain timing actuation instead of helical gear drive. Car bodies were built by Biddle & Smart- many with aluminum panels, which kept the weight down considerably. In 1925 a new brougham body from this company became an overnight success because at a mere $1450, it looked every bit as good as a $3000 Cadillac!

The popular Coach model was also redesigned in this year, and was given slimmer side pillars and a curving windshield base. Tires were changed from 33 x 6.2 to 33 x 6.00. The overall appearance of the car is that of a solid and well-engineered vehicle, which was no slouch either, with its smooth six-cylinder engine.

Engine: L-Head Six-cylinder Inline

Displacement: 289-cid

Horsepower: 76

Base Price: $1165

Number Produced: 109,840

1925 Hupmobile Series R Touring

Engine: Four-cylinder in-line	
Horsepower: 39	
Transmission: Three speed selective	
Base price: $1225	

In 1909 Robert Craig Hupp began his company with a small cute runabout, which was surprisingly well equipped with a four-cylinder engine of 16.9 horsepower and two-speed sliding gears. Priced at only $750 it was designated the Model 20. Thus began a marque that was to last for thirty years and always provide interesting value-for-money cars. Sadly, Hupp himself parted from the company quite early on in its history. He was gone by 1911. Like Ransom Eli Olds, who had parted early on from the company he had named, Hupp had to resort to using his initials for future car exploits but RCH doesn't really have the same resonance as REO and after a few years he disappeared from sight. The two board members he quarreled with were Charles Hastings and J. Walter Drake. Both men would be involved with the company for a good many years to come. Early sales success was credited to Hastings hard work in establishing the car's reputation. Sales rose quickly, going from 1,617 in 1909 to 6,079 in 1911.

A 1910 car straight off the production line took part in a round-the-world promotional trek covering 48,600 miles and traveling through 26 countries. The man responsible for engineering the car following Hupp's departure was Emil Nelson, who left the company in 1912 and who formed his own manufacturing company - Nelson in 1917. Following Nelson's departure, Frank E. Watts took over and remained in charge of engineering for the next 26 years - virtually to the end of the company's life. By 1916, Hupp was in some disarray. Drake had led the company into developing an over-complicated model range and some large, ambitious vehicles that were moving away from the company's founding ethic of value for money cars. A parallel to what happened at Olds following R. E.'s departure.

The only man who could stand up to Drake was his former colleague Charles Hastings. Hastings had retired to California some three years earlier. He was recalled by the board and invited to serve as President from 1917 onward. In that first year he was responsible for a new, smaller model – that put the

Below: Artillery style wooden wheels were pretty old hat by 1925.

Below right: The radiator mounted H for Hupmobile badge.

company right back at the popular end of the market. This, 'Series R' car was given the familiar four-cylinder power unit that had been the mainstay of the company all along. The model revived the company's fortunes, reaching an apogee in 1923 with 38,279 cars sold.

The series R was available in 5 body styles: a touring model, roadster, coupe, club sedan and regular sedan. It was well finished, with two-piece bumpers, nickel-plated drum headlights and radiator shell, and had a louvered hood. But by 1925 the body style was looking dated.

Above: The last of the Four-cylinder Hupmobiles. The following year the company went straight-eight!

1927 Chrysler Series 70 Royal Coupe

Engine: L-Head Inline Six-cylinder
Displacement: 218.6-cid
Horsepower: 68 at 3000 rpm
Transmission: Three-speed manual
Body Style: Two-door Coupe
Weight: 2935 lbs
Base Price: $1695
Number Produced: 72039

Walter Chrysler lived the original American rags to riches story. The one time farmhand, grocery boy, silverware salesman and roundhouse sweeper for Union Pacific was 'discovered' by Billy Durant in 1910 when he put him in charge of Buick as works manager. He learnt his craft well. When he left General Motors just nine years later, he was already a millionaire. Durant reportedly paid Walter Chrysler $10 million to buy out his share of Buick. By 1919 Chrysler had had enough of Durant, and his next job was at the Chase National Bank. The bank had many big customers in the automotive industry and by the early '20s many of these companies were ailing. It was Chrysler's knowledge of the industry that got him the job to help straighten these out. Firstly, he was asked to try to put Willys-Overland back on track. In the course of this job one thing – a new six- cylinder power plant – caught his eye. It was lying fallow at the Elizabeth, New Jersey plant that Willys were forced to sell. The creators of this interesting piece of engineering were a team we have encountered already at Studebaker, Fred Zeder, Owen Skelton and Carl Breer.

In the meanwhile, Chrysler had moved on to sort out the equally unstable Maxwell company, and it was this that he used as his final platform to launch Chrysler. He hired the Maxwell team and set them up at the Chalmers-Maxwell factory in Detroit to further develop the engine that he had decided would power his first car - the Chrysler B 70. He launched his first model at the Hotel Commodore in 1924.

Chrysler's marketing strategy was simple –to sell a car for $1500 that provided the same kind of motoring pleasure for which people were willing to pay $5000. This strategy worked well and the car soon evolved into 'Series 70' with nine different body styles. These included a roadster model, phaeton, the Royal Coupe, a sedan, brougham and Crown Sedan. All the models were all based on a single 112-inch wheelbase and the same basic chassis. Four-wheel hydraulic brakes were standard on the entire series.

Below: The new 216 cid six-cylinder power plant designed by Zeder,Skelton and Breer for Chrysler gave out 68 horsepower.

Below right: Three-speed manual gearshift and neat dash characterize the first Chrysler.

The engine was an impressive six-cylinder, with 218.6 cubic inches and a high compression rating (for the time), with cylinder heads giving a 4.7:1 ratio. This developed a respectable 68 brake horsepower. Quite a package!

Above: The aristocratic prow of the Royal Coupe decries its comparatively modest price of $1695. This closed car retains the drum-style headlights that were soon to be replaced with the bullet type.

1927 Ford Model T Pick-up

Engine: L-Head Four-cylinder
Displacement: 176.7-cid
Horsepower: 20 at 1600 rpm
Transmission: Two-speed planetary
Body Style: Roadster Pickup
Weight: 1736 lbs
Wheelbase: 100 inches
Base Price: $381
Number Produced: 75,406

Opposite page: A Roadster Pick-up from 1927 – the final year of Model T production. In this year alone over 75,000 units found buyers.

Henry Ford was in an unassailable position within his company by 1925, he owned it, he was President and he decided what went on. He certainly wasn't going to listen to anyone criticizing his revered Model T. When critics called it old fashioned he had the company's leading position in the car sales chart to reassure him. That, and a few million dollars in the bank. But his management team recognized the threat posed by Ford's competitors, particularly Chevrolet's Superior range. The defection of the Head of Ford Production, William S. Knudsen, over to Chevrolet was also worrying. But instead of making any radical changes, Henry Ford introduced cosmetic adjustments in an attempt to silence his critics.

The tires for the 'T' were upgraded to the balloon type in 1925. The most radical departure for '26 was the reverse of the 'black as the only color option' diktat. Other minor changes to the Model T that had been announced in 1925 as part of the 'Improved Ford' promotion, included new fenders, running boards and a modified chassis. Extra models were also added to the range. This was probably the most significant number of changes that anyone could remember in the car's history. But such measures were actually too little, and certainly too late.

By 1927 the Model T line was shut down with no replacement planned. There were six long months of feverish speculation before the line began rolling again. However, some smart marketing had come out of the 'Improved Ford' package, in the shape of the Roadster Pickup. Because of their rugged practicality and low price, people had been using Model Ts as utility vehicles all along. The option had been to buy the chassis-only version, which gave customers the option to construct their own buckboard style back, modified to carry fence posts, livestock or whatever was required. Ford now marketed a ready-to-buy package for practical hauling. This was a great product that, in years to come, would be marketed as a vehicle in its own right, and become an essential part of American transport culture - the pick-up.

Below: the front suspension and braking on the 1927 Model T was basic but worked.

Below right: Ford's L-Head engine had remained virtually unchanged since the car's launch in 1909 whilst other competitors had moved on.

Ford 1927

1927 Hudson Roadster

Engine: F-Head Inline six-cylinder	
Displacement: 288.5-cid	
Horsepower: 92 at 3200 rpm	
Transmission: Three-speed sliding gear	
Body Style: Two-door Roadster	
Weight: 3480 lbs	
Wheelbase: 127 3/8 inches	
Base Price: $1500	
Number Produced: 66034	

The late '20s were a good time for Hudson. Adding together Hudson and Essex sales, the company was always in the top five producers. Essex had been created specially to pioneer models for the budget end of the closed car market, and it had been recognized as such by no less a force than the New York Times. In an article of January 1926, the paper expressed a view that: 'the flood of new, small closed sixes is one of the outstanding features of the year... That the light economical six makes a definite and potent appeal cannot be doubted. Indeed the Hudson and Essex organization has one of the most remarkable achievements of 1925 to its credit in the production of 250,000 cars of both makes, almost entirely closed models'. That year, Hudson invested a massive $3 million in a new production plant. Hudson's Model 'O' of 1927 demonstrated the same commitment to strong investment in the company's future. This was a completely new model with a host of improvements on its predecessors.

Astonishingly, the holy of holies - Hudson's trusty L-Head Super Six engine was also revamped, with a tasty new F-Head, better manifolding and improved valve design. Whilst it was of identical bore and stroke to the old unit, the improvements took the power up to 92 horsepower. This was transmitted to the rear wheels via an improved single plate clutch. In fact, many experts regard the horsepower rating as a conservative estimate of the engine's output. Claims of 100 mph plus were not unusual from drivers of the car. The brakes were also beefed up. These were four-wheel mechanical. A new rear suspension, 18-inch wheel rims, full-crown fenders, reshaped hood profile, a

Right: 1927 was the first year for the 288.5 cubic-inch Super Six F-head unit which featured an overhead intake and developed 92 horsepower.

four-inch lower bodyline and bullet-shaped headlights all contributed to the truly racy look of the new car.

But why was all this undertaken? Hudson already had a pretty good market share with a proven car. The explanation may lie in Chairman Roy D. Chapin's mission statement of the same year. He noted that 'buyers are now insistent that cars shall excel in appearance and convenience as well as in the fundamental qualities. The demand for improved performance is widespread and is being met by better design, materials and workmanship'.

Above A hot rod in the making. The sporty looking Hudson Roadster just invites you to climb in!

1927 Studebaker Big Six President

Albert Erskine, President of Studebaker was worried. He just didn't feel that South Bend, Indiana, the location of the company plant, had enough style and class for his enterprise. He had designs on the prestige car market. Erskine's solution was to book space at the Plaza Hotel in New York that summer, in order to present the company's range to the socialites of Manhattan.

But the presentation wasn't a great success. So instead, he launched a new model, the Erskine, styled with so-called European flair. This model also failed to impress. As a last resort, he decided to rely on the Studebaker product speaking for itself, at least for the time being.

The six-cylinder cars had made Studebaker famous, and the company planned an impressive, all new 'six' for 1927. The President Big Six model was restyled to include a lower body line, sloping rounded roof, double-side body moldings, a two-tone paint job, steel disc type wheels, 'acorn' shaped nickel-plated headlights and nickel-plated double bumpers. The cars also had 'Atlanta-style' hood mascots and 'French' windshield visors.

Two wheelbase lengths were offered, 120 and 127 inches. A host of extras could also be ordered, which could put the car firmly in the luxury category. These included a no-draft ventilating windshield, Watson stabilizers, engine thermometer, clock, hydrostatic gas gauge, coincidental lock, oil filter, Alemite chassis lubrication, automatic wiper, double rear vision mirror, armrests, toggle grips, auto dome light, rear signal light, emergency lamp on an extension cord, four wheel mechanical brakes, balloon tires and even a Vanity case smoking set!

The famous six-cylinder engine was boosted to 354 cubic inches, giving 75 horsepower at 2400 rpm. Three body styles were now available: a four-door custom sedan, four-door limo and a four-door duplex phaeton. The entire range could accommodate seven passengers in great luxury.

1928 Cunningham Series V7

Just like Studebaker, Cunningham was another auto producer that started out by manufacturing high quality horse-drawn vehicles. The company's origins went back to the mid-nineteenth century. Joseph Cunningham inherited his fathers carriage business in 1886 and consolidated the company's worth by concentrating on a range of superbly handcrafted vehicles, from dogcarts to Berlins. But like many smart businessmen at the turn of the 20th century, Joseph could see that there was a wind of change sweeping through society, and that horseflesh would soon be replaced by horsepower of a different sort. His intention to produce a horseless carriage was realized in 1907 when he completed a vehicle using a Continental engine. By 1911 the Cunningham full production model was ready. The 'Model J' had its own four-cylinder engine, which developed 40 horsepower.

Joseph Cunningham obviously thought big. Only two years later, he graduated to a large, 442-cid V-8 capable of a modest 45 horsepower. The cars themselves were also big, with no wheelbase smaller than 130 inches, and were to grow even larger in subsequent years.

By 1919 Cunningham cars were winning speed trials, and the famous racing driver Ralph De Palma, drove one at a reputed 98 mph in a speed trial at Sheepshead Bay. As well as performance ability, the cars also had a reputation for the immaculate style and finish of the coachwork, a reflection of their pre-engine heritage. The cars were long and sleek and their good looks and classy reputation attracted a rich and famous clientele. Hollywood movie stars and celebrities, such as William Randolph Hearst, Mary Pickford, Cecil B. DeMille, Fatty Arbuckle, Blanche Sweet and Harold Lloyd all drove Cunninghams in the '20s. Unsurprisingly, the price of the cars was very steep – buyers had to pay between $5,000 and $9,000 for the privilege of ownership. Such lavish cars

had their time and place. When the economic downturn arrived in the 1930s, conspicuous consumption was no longer considered smart, and the company was forced to retrench into ambulance and hearse production.

1928 Cunningham models included a touring car, cabriolet, roadster, two-door coupe and Town car. A choice of wheelbase was offered, of either 132 or 142 inches.

1928 Ford Model A

Engine: Inline L-Head Four-cylinder Inline	
Displacement: 200.5-cid	
Horsepower: 40	
Base Price: $550	
Number Produced: 633,594	

Opposite page: Model A Tudor, Walton-style outside a lumber mill.

Below: The Ford Model A two-door roadster.

So here it was at last! The first all-new Ford model for nineteen years. Years in which Ford had completely failed to develop a replacement for their one and only model. Instead of having a new lineup ready and waiting in the wings, Ford were left with a gaping lack of product, and were forced to shut down their entire production line in May 1927. This allowed Chevrolet to take over as the number one-selling manufacturer. In the ensuing six months, rumors circulated about the introduction of a Linford model - a cross between a Lincoln and a Ford. This was not entirely surprising. Ford had purchased the Lincoln Company out of receivership in 1922 from the Leland Family, and Henry Ford's son, Edsel, had run the company very successfully from that time, developing a model range that Ford now desperately needed to access.

The new car was inevitably compared to its predecessor, and its heritage was recognizable in some regards. The same four-cylinder L-Head engine was present, but this time it displaced 200 cubic inches and generated twice the horsepower. The 'A' also shared the semi elliptic front and transverse leaf spring rear suspension that Henry Ford so passionately believed in. He defended the re-cycled suspension by saying 'we use transverse springs for the same reason that we use round wheels - because we have found nothing better for the purpose'. However, beyond these similarities the two models had little in common.

In place of a magneto, the Model A had a modern electrical system, complete with battery and coil type ignition. Nickel-plated electric headlamps were standard. The car also had four-wheel mechanical brakes and double action shock absorbers. In place of the old two-speed planetary gears, a three-speed selective sliding type gearbox was now fitted. The wheelbase was also up by 31/2 inches. The new bodyshell had an attractive longer, lower look that the Model T, with sleeker windows and a high beltline. The fenders had arched cross sections and welded wire wheels replaced the old wood spoke artillery type. Two-piece nickel-plated front and rear bumpers were added, and the radiator shell now had a distinctive 'V' shaped dip over the top front of the radiator core. Nine body options were offered, something for everyone, including a two-door roadster, phaeton, Tudor and fordor sedan, sport coupe and even a taxicab. Strangely, the taxi was the most expensive car that year, priced at $600. Maybe Henry figured they made up the money with the extra tips.

All in all, it is fair to say that the Model A was in every way a more than worthy successor to the crown of America's number one selling car.

1928 Pontiac 6-28 Roadster

Engine: L-Head Six-cylinder inline
Displacement: 186.5-cid
Horsepower: 48
Base Price: $775
Number Produced: 80,000

Opposite page: The Two-door roadster in blue with later type Jaxon ten-spoke wood wheels and a standard front bumper.

Depending which way you look at it, Pontiac is either a suburb of Detroit or an Indian Chief. Either way, it's a good name for a car. The Pontiac Buggy Company was originally part of the Oakland Motor Company, started in 1907 by Edward Murphy and Alanson P. Brush. When Oakland was bought by Billy Durant in 1909 to join the General Motors empire, the company continued to produce mid-priced models. The General Manager, George Hannum, decided that the company needed a lower priced market entry model and code-named his idea 'Relot'. Alfred P. Sloan who was managing GM, after the departure of Billy Durant, recognized the sense of this. After all, Hudson had started Essex for the same reason a couple of years earlier. Having two car marques at completely different price points made good economic sense. This strategy had pole-vaulted Hudson several rungs up the sales ladder. Experimental engineers Ormond E. Hunt and Henry M. Crane were asked to develop the idea. Their proposal was to graft a six-cylinder engine onto a Chevrolet chassis. At this time, Chevrolet's engines were almost universally fours. Essex's success had been won by six-cylinder 'closed cars' and this was the target to beat. Hannum's leadership wasn't to last. He was succeeded by Alfred Glancy, who hired his own team to see the task completed. The 'four horsemen' as they were known Ben Anibal, Fenn Holden, Herman Schwarze and Roy Milner were all seasoned professionals, having spent a decade at Cadillac learning their craft. The Pontiac name was already part of the company's heritage, and it provided the excuse for an Indian head hood ornament to hang the publicity on.

The sales launch was designated 'the Pow Wow with heap big eats at lunch-time'. This took place in 1926 at the New York Automobile Show. Although Series 6-27 was hardly remarkable, it was well designed for its purpose, and in its first year it set a sales record by finding 76,696 buyers.

Below: The rear view of the Pontiac Roadster with its boat tail style back concealing the dickey seat.

The car's chassis was conventional and its L-head six was decidedly unexciting (with a meager forty horsepower developed from 185-cid), but it was well targeted to compete against other low-priced sixes. At $825 only the Essex was cheaper. The sales just kep on coming - by March 1929 100,000 Pontiacs had been sold. The formula was obviously working.

By 1928 Pontiac had a fully fledged range, with seven models on offer. Its body styling now had more sweeping lines, a raised panel on the top of the hood, deep crowned fenders. Mechanically, other improvements included four-wheel mechanical brakes. Pontiac was here to stay.

3013

1929 Auburn Speedster 120

The origins of Auburn lie in a small carriage company started by Charles Eckhart back in 1874. Eckhart had previously worked for Studebaker in South Bend where he picked up the skills of his trade. His sons Frank and Morris took over the family firm in 1893 and started the Auburn Automobile Company. A single-cylinder, chain drive runabout soon gave way to a two-cylinder car in 1905, and in they manufactured a four-cylinder version in 1909. Sales continued throughout World War I, but by 1919 the company had began to lose direction. A consortium of Chicago businessmen led by William Wrigley Jr. (of chewing gum fame) bought the Eckharts out.

Despite restyling the range to add the Auburn Beauty Six, the company got caught in the post war sales dip. In four model years, just over 15,000 cars were sold. By 1924 the company was producing fewer than 10 cars a day. Enter hotshot salesman Erret Lobban Cord, who had made over $1,000,000 selling Moon cars for the Chicago based Quinlan Car Company. He was now looking to invest. First he took a lowly paid position as General Manager of Auburn, on the understanding that he could acquire stock if he could get the company 'off the bottom'. He authorized some cosmetic restyling, adding a touch of nickel plate here and a custom paint job there. Thus beautified, the backlog of cars on the Auburn lot finally began to sell. In 1925 he began buying straight eights from Lycoming and he had James Crawford (Auburn Chief Engineer) bolt these into the existing six-cylinder chassis. He then commissioned an outstanding 'boat tail' design by Al Leamy. A combination of the two factors, an improved engine plus and eye-catching looks, really made Aurburn sales take off.

At about this time, the demise of Mercer left a gap in the market for somebody to challenge Stutz, and the new Auburn wanted the slot. The car was well engineered for the times, with four-wheel hydraulic brakes and Bijar lubrication.

Seven models were on offer for 1929: a speedster, cabriolet, phaeton, Victoria, sedan (both five and seven seat), and sport sedan.

Engine: Eight-cylinder inline

Displacement: 298.6-cid

Horsepower: 125

Transmission: Three-speed manual

Body Style: Two-door Roadster

Number of seats: 4

Wheelbase: 136 inches

Base Price: $1895

1929 Chrysler Imperial L Sedan

Walter Chrysler had been busy. To consolidate his position in the motor industry, he had improved his model range dramatically, widening the number of models available and moved toward the upper end of the market, while selling at prices that undercut those of his rival Cadillac. He had bought Dodge in 1928, not for cash, but paid for in Chrysler stock. This move would provide him with the distribution outlets he needed to crack the volume market.

He then created Desoto and Plymouth to give him access to all other market sectors. This meant that, as Chrysler approached the '30s he was in a very strong position to weather the storm of the Depression. His combined companies thrust him high in the industry sales chart taking him to third place in 1928, behind only Ford and Chevrolet.

The Chrysler Imperial was Walter P's prestige car, which had been originally launched in 1927. Styling was long and low based on a 136-inch wheelbase chassis. The car had a scalloped hood and radiator shell, a style some say was 'borrowed' from Vauxhall of England. In fact, Chrysler was not averse to incorporating all kinds of elements from other cars in his designs.

It was hoped that here was a series of cars that could take on Cadillac head-to-head. In fact, although the Imperial worked out about $600 cheaper than an equivalent Cadillac model, in many ways the Chrysler was more technologically advanced. For example, the Imperial had four-wheel hydraulic brakes, while the Cadillac was still using mechanical brakes. Also, the beefed up 309 cubic inch version of the Chrysler six engine block, complete with the high compression Red Head (5.2:1) was churning out a stunning 110 brake horsepower, whereas the Cadillac's 314-cid V-8 scraped together a mere 85 horsepower. The Cadillac probably would have had the better torque of the two cars, but hey, not everybody drove like their granny even then. Chrysler even offered the same kind of options on special custom-built bodies that Cadillac did: versions by LeBaron, Dietrich and Locke were all available (at a price) in addition to the eleven factory-listed bodies.

Last but hardly least, the car looked great, with its cadet-style windshield, its long lines accentuated by the new two-tone color scheme, bullet style headlamps, body-colored wire wheels and distinctive Viking's head winged radiator cap.

Engine: L-Head Inline six-cylinder

Displacement: 309.3-cid

Horsepower: 110 at 3200 rpm

Weight: $4460

Wheelbase: 136 inches

Base Price: $3095

Number Produced: 2506

1929 De Soto Series K

Engine: L-Head Six Cylinder in line	
Displacement: 174.9 cid	
Horsepower: 55	
Transmission: Three-speed selective sliding gears	
Fuel System: Stromberg IV carburetor	
Body style: Four-door Phaeton	
Number of seats: 5	
Weight: 2445 lbs	
Wheelbase: 109.75 inches	
Base Price: $845	
Number Produced: 81,065	

At the same time that Walter P. Chrysler was trying to buy Dodge Brothers from bankers Dillon Read and Co., he thought it would be smart to hedge his bets by creating his own division to bridge the gap between the top-end sales of Chrysler and his budget priced Plymouths. This was the inspiration behind De Soto. De Soto was named for the Spanish explorer Hernando De Soto who had explored the Mississippi Valley some centuries earlier, along with La Salle. But by the time the first De Sotos were ready in 1929, Walter Chrysler had managed to buy Dodge and it could have been thought that De Soto was surplus to requirements. But first year sales for the line set an all-time record, one that wasn't even approached for another 30 years. A massive 81,065 cars were sold in the model year, beating Chrysler's own sales and those of Pontiac.

The Series K was not the most outstanding car around, but at $845 it offered a lot for the cash. Seven models were available all with the L-Head six of 174.9 cubic inches, giving a respectable 55 horsepower. This was enough to give the Series K models a lively and enjoyable performance. Four-wheel

hydraulic braking was by manufactured by Lockheed. Wood spoke wheels were offered as standard, but wire wheels were an option taken by most customers. Lovejoy shock absorbers and rubber engine mounts contributed to a smooth ride quality and passenger comfort. Externally, there were also some nice styling touches. Chrome-plated bullet headlights were mounted on an upward curving tie-bar, triple groups of hood louvers, built-in cowl lamps and a host of special paint jobs all went to suggest a car of quality. The fact that the car was on sale at under $1000 was a real incentive.

Above: The De Soto Series K sedan was good value for money.

Opposite page: The L-Head six-cylinder engine produced a respectable 55 horsepower which kept De Soto in the running.

1929 Dodge Senior Six

Engine: Six-cylinder L-Head inline
Displacement: 214.6-cid
Horsepower: 78 at 3000 rpm
Transmission: Three-speed selective sliding gears
Fuel System: Stromberg one-barrel carburetor
Body Style: Two-door Roadster
Number of Seats: 4
Weight: 2695 lbs
Wheelbase: 120 inches
Base Price: $995

After the untimely end of both of the roistering Dodge brothers in 1920, the famous company was left to the two widows. Uncertain how to continue the company and concerned by a marked downturn in sales, the two women decided to sell out to New York bankers Dillon Read and Company. Dillon Read paid $146 million for the company, which was then the largest transaction ever known in auto trade banking.

Under Dillon, Read & Co, 1926 sales jumped to over $21 million. Four new models were introduced in 1927 for the 1928 year: the Fast Four and three lines of six-cylinder cars, the Senior Six, the Standard Six and the Victory Six. Dodge decided to improve the way the vehicles were made, effectively pioneering the industry standard of welding steel panels together. Unfortunately, the amount of money needed to develop new manufacturing technology coupled with increased vehicle prices saw Dodge sink even further down in the US market to 7th place. At the end of 1927, sales had dropped a huge amount to just over $9.6 million.

The bank put accountants in charge of the company, who had no knowledge of the auto business. The company's fortunes continued to wane. Walter P. Chrysler was finally able to gain control of the company in 1928, having been rejected on his earlier attempt. Chrysler's vision for Dodge was to allow the company to operate without any drastic change of image. Dodge had good customer loyalty in the lower end of the mid-priced market. More importantly, the company had a great dealer network. This asset would help Chrysler increase his penetration into a wider market segment.

The Chrysler Corporation was only three years old at the time of acquisition and the purchase of Dodge made Chrysler the third largest automobile manufacturer overnight. Dodge Brothers, Inc. became Chrysler's new Dodge division. Production of the Fast Four was quickly ceased and vehicle prices were increased slightly. Even with the price increase, however, the basic Dodge Six was offered for $765 in 1929, while other makes sold close to $1000.

Dodge had always abreast of current trends. The Senior Six had had four-wheel hydraulic brakes by 1927. Further improvements for our featured year of 1929 included a restyled radiator shell, vertical hood louvers, automatic radiator shutters, a cadet-type sun visor, frosted silver-finish dash panel with full instrumentation, automatic windshield wipers, front and rear bumpers, interior courtesy lights, velvet mohair upholstery, smoking cases and

Below: The unremarkable Dodge L-Head six engine was boosted in 1929 to give 78 horsepower.

mahogany finish steering wheel. The option list is very extensive.

More substantial changes were made in the lengthened chassis, which gave a four-inch increase to the length of the wheelbase. A more powerful six-cylinder engine was now fitted. Lovejoy brand shocks were also added to improve the ride and handling.

No less than nineteen different body options were now available, too many to list. In short, the car was an astonishing package for under $1000!

Above: This dark red two-door Roadster Senior Six looks like a car that should cost over $2000. In fact, it was available for just $995.

1929 La Salle

Engine: Ninety-degree L-Head V-8	
Displacement: 328 cid	
Horsepower (Estimated): 80	
Body Style: Two-door Roadster	
Number of Seats: 2/4	
Weight: 3990 lbs	
Wheelbase: 125 inches	
Base Price: $2345	
Sales: 22,961	

Cadillac was named for the French explorer who discovered Detroit. So what was more natural than to dip back into history once again to name Cadillac's new brand for 1927. Another Frenchman this time, a Monsieur La Salle, who explored the Mississippi valley in the 1600s gave his name to a model that was to turn out to be a flop for Cadillac and GM.

Created to fill the price gap between Buick and Cadillac, the La Salle brand should have worked well. A young, up and coming designer, J. Harley Earl, was working at the Don Lee Corporation in California when he came to the attention of Cadillac's President Lawrence P. Fisher. Fisher had taken over as President in 1924. He quickly recruited Earl to design the new cars.

Cadillac knew that they had to increase sales. Adding a division with more cutting-edge design and less staid looking models seemed a good idea. To achieve this, Fisher founded a new department 'art and color' which was an attempt to disentangle the design function from engineering considerations. Earl always hated the department name, which he felt sounded superficial. When American auto manufacturers want their product to look better, they often turn to Europe for inspiration. Earl did exactly this, and was impressed with the looks of the Hispano-Suiza. But Earl's designs far exceeded anything that Europe could offer, and the La Salle styling was always one of the strongest aspects of the range.

Below: The handsome four-door sedan version of the 1929 La Salle.

Because La Salles were engineered to Cadillac's high standards, their customers were buying into a high-quality range. Indeed it has been suggested that if the cars had been badged as Cadillacs, theirs could have been a success story. The 303 cubic inch V-8, capable of 75 horsepower pushing along Harley Earl's elegant body very nicely. Test ace Bill Rader drove a stripped La Salle stock car at the GM proving track for 951 miles at an average speed of 95.3 mph.

The La Salle experiment initially looked good. Combined sales took Cadillac over 40,000 units for the first time ever in 1928. By 1929 sales were at over 50,000 cars.

Above: This dark red 1929 La Salle Roadster is an example of classic Harley Earl styling.

1929 Lincoln Model L

Engine: 60-degree V8 L-Head	
Displacement: 384.8 cid	
Horsepower: 90 at 2800 rpm	
Fuel System: Stromberg 03 Updraft carburetor	
Base Price: $4800	
Number Produced: 7,641	

Henry Martyn Leland, the founder of Cadillac was already 74 when he stormed out, leaving the company after a quarrel with Billy Durant.

Durant insisted that he had fired the Lelands. They maintained that Durant had refused to do his patriotic duty by producing Liberty Aero Engines for the War effort. Whichever version was true, Leland subsequently set up a company to make the engines, receiving a government contract and a ten million dollar advance to do so.

But towards the end of the war, the Lelands ended up with a huge factory, 6000 workers and mounting debts. To power their way out of these financial problems, they did what they knew best - manufacture cars. Within three hours of announcing his return to auto manufacturing, Henry Leland had raised $6.5 million worth of stock.

Naming the new company 'Lincoln', for the president Leland had voted for, might have indicated the era Henry Leland was most comfortable in. The auto market had moved on by this time, and the first Lincolns were very old-fashioned looking. Precision engineering was his watchword with the new cars and due to Leland's attention to detail, the cars were also late in getting to the market. Two demerit points already! After 17 months of trading, only 3400 cars had been sold against projections of 6000 per year. A panic-stricken board of directors called in the receivers.

Henry Ford was immediately attracted to Lincoln for three interesting reasons. Firstly, the company would give him instant access to the upper end of the market with a good basic product line that could be quickly overhauled. Secondly, he was looking for a task to occupy his son Edsel. Edsel had a strong instinct for style and taste, which was unsatisfied by his work on the Model T. Thirdly, was Ford's long-standing enmity for Henry Leland .

Cadillac had sprung from the ashes of one of Henry Ford's first companies. It was also Leland who had persuaded the shareholders to desert Ford in his favor. Buying Lincoln out of trouble cost Ford $8 million, but he probably enjoyed spending the money. A picture survives of the signing of the agreement. White-bearded Leland and his son Wilfred both look thoroughly dejected.

Edsel Ford stares confidently at the camera and even Henry wears a tight-lipped smile. Appropriately, a portrait of President Lincoln looks down on the scene.

Below: The 60 degree V8 L-Head Lincoln unit remained in production until 1930.

Above By 1929 Lincolns had become handsome and tasteful, thanks to Edsel Ford. He chose the greyhound hood mascot personally.

Left: The body plate of the Lincoln clearly proclaims Ford's ownership of the company.

1929 Plymouth Model U

Engine: Four-cylinder Inline, Valve in block	
Displacement: 175.4 cid	
Horsepower: 45 at 2800 rpm	
Transmission: Three-speed sliding gears	
Fuel System: Carter 103 S Carburetor	
Body Style: Two-door Roadster	
Number of Seats: 2/4	
Weight: 2285 lbs	
Wheelbase: 109.75 inches	
Base Price: $675	
Number Produced: 108,345	

Plymouth was named for the Plymouth Rock, an enduring symbol of New World spirit, and associated with the Pilgrim qualities of integrity, simplicity and honesty. The name exemplified the ambitions of the brand. Walter P. Chrysler designed the Model U as a no-nonsense entry into the budget end of the car market. It was launched to go head-to-head with Ford and Chevrolet. The model's first year sales were respectable at 50,000, but this hardly swamped the opposition. The second year was better and the 100,000-unit mark was crossed. Times were tough in the ensuing years, with the Wall Street Crash and the economy-minded Plymouth came into its own, entirely justifying Chrysler's strategy.

The Model U was the first real Plymouth, as the earlier cars bore the Chrysler prefix. But both models are virtually interchangeable. The 'U' offered no breathtaking surprises but was well appointed for a car in this class. Power was provided by a four-cylinder unit, equipped with aluminum pistons and full-pressure lubrication. The 175.4 cubic inch displacement generated a steady 45 horsepower. Four-wheel hydraulic brakes were fitted as standard,

and the car had an independent handbrake. The competition didn't offer these specifications for a few years. The model prices were also competitive, at between six- and seven-hundred dollars. The car came with fitted wood spoke wheels as standard but wires were an option, with two-piece rounded bumpers and Twolite headlamps. Seven body options were offered, including a roadster and coupe with rumble seats, two- and four-door sedans, plus a touring car.

Above: The Two-door Roadster with Rumbleseat in two-tone brown and yellow on a day at the beach.

Opposite page: A four-door sedan version of the Plymouth Model U.

1930 Duesenberg Model J

Engine: Eight cylinders
Displacement: 420-cid
Horsepower: 265 at 4250 rpm
Transmission: Selective manual, floor mounted lever
Compression Ratio: 5.25:1
Fuel System: Schebler carburetor
Weight: 5200 lbs
Wheelbase: 142.5 inches

Was the Model J the finest American car ever built? E. L. Cord described it as "the world's finest". This was a typically modest remark from the showman empire builder. The Duesenberg Company was formed by brothers Fred and August Duesenburg in 1913. They quickly achieved a reputation for brilliantly engineered race engines. In 1921, their Model A arrived – a good, solid car but no style-setter. The lack of business success resulted in Duesenberg Motors being sold to the charismatic Cord who immediately put the company to work. In 1928, the stupendous Model J emerged, to countrywide acclaim. The Model J came with a 420-cid straight eight designed by Fred Duesenberg, and built by Lycoming: it was an engine capable of 265 horsepower – truly exceptional for the time.

But it wasn't only their sublime power, which gave these cars their reputation. The specifications both interior and exterior were something else and aimed at only the rich. The engine had overhead camshafts driven by chains to four valves per cylinder. Engine fittings were finished in exotic, shiny nickel, stainless steel or chrome. Aluminum was used throughout to keep the weight down – even in the gas tank. Models J's were massive with a standard wheelbase of 142.5 inches but they weighed in at only just over 5,000 pounds. In high gear, they could reach 116 mph and could achieve almost 90 mph in second. For the comfort of the driver, the cars were equipped with a range of instruments to satisfy any enthusiast: a speedometer calibrated to 150 mph, ammeter, water and oil pressure gauges, split second stopwatch, tachometer, barometer and break pressure gauge. More detail was found in the car's warning systems

Below: The Model J engine was a straight eight unit built by aero engine manufacturer Lycoming to Fred Duesenberg's design.

designed by Fred – reminders to add or change oil, or top up battery water. And the cost of this luxury and attention to detail? Well, you didn't buy a finished car, but only the chassis, which cost a stupendous $9,500 in 1930. The bodywork was added later. This all added up to a finished price of around $17,000. A few cars were as expensive as $25,000, crazy money at the time. Between 1929 and 1936, only 470 chassis and 480 engines were built. This exclusivity was another selling point. The Model J was a rare and fabulously opulent car.

Above: A king among cars: the mind-blowingly expensive 1930 Duesenberg Model J.

1930 REO Flying Cloud Model 25 sport sedan

Engine: six cylinder
Displacement: 268.3 cid
Horsepower: 85 at 3200 rpm
Body Style: Sports Sedan
Number of seats: 5
Wheelbase: 125 inches
Base Price: $1845

Mr. Olds left his eponymous company in 1904, after a major fall-out with the controlling Smith family, who wanted to phase out the beloved Curved Dash and bring in some more up-market models. They were furious when he tried to use his name as a moniker for a rival company, and it was this argument that resulted in Old's new company being named REO, after this initials, Ransom Eli Olds. Ultimately, REO was a victim of the Depression, but not before the company produced the Flying Cloud Sixes. The 1930 versions were little changed from the previous year - the Series 25 was the Flying Cloud of 1929. It had a six-cylinder 268.3 cid engine capable of producing 80 horsepower at 3200 rpm and this was fed fuel from a 21-gallon tank. The wheelbase was 125 inches and the car's braking system was the Lockheed hydraulic brake system. In one respect the REO 25 stood out, it had outstanding styling and a dynamic, desirable profile. The body design was in the hands of the talented Amos Northup. He turned out a style that was both formal and as a result, classic in line and proportion.

The REO cars were beautifully finished. The workman-ship on the car was solid and all the furnishings and appearance detailing were of the highest quality. The Model 25 was the upscale version of the three models and drove quietly and smoothly not least due to the silent second transmission. The junior REO car for the year was the 115-inch wheelbase model 15, complete with a 65 horsepower, 214.7 cid continental engine sitting on a shorter 115-inch wheelbase. Then came the first of the senior models, the model 20 and the model 25. The catalogue for the year lists seven body types ranging in price from $1,125 for a small series 15 Coupe, to $1,845 for the top of the range model 25, a five-passenger sedan.

The most fabulous REO of all arrived in 1931: the Royale, a 125 horsepower straight eight with brilliant coachwork by Northup. In 1936, REO announced it would cease building cars, Olds had given up control two years earlier.

He died in 1950.

Below: The Reo two-door coupe version of the Flying Cloud.

Above: The fender mounted, rather than on the cowl, parking lights were a feature of the 1930 REO Model 25s - a detail from Amos Northup's design.

1931 Buick Model 54

Buick's Marquette, a basement priced car rose with the dawn of 1930... and rapidly disappeared from sight – gone after a fruitless year. Not the greatest start to a decade. By 1931, most autos were powered by straight-eight engines. The company was hard hit by the Depression, but owners General Motors employed Harlow H. Curtis to bring Buick back from the edge and five years later, he had built up a strong line. The decision to go for the eight came just before the Crash. The mid-range series 50's were built on a 124-inch wheelbase, and used a 331.3-cid engine, which developed 99 horsepower. The 50 series offered just a four-door sedan and a four-place sports coupe.

The line up in 1931 was expanded and saw the introduction of the first eights, which were possibly the most advanced power unit of the day. They were smooth and reliable engines designed by division chief engineer F. A. Bower. The 50 included a 77 horsepower, 220.7 cid power plant, and was now the cheapest model in the range. The top of the line offerings, Series 80 and 90 had a lengthy roster of models that including roadsters and a seven seat 90 limousine. The 50 line included a second series group announced early in the year, which rode on a 114-inch wheelbase taken from the deceased Marquette. The straight eight remained Buick's mainstay power plant for the following 22 years.

The new 1931 engine proved itself at the Indianapolis 500 that year, powering racer Phil Shafer to qualify at a speed of 105.1 mph averaging 86.4mph during the race. All the showroom cars of '31 were quick for their day: 0-60 mph took about 25 seconds and a top speed of 90mph was possible. The canny Scot David Dunbar Buick had evolved a successful series of cars that put him back into the auto business. Buick had come a long way from the his origins in bathtub enameling. Even his first car of 1903 had had the well-known Buick feature of overhead valves, then a rarity, but still uncommon even in the 1930s.

Engine: Eight-cylinder

Displacement: 220.7-cid

Horsepower: 77

Body Style: Two-door Sedan

Number of seats: 4

Weight: 2935 lbs

Wheelbase: 114 inches

Base Price: $1055

Number produced: 907

1931 Cadillac Fleetwood Phaeton

Even in the thirties, Cadillac kept its blue chip image and continued to build outstanding luxury cars. The surprise for 1930 was the introduction of the sixteen-cylinder, a magnificent brute of a power plant. It was only nine months old when Cadillac introduced another multi-cylinder engine, a 368-cid V-12. Essentially, this was a V-16 with four fewer cylinders generating 135 horsepower, used to power slightly smaller cars such as the Fleetwood phaeton. The V-12 was fitted onto the Cadillac eight's 140 inch wheelbase. The twelve's engine was free revving, and delivered smooth and even power. Most twelve's could cruise all day at 70 mph. However, in the market the cars fell short of being best sellers. They were staggeringly expensive, although beautiful to look at. At this time, cars with more than eight cylinders seemed like too much of an extravagance.

Cadillac designed and built most of its bodies in house through either Fisher or Fleetwood. These were two highly respected coachbuilders that GM had acquired. Classic Cadillac styling was a big feature of all the 1930-31 models with the distinctive and grand thick collar, vertical radiators, stunning rounded hoods and crisp but flowing lines.

There were fourteen Fleetwood custom models of this period with the most elegant and breathtaking being the Madam X models (fitted with the V-16). These cars were named after a stage play of the time. They had sleek chrome windshield and door mouldings. The design of both the V-12 and V-16 engine was credited to Owen Nacker. Cadillac's early '30's V-8 was based on a 341-cid unit, which was first introduced in 1928. It was resized to 353-cid for 1930-35 and delivered between 95-130 horsepower. It continued to be used up to 1949 in one form or another. It was the best selling line for the fortunes of Cadillac though the great V-16s and V-12s remained low volume after an initial sales spurt.

Engine: Eight cylinders

Displacement: 353-cid

Horsepower: 95 at 3000 rpm

Transmission: Selective, synchromesh

Compression Ratio: 5.35:1

Body style: Four-door Fleetwood Phaeton

Weight: 4395 lbs

Wheelbase: 152 inches

Base Price: $2945

1931 Oldsmobile F-31

Engine: Six-cylinder	
Displacement: 197.5-cid	
Horsepower: 65 at 3350 rpm	
Transmission: Synchromesh	
Compression Ratio: 5.06:1	
Fuel System: mechanical Stomberg downdraft carburetor	
Body Style: Two-door Roadster	
Number of seats: 2	
Weight: 2965 lbs	
Wheelbase: 112 inches	
Base Price: $995	
Number produced: 3,500	

Oldsmobile, who was America's leading automaker by 1903-4, hadn't lost their desire to meet a challenge. During the Depression year of 1931, they introduced the pioneering synchromesh transmission. Many of Oldsmobile's technical advances in the '30's were down to Charles McCuen, who was first the chief engineer and then general manager of the company. The Olds F-31 followed the general styling fashion of the time, with a typically squared shape. The '30 and '31 lines offered popular styles in the Standard, Special and Deluxe trim and all sixes came on a 113.5-inch wheelbase. The cars were all powered by the tried and true six-cylinder power plant with 197.5-cid producing 65 horsepower at 3350 rpm. Both Standard and Deluxe models were offered in a variety of open and closed body styles, and the Deluxes came with front and rear bumpers, twin side mounts and a rear platform designed to carry a trunk.

Prices started at around $900 and rose to $1000. Oldsmobile suffered less from the economic effects of the Depression than many carmakers. This relative success was attributed to conservative and recognizable styling cues, which the customers could identify with and even more importantly, trust. The performance was adequate, and reliability was just as good as any other make. Perhaps it is no accident that this solid profile for the manufacturer remains intact, as it is now America's oldest car nameplate. The six-cylinder engine served the cars well, and it was only in 1932 that the first straight eight arrived while in the same period the six grew to 213.3-cid, developing 74 horsepower for 1932.

Opposite page: A 1931 Oldsmobile F-31 two-door Roadster, with bright-eyed headlamps. Over 3,000 were built, selling for just below $1000.

Right: The Six-cylinder L-head power plant produced 65 horsepower with its new designed porting and downdraft Stromberg carburetor.

1931 Stutz SV 16 Model MA-23

Engine: Eight-cylinder
Displacement: 322-cid
Horsepower: 113 at 3300 rpm
Wheelbase: 134-1/2 inches
Base Price: $3495

Top right: A stunningly elegant, customized 1933 two-door Stutz SV-16 Model MA-23.

Bottom right: An immaculately kept Convertible Coupe by Derham.

Below: The Stutz straight-eight employed twin overhead camshafts and two valves per cylinder unlike his earlier multiple valve options hence 16 SV. Despite being technologically advanced for the time the engines were complicated and expensive to produce.

Harry C. Stutz left the company in 1926, he was replaced by European born and influenced Frederick E. Moskovics. The new owner had introduced the 'Safety Stutz' models in the same year. These cars displayed beautiful styling. The cars were powered by Stutz's first eight-cylinder engine, an inline configuration with a single overhead camshaft and dual ignition with two plugs on every cylinder. It was known as the Vertical Eight. Moskovics resigned from Stutz in 1929. From this time, it was Charles 'Pop' Greuter who engineered power plant development for the company. Stutz couldn't afford to invest or build a 12- or 16-cylinder engine they needed for 1931 so they experimented with a supercharger that lifted the Vertical Eight to 143 horsepower. It did the job, but was ungainly and had fuel feed problems. So, after looking around the industry, Stutz decided to follow the Duesenberg solution by introducing a 32-valve twin-cam cylinder head. This engine abandoned the dual ignition system, and was named the SV-16 (single value). The company also produced a DV-32 (dual valve) assembly, which could produce 161 horsepower at 3900 rpm. Stutz attempted to enter the low price car market, but did not have the product so they were left trying to market cars that were high-priced and uncompromising with a sports-like finish.

This made Stutz an oddity in the marketplace and an expensive option. Standard body models sold for around the $3,500 level. There were at least thirty custom styles available on the SV-16 chassis, all from high-end coachbuilders such as LeBaron, Rollston, and Waterhouse. One customizer was Waterhouse who offered fabric bodies, which were strong but light, safe and durable. Compared to the steel shells, the fabric bodies absorbed the road noise well, lasted longer and were easier to repair.

The MA series used a 322-cid, eight-cylinder power plant, which developed 113 horsepower at 3300 rpm. The wheelbase was 135 inches and the engine was fueled from a twenty-gallon tank. Sales for the year and the series were both very low. By 1935, the company had discontinued car production.

STUTZ
1931

1932 Buick 96C Convertible

Engine: Straight eight-cylinder	
Displacement: 344.8 cid	
Horsepower: 104 at 2800 rpm	
Transmission: Sliding gear synchromesh	
Compression Ratio: 4.5:1	
Fuel System: 2-barrel Marvel updraft carburetor	
Body Style: Convertible Coupe	
Weight: 4460 lbs	
Wheelbase: 126 inches	
Base Price: $1805	
Number produced: 289	

The company had big news for 1932. They introduced 'Silent Second Synchromesh' transmission and included more horsepower for all their engines. But sales were not so clever, and by 1931 they had a new president in Harlow H. Curtice. He demanded 'more speed for less money' - not a bad start for a consumer driven business. He backed up his dictum with an all-new 117-inch wheelbase series 40 for 1934. The result was an upturn in sales and a modern streamlined styling that broke with the now dated square designs of the '20s that beleaguered the early 30's models. Also featured in the line up was GM's new knee action independent front suspension, which was then a great step forward. Curtice also launched a $64 million factory modernization program.

The 1932 cars came in four series: 50, 60, 80 and 90. The 90 series was Buick's largest and had nine models and horsepower was rated at 104 mph at 2900 rpm. The 96C convertible coupe was powered by a straight eight, 344.8-cid engine. Innovations in 1932 included adjustable shock absorbers, 'wizard control' freewheeling and an automatic clutch. By then end of the year they had discontinued the use of wooden spoke wheels on the 50 series. But by 1932

Right: The Buick is especially imposing head-on, the twin spare wheels mounting for the rear view mirrors, the notched radiator grill , and a respectable array of driving headlights.

wooden spokes were not the only things being dumped. The economy was also dumping on the people and with almost everything at a standstill, Buick's sales slumped by more than 50 per cent with only 40,000 plus units built for their 2,000 odd outlets. This made them the third largest auto dealership of GM but there were very few buyers around, and sales slipped to seventh spot nationally. But there was always the glamorous and the imposing sight of a stunning 96 convertible to raise the spirits and drive the company on. It was an outstandingly comfortable and powerful car to drive, with all the qualities expected from a top of the range model.

Above: Graceful and imposing, this convertible 2 door Buick 96C has a 134 inch wheelbase length, a top speed in excess of 100 mph, brakes on four wheels, painted wire-wheels and 18 x 7.00 tires.

1932 Chrysler Imperial Speedster

Engine: Eight-cylinders inline	
Displacement: 384.84-cid	
Horsepower: 160 at 3500rpm	
Transmission: Three-speed manual	
Compression Ratio: 6.2:1	
Fuel System: Stromberg Model DD 3	
Body style: Two-seat Sports Convertible	
Wheelbase: 135 inches	
Number Produced: 1	

Right: The beautiful 1932 Speedster remained in the ownership of Walter P Chrysler Jr. for thirty years.

Below: The straight-eight engine used in the Speedster was a stock Chrysler Imperial part but utilized an high compression alloy head to boost power from 125 to 160 horsepower.

Walter P. Chrysler was not a man to have anything get in his way. Having jumped into the luxury car market in 1925 with the Chrysler Imperial E80 he wasn't going to let a little thing like the Depression of the early thirties stop him from going ahead with his plan to develop luxury cars.

Chrysler's stability was underpinned by solid sales of yeoman models like Plymouth and Dodge, which allowed Chrysler to continue development when other manufacturers were retrenching. Walter Chrysler established the experimental Custom Body Shop in 1931 with engineer Carl Breer at its head. This was a move in keeping with the company's dedication to research and development. Even so, some of the department's projects seemed somewhat self-indulgent. A special car was created for the boss's son, Walter P. Jr. The design for the car was credited to Herbert Weissinger, head of the Chrysler Art department. Although it has been mistaken for a Le Baron of the same vintage, it is in fact a totally redesigned one-off model, using stock Chrysler parts of the day. Its long low sweeping lines, accentuated by the long uncluttered hood stretching from radiator shell to windshield, gave the car the sleek and powerful appearance of its twelve- and sixteen- cylinder contemporaries, such the Cadillac V-16. A cut-down, swept-back windshield added to the illusion. The body panels were crafted in aluminum to keep the weight down.

Beneath the skin lurks the Chrysler Straight-Eight of 384.84-cid, fitted with a cast-alloy, high-compression head, which boosted the power from a standard 125 to 160 horsepower. Bringing this powerful cast-iron monster to life was a gas pedal actuated starter with an automatic stall restart system exclusive to Chrysler. The whole rig sat on a reinforced double-drop boxed frame for rigidity. With its collection of experimental gadgetry, combined with tried and tested mechanicals, the car is every bit as impressive on the road as it looks parked.

The result is probably one of the most beautiful cars of all time.

1932 Packard Dietrich Body

Engine: V-12, modified L-head

Displacement: 445.5 cid

Horsepower: 160 at 3200 rpm

Compression Ratio: 6.0:1

Fuel System: Stromberg-Duplex carburetor

Wheelbase: 142 1/8 inches

The co-operation of a luxury producer like Packard with a coachbuilder of the stature of Dietrich was always going to result in the creation of a very special car. Leading the Packard line-up for 1932 was the new twin six, which later became the twelve. The car was equipped with a new V-12 engine that had been specially developed for a front wheel chassis model that had never actually materialized. This new 445.5-cid engine purred. It was designed to produce a smooth and sustained pace rather than high-speed motoring. The Packard factory boys claimed that the car could cruise at 100 mph, but that was in test conditions.

A more normal driving experience was for the car to run beautifully at about 90 mph, if it was pushed much harder than that, it ran out of breath. The optimum speed for comfort was between 60-70 mph, when the car almost seemed to glide along. The 1932 cars were designated the ninth Packard series. Although these cars were more streamlined in their styling and had a lower profile, they were essentially similar to the previous year's models. The 1932-34 V-12s shared the same wheelbases and most of their body features with the Eighth series, now designated 'Deluxe' for 1932. The 160 horsepower engine produced a level of performance similar to that of KB, another twelve-cylinder car. Lincoln was Packard's head-to-head competitor throughout the 1930s.

Packard had not planned to build many of these cars, but the Packard twelve became a classic model, and was built in various, ever more impressive, configurations from 1932 right through to 1939. Sales of the twelves began to really take off in 1934 when almost a thousand cars were built. This was double the production total for the previous two years. By 1935 the engine was stoked up to 473 cid, and developed 175 horsepower. The car became more and more powerful and spectacular, both to drive and be seen in. Among its many distinguishing features was the mascot that perched on the hood. Dual rows of hood louvers aided engine cooling and were a styling addition. Together with these features, the bullet headlights and further streamlining characterized the Packard style of this vintage.

Right: A 1932 Packard Twelve with a custom Dietrich body found fewer than 500 buyers during the Depression years.

Below: The 1932 Packard featured this intricate Flying Lady hood mascot. Many other marques of this vintage had similar decorations.

STANDARD
OIL
PRODUCTS
SERVICE
JONESY'S
STANDARD OIL PRODUCTS
STANDARD OIL PRODUCTS
REGULAR
22 9/10
ETHYL
27 9/10
'32 PACKARD

1932 Pierce-Arrow Model 54

Engine:	straight eight
Displacement:	366-cid
Horsepower:	125 at 3000 rpm
Transmission:	Three-speed
Wheelbase:	137 inches
Base Price:	$2,385

Right: The distinctive radiator of the 1932 Pierce-Arrow Model 54.

Below: The Model 54 used the more mundane 366-cid straight-eight unit, giving 125 horsepower. The engine received an eight point mounting in 1932 for extra smoothness and rigidity.

The Pierce-Arrow Company always had a sense of drama associated with its cars – epitomized by the single, radiant red sports phaeton trimmed in stainless steel that was shown and stood out even against the pomp and ceremony of the annual Princes' Convention in New Delhi, India, in March 1931. A typical act of flamboyance and panache that was the spirit of Pierce-Arrow came in the 700 shades of beige offered on the 1930's car. The 1932 cars were significant. First, there were two lines of twelve-cylinder cars, Pierce decided to cut back on the eights.

The flagship of the fleet was the model 52, offering five body types on 142 and 147-inch wheelbases. Both cars were powered by the new 150 horsepower 429-cid V-12, designed by the chief engineer, Karl M. Wise. It was a super smooth engine with an 80-degree cylinder bank angle and dual downdraft carburetors. A smaller 140 horsepower engine was used in the model 53. Anchoring the line was the model 54. Effectively, this was Model 53 with an improved 366-cid straight-eight power plant generating 125 horsepower. Common to all '32 models was solid, eight-point engine mounting, stronger frames and adjustable shocks for a fantastic ride.

The first common model produced with Studebaker also dates from this time. Studebaker now owned Pierce-Arrow and had successfully managed to expand the market for the company's product, despite the Depression. The car they produced jointly was a two-door sedan sold as the Club Brougham. The car incorporated the first recognizable exterior design changes for three years. Looking at the car's profile you can see that the radiator shell was now V-shaped and the side windows were more curved. As three-dimensional body moldings were now used in manufacturing, a more pronounced curve was put into the rear fenders. When all of these changes were put together, the result was the transition to the aerodynamic styling that would be at its zenith in the 1934 cars.

1932 Studebaker Dictator Eight

The Dictator line-up for 1932 was the same as in the previous year. The president of the Studebaker was now the company's former accountant A. R. Erskine, who brought with him all the imagination that accountant's tend to have when faced with a product requiring vision and style. None, in other words. Erskine led the company from one production disaster to another. The first car he launched was named Erskine. Unsurprisingly, it failed.

Then came the exquisite and much-sort-after name-choice of the 1930s, the Dictator. Erskine oversaw the design and production of the model's eight-cylinder engine. Studebaker engineers Max Wollering and Guy Henry objected to this strategy, and they were, therefore, bypassed. Erskine brought in Delmar G. 'Barney' Roos who was happy to follow orders. The result of their combined endeavors was the Studebaker straight-eight power plant, capable of delivering 100 horsepower. The cars were styled by Ray Dietrich.

In 1928, Erskine sent the new cars to the races, where they cleaned up and caught the eye of the public. The 1932 Dictator eight-cylinder engine had 230-cid that produced 81 horsepower mounted on a 117-inch wheelbase. Under the hood Delco Remy ignition was fitted, as were mechanical Bendix brakes, a Synchomesh transmission and a vacuum advance distributor. Fuel was fed to the engine by a Stromberg carburetor. Exterior styling included a slanted windshield, oval headlights and one piece fenders.

The big brother of the Dictator was the President (a more inspired choice of name) eight-cylinder model, which took on the 1930 Indianapolis 500 course in two private specials. These cars went well enough for Studebaker to enter a factory team in the 1932 Indy – a rare event for an American manufacturer. The Studebaker was driven by Cliff Bergere, and finished third.

Engine: straight eight

Displacement: 221 cid

Horsepower: 81 at 3200 rpm

Compression Ratio: 5.0:1

Fuel System: Stromberg carburetor

Wheelbase: 117 inches

1933 Ford Deluxe Model 40-720 Coupe

Ford's first V-8 was introduced in 1932, the famous cast-iron flathead first delivered 65 horsepower from 221-cid. The new V-8 was a great bargain and had a top speed of 78 mph from 75 horsepower and caused a storm of interest. Millions were drawn to its March unveiling but it was launched too quickly for the engineers and without an ideal engine-testing program because of the need to get the V-8 into the market as quickly as possible. There were durability problems, such as cylinder head cracks and a rapid oil burn off. In some cars, the engine mounts worked loose and ignition problems came to light. Ford ended up having to replace thousands of these, and the engine problems were reflected in lower sales for the year. But Ford didn't allow themselves to be deflected by these teething difficulties. They were soon resolved, and the V-8 gained a reputation as a solid power unit that could that could withstand a considerable assault from being driven hard and fast. Everything was sweet by 1933. The generic Detroit style now consisted of streamlined models. Edsel Ford was now the force behind the company's design, and came up with a balanced new look for the 1933 Ford line-up that was very well received. The hood extended back to the windshield and the sharp corner on the fender was rounded off. The V-8 durability kept improving and the line was now named the model 40. The speedy new V-8 was attracting many new enthusiasts, amongst them Public Enemy Number One, John Dillinger. Dillinger actually had the cheek to write to Henry Ford, praising the new engine.

The '33 wheelbase was 112 inches, now increased from the short 106.5-inch version. All models were delivered with black fenders and 17-inch wire spoke wheels. With improved reliability and aluminum cylinder heads, the engine now delivered a performance comparable with the detailing of the interior, which featured a new dash arrangement with the gauges arranged directly in front of the driver. Ford was finally getting in tune with its customers.

Engine: eight-cylinder

Displacement: 221 cid

Horsepower: 75

Transmission: Sliding gear

Compression Ratio: 6.3:1

Fuel System: Mechanical, Detroit Lubricator Downdraft, single-barrel carburetor

Weight: 2538 lbs

Wheelbase: 112 inches

Base Price: $540

Number produced: 15,894

1933 Lincoln KB Convertible Coupe

Engine: V-12

Displacement: 447.9 cid

Horsepower: 150 at 3400 rpm

Transmission: Sliding gear

Compression Ratio: 5.25:1

Fuel System: Stromberg DD downdraft carburetor

Number of seats: 2

Weight: (estimated) 5500 lbs

Wheelbase: 145 inches

Base Price: $4,200

Right: The elegant, head-turning, twelve-cylinder 1933 Lincoln KB convertible coupe with custom bodywork.

Lincoln was founded by Henry Martyn Leland, the man who gave America the Cadillac at the end of the Great War. But by 1922 he had to sell up, to Henry Ford. Henry and his son Esdel brought out the Model K in 1931. It had a new 145-inch wheelbase chassis with an upgraded 120 horsepower V-8. The colossal chassis had side rails that were nine inches deep, and six cross members with cruciform bracing.

The K also had a floating rear axle, torque tube drive, Houdaille hydraulic shocks and Bendix mechanical brakes. It accelerated well and had a good top speed. The K chassis was designed for the new V-12 that arrived in 1932 for the KB series. This new engine was a 448-cid with 150 horsepower. The KB was a machine that stood apart from its rivals, both as an efficient town car and as a fast open road tourer. On the outside of the car, Esdel had been to work. He transformed the Lincoln, updating the standard factory built bodies and securing many custom or semi-custom styles from the very best of America's coach builders - Brunn, Dietrich, LeBaron and Willoughby. Between them, they styled, designed and produced some of the best examples of shapes from this classic era. A cautious approach to streamlining had begun in 1932, but by 1933 the KB had a distinguished radiator with a chrome grille, hood louvers, skirted fenders, vee'd front fender and new truck racks.

Sales for the KB, as for so many models of the period, were difficult because of the effects of the Depression. Lincoln consolidated power in '34 by boring out the old 414-cid KA engine to produce the same 150 horsepower as the KB. They also introduced aluminum cylinder heads, which were quite revolutionary for the time. This was made possible by the arrival of 70-Octane fuel, which was almost as refined as aviation fuel. The top speed for the Lincoln was 95 mph.

Right: The meaty V12 was at the heart of the Lincoln's 95mph performance.

1933 Plymouth PD

Plymouth was a success from the outset. The company prospered throughout the 1930's with a strong marketing combination of low prices, eye-pleasing styling, and engineering that was often more advanced than that of its competitors, such as Chevrolet. By 1932, Plymouth was American's number three auto manufacturer. 1933 was one of the roughest industry years on record, but Plymouth came through unscathed. This was no accident, the cars had features customers wanted, such as all-steel construction, at a time when wooden framed bodies were still in production, four wheel hydraulic brakes and rubber engine mounts that gave 'the economy of a four and the smoothness of an eight'.

The 1933 PD was launched in the second period of the year as a two series line of standard and deluxe Sixes on wheelbases of 108 and 112 inches. A convertible coupe came in the Deluxe line and the cars began to follow the industry trend by moving towards a streamlined profile, with skirted fenders and a rounder hood assembly. On top of the hood came the new winged-lady mascot.

The PD grew out of the not-so-popular PC launched earlier in the year. Plymouth engineers took the Dodge Model DP frame, modified the wheelbase, added a redesigned front end and then manipulated the hood panels, front fenders, running boards and splash aprons. In this way, they ended up with a new car without too much retooling or expense.

The PD on its longer wheelbase corrected all the perceived faults of the PC. Some of the design details included: a painted or chrome grille with the hood stretched back over the cowling to the windshield. Bullet shaped, chrome plated headlamps helped the look and wire wheels gave the car a styling image. Many were fitted with trumpet horns to give the car's profile a lift. Not surprisingly, sales rocketed, not least because of the lower $495 price.

Engine: Six-cylinder

Displacement: 189.8-cid

Horsepower: 70 at 3600 rpm

Transmission: Sliding gear

Compression Ratio: 5.5:1

Fuel System: Carter C6A-C6A2-C6A3 or C6A4

Weight: 2545lbs – 2645lbs depending on body type

Wheelbase: 112 inches

Base Price: $525

Number produced: 49,826

1933 Pontiac Eight

Pontiac veered towards collapse in 1932 but was saved by the effective leadership of GM President Alfred P. Sloane. He integrated Pontiac's manufacturing with that of Chevrolet early in 1933, which resulted in shared bodies, chassis and other main components. He achieved economy of scale. The resurgence of the company was overseen by Pontiac general manager William S. 'Big Bill' Knudsen, chief engineer Benjamin H. Anibal and chief designer Franklin Q. Hershey. Anibal became known as the father of the Pontiac Eight, whilst Hershey was responsible for its unique styling. The engine for this year was a 223.4-cid inline-eight, generating 77 horsepower. The car rode on a 115-inch wheelbase. Its body styling was in a transitional phase - moving from the boxy shape of the early 30's towards the streamlining that followed in the mid 30's. In fact, the new Pontiacs were amongst the most attractive looking cars of the year.

The Economy Eights looked unique due to the design skills of Hershey and chief body engineer Roy Milner. Both were strong-willed men who were looking to make a lasting mark. Hershey collaborated with famous styling director Harley Earl to achieve more streamlined and showy Pontiacs. They subsequently designed a distinctive new radiator and added skirted, front fenders with horizontal go-faster speed streaks. Milner also gave the open models a smooth deck. The car was equipped with a Fisher body with beavertail rear, plus a slanting V-shaped radiator with vertical bars. The hood had four wide louvers slanting backwards towards the cowl. Inside the car, airplane style instrumentation appeared on the dashboard.

This excellent style and profile package, combined with the powerful straight-eight and knee action independent front suspension gave Pontiac a winner that was to pull the company through the effects of the Depression and secure its future.

Engine: Inline eight-cylinder
Displacement: 223.4 cid
Horsepower: 77
Transmission: Muncie synchromesh
Compression Ratio: 5.7:1
Fuel System: Carter IV carburetor
Body Style: Four-door Sedan
Number of seats: 5
Weight: 3020 lbs
Wheelbase: 115 inches
Base Price: $695
Number produced: 90,198

1934 Nash Ambassador

Engine: Eight-cylinder
Displacement: 322 cid
Horsepower: 125 horsepower at 3600 rpm
Transmission: Selective sliding gear
Fuel System: 5.25:1
Wheelbase: 133 inches

Automaker Charles Nash believed in offering a lot of car for your money. His Nash cars were

full of innovations and extras. For example he produced aircraft-type instruments for his customers in 1934. But, like most auto companies, Nash was hit hard by the Depression and in the previous year had recorded his lowest ever output, with less than 15,000 units produced. The company needed a new approach, look and improved performance. However, difficult economic conditions delayed the planned 1934, which was postponed for a year. Despite their problems, Nash produced its millionth car in 1934, but also recorded losses of over $1.5 million...

The Ambassador model continued with a revised twin-ignition eight-cylinder engine delivered between a 100 and 115 horsepower. By 1934, these power plants had evolved into two engines of 260.8 and 322 cid delivering either 100 or 120 horsepower. The styling of the model remained classically upright even though the industry was turning towards rounder and more streamlined shapes by this time. Some other innovations that Nash introduced in his cars

Below: A fairly sparse interior for a luxury car.

during the early '30s included sports cowl ventilation, a dashboard starter button, shatterproof glass, downdraft carburetors, and a Syncro-Safety Shift. The latter featured a dash-mounted gear lever in 1931. Freewheeling was introduced, followed by an ignition and steering wheel lock. The Ambassador body was designed by Count Alexis de Sakhnoffsky and featured a high styled body, deep-skirted fenders and chrome, bullet shaped headlamps with a full beavertail back end. The model ran on the larger Nash engine, which produced 125 horsepower at 3600 rpm. It had mechanical brakes on all four wheels and Baker Axleflex independent front suspension. The wheel size was 17- inches and the wheelbase was either 133 or a massive 142 inches.

Above: In a troubled year, the 1934 Nash Ambassador held its own in terms of stature and comfort even though the styling was looking its age as most cars entered the slipstream age.

1934 Packard Model 1104 Super Eight Phaeton

Engine: Straight-eight	
Displacement: 384.8 cid	
Horsepower: 145 at 3200	
Transmission: Selective synchromesh	
Compression Ratio: 6:1	
Fuel System: Stromberg carburetor	
Weight: 4645 lbs	
Wheelbase: 142 inches	
Base Price: $3090	

The 'Super' designation had grown out of the previous Deluxe model for 1932. The Super Eight Series offered eleven body styles ranging from a five-passenger sedan through to the top of the line limousine, with seating for seven passengers to travel in comfort. The Phaeton was built on the 142-inch wheelbase, and had a straight-eight power plant putting out 45 horsepower. There were three forward gears and a single reverse. The car used mechanical brakes on all four wire-wheels. Amongst the options offered on the Phaeton were a radio, clock, cigar lighter, spotlights and mirrors. During the year, 2,000 cars were built across the range. The opening price for the range was $8,700 for the model 1104 range. Models 1103 and 1105 were also available for 1934. Wheelbases for the Super Eight series ranged from 135 to 147 inches. Throughout this period, custom coachbuilding was an important part of the Packard package. Although the car had an upright appearance, the custom cars were built on the longer wheelbase chassis.

Packard had trouble selling its grand statesman-like cars in the Depression years, so had tried to develop a medium-priced car – and to get it to the marketplace ahead of Cadillac and Lincoln. The company launched its Light Eight in 1932. It was a quality product built with the same meticulous care as the larger models. True to its name, it was smaller and lighter than the senior models, but looked chunkier because of its short wheelbase. Unfortunately, the car cost as much to build as the standard Packard, and the model quickly developed into more of a liability than an asset. It was dropped after a single production year. Packard buyers wanted substance and weight in the cars they bought.

1935 Auburn 851 Salon Coupe

Engine: Eight-cylinder	
Body style: Coupe	
Weight: 3,105 lbs	
Wheelbase: 120 inches	
Base Price: $1221	

Auburn's offer was a car with value and amazing style. They was one of the few automakers that produced higher car sales immediately after the 1929 Wall Street Crash . They had beauty and power, and great prices matched to outstanding performances from the eight- and twelve-cylinder engine configurations. But this success did not last, and sales plunged in 1932.

The 1933 model eights and twelves continued with little change into 1934. Streamline styling was then added by Alan Leamy of Cord L-29 fame. But little changed – nobody wanted the twelve. After Auburn's owner, the charismatic E. L. Cord, fled to England in 1934, the marque was handed over to Duesenburg president Harold T. Ames. His arrival also bought in the designer Gordon Buehrig and engineer August Duesenburg. By now both men were legends in the industry. Buehrig was given the brief and a limited budget to improve the Auburn styling for 1935. Augie was briefed to design a new eight working with Lycoming and supercharge specialist Schwitzer-Cummins. And what came out of this partnership? A beautiful line of cars, with the superb 851 Speedster at the helm. It turned out to be Auburn's final moment of glory. The new design by Buehrig was a reworked 1934 twelve giving the old bodies a new radiator, sleek hood, eye-catching curved fenders and flamboyant exhausts. The engineer Duesenburg added two cylinders onto the '34 six to come up with a 280-cid eight at 115 horsepower, or supercharged with the blower to 150 horsepower. It could hit 100 mph straight from the showroom to the road. As ever the price was great – just $2245. Fifteen models were available starting with the bargain priced five-seater, six-cylinder Brougham for a modest $695 through to the two-seater Speedster. The year turned out to be a poor one – just over 5,000 units were built, putting the company in 18th place on the industry sales ratings, little return for the creative skills harnessed to build this line of inspiring cars. These breathtakingly beautiful cars were duds in the market. They were also Auburn's final flourish. The fleet was brilliant in concept – the cabriolets, broughams, phaetons, sedans and the magnificent supercharged Speedsters – but production was very low. Cord returned from exile to face a crumbling empire and personal financial investigation. The 1937 range was cancelled – the make was now dead in the water.

1935 Chevrolet Suburban

Engine: Six-cylinder OHV inline
Displacement: 206.8 cid
Horsepower: 80
Body styles: Suburban Carryall
No of seats: 8

Chevy introduced America's very first SUV in 1935. This foray into fun and functionality was such a success that it allowed the model to remain at the cutting edge for over sixty years. In fact, Suburban is one of Chevy's best-loved nametags to this day, and is the longest-running model name in automotive history. The vehicle was extremely versatile, with both off-road and hauling capability, while having a comfortable ride for long-distance travel.

Chevy's core engine product for 1935 was based on an original design from 1929 when it had been introduced as the 'Stovebolt Six', better known as the cast-iron wonder. The names were drawn from the engine's cast iron pistons and slotted head bolts. It was not pretty but it was effective. The Stovebolt was designed and built by Ormond E. Hunt. By 1930, it produced around 50 horsepower from 194 cid. The engine then underwent year on year improvements and as a result remained as Chevy's only power plant for almost thirty years. The solid, overhead-valve engine came in two versions for 1935. The less powerful version was the 181 cid, which developed 60 horsepower, the other was a 206.8 cid which blew out 80 horsepower. It was a heavy engine but in the following year it was redesigned as a six that was both shorter and lighter than the original. At this point, the public relations people stepped in.

They wanted to change the public's perception of the 'Old Stovebolt' and attempted to rename this venerable power plant in 1934. First they tried to get 'Blue Streak' to catch on, then Blue Flame Six in 1935. The thinking was of Ford's publicity success of that year when they introduced their 90 horsepower

Below: This immaculate red Suburban has well –preserved details like the hood mascot and spare wheel.

V-8 engine to the market with great pizzazz. But the Stovebolt name resisted the change. The Master Deluxe of the year had an extra inch on the wheelbase and a much slimmer body with a vee'd windshield and streamlined fenders. The radiator was a new raked back design with the cap positioned under the hood – innovative for the time.

This was also the year of the steel Turret Top construction, so not all the glory went to FoMoCo that year. The 1935 Chevrolet models were the last models with any styling throw back to the classic era of the early 30's

Above: A two-door 1935 Chevy Suburban in a brilliant red livery with the silver hood-mounted mascot gleaming in the sun.

1935 Chrysler Airflow C1

Chrysler came of age in 1934 when it made a major production and design decision. It looked into the future and saw the time was right for a big step up. Chrysler attempted to build the most radical car ever attempted by a US auto manufacturer. In 1934 came the Airflow, an engineer's car that the conservative Walter Chrysler had approved without a care as to what the public would think of it. Wind tunnel tests on the concept, built by the design triumvirate - Fred Zeder, Carl Breer and Owen Skelton - had suggested a modified teardrop shape. An eight-cylinder engine placed over the front axle, rather than behind it, freed up passenger space. The seats were a generous 50 inches across. A beam and truss body engineered along aircraft design principles provided strength with less than normal weight.

Body styling was directed by Oliver Clark who had designed the first Chrysler and although the aerodynamic considerations really styled the Airflow. It had a hood that cascaded into a grille and looked like a waterfall with flush mounted headlamps on each side. 'The beauty of nature itself' was the description given to it by Chrysler, 'breathlessly different-looking' was a comment from Harper's magazine. Then came the outstanding performance. On the Bonneville Salt Flats, an Imperial coupe motored the flying mile at 95.7 mph. Over 500 miles it clocked a constant 90 mph and set 27 new American speed records. Sadly, 1934 sales were under-whelming.

But the cars lost money. Unfortunately, the public found the model too bizarre or just plain ugly. The 1935 Airflow C1 was powered an eight-cylinder 323.5 cid engine developing 115 horsepower. On the outside, the car had a new hood that extended forward in the shape of a Vee, and single broad bumpers. The new grille had a great top to bottom slope. Standard equipment included Autolite ignition, hydraulic brakes, Floating Power, automatic choke and a stabilizer in the back of the front axle.

Engine: inline eight-cylinder

Displacement: 323.5 cid

Horsepower: 115 at 3400 rpm

Transmission: Manual

Compression Ratio: 6.2:1

Fuel System: Stromberg IV model EX-32

Body style: Coupe

Weight: 3883 lbs

Wheelbase: 123 inches

Base Price: $1245

Number Produced: 379

1935 DeSoto SG Airflow Coupe

Engine: Six-cylinder
Displacement: 241.5 cid
Horsepower: 100 at 3400 rpm
Transmission: Manual synchromesh
Compression Ratio: 6.5:1
Fuel System: Ball & Ball IV E6F2
Body style: Coupe
Weight: 3390 lbs
Wheelbase: 116 inches
Base Price: $1015
Number Produced: 418

By 1933, DeSoto sales were in decline but a new engine was planned to boost performance for the 1934-36 models. The original six had never been exciting. It was now upgraded to 100 horsepower at 3400 rpm and an increased displacement of 241.5- cid. It was, however, sturdy and reliable and would run for long spells with little maintenance other than oil and water. It was also good with gas consumption, turning in 22 mpg with steady use.

By the end of 1934 the Chrysler Airflow was in retreat and although DeSoto had their version, they also produced an Airstream model for 1935 to go alongside the unpopular Airflow design. The Airstream proved to be more saleable. The DeSoto Airflow disappeared after 1936, a year ahead of the Chrysler model. The Airflow series SG was a six and in an attempt to make it appear less radical, the DeSoto Airflow was given a new front end with a V-shaped radiator. It followed that the grille would also change and it became more sloping than rounded which served to increase the length of the hood and give the car a more acceptable look. This meant that there were now only three horizontal hood louvers instead of eleven. Freewheeling was no longer used but the car had the anti-sway stabiliser bar moved from the rear of the car to the front.

There were now more standard features such as the centifuse brake drums and Autolite ignition. By 1935 DeSoto was America's 13th largest car producer and for the second year in a row, the Airflow won the Grand Prix Award for aerodynamic styling at the Monte Carlo Concours d'Elegance. DeSoto also provided dealers with a kit that could upstage the 1934 Airflow to give them the same general appearance as the 1935 models. Over 400 SG coupes were sold at around $1015 each.

1935 Duesenberg Model JN coupe

Engine: eight-cylinder	
Horsepower: 256	
Body style: Rollston	
Wheelbase: 142.5 inches	
Base Price (Chassis only): $9,500	
Number Produced: 10	

Fred Duesenberg died after a car crash in 1932. Before this tragic event he had supercharged the Model J to create the spectacular, powerful model SJ. His brother, Augie Duesenberg, wanted still more power and fitted a set of ram's horn manifolds. He was astonished to see the horsepower reach 400. In late 1934, A. B. Jenkins drove a special Roadster on the Bonneville Flats, and averaged 135 mph over 24 hours, clocking 160 mph for one lap. Yet again, it was an incredible car.

Only two Model Js could be described as sports cars. These were the Specials, designated Duesenberg Model SSJ. Both of these cars were first owned by Hollywood stars: Gary Cooper and Clark Gable. Then came another development in the life of the J, the 1935 JN. Only ten of these were ever built, all with Rollston coachwork. They had a 153.5-inch wheelbase and 17-inch-diameter wheels. The bodies were set below the frame rails for a lower profile. Superchargers were added to two of the cars but were removed later. This set up some confusion. Not all cars with Duesenberg plumbing on the outside had blowers and some superchargers were added to originals, so confirming specifications and accurately cataloging the cars is difficult. The face of the JN also changed. Some versions had a shorter running board apron, which meant that the battery carrier and toolbox were redesigned to give a lower door line. The bullet-shaped tail lamps were finished with the fender color. The headlamps, designed in the same recognizable style, were factory fitted, rather than dealer installed. On several JN's, the cowl lamps were absent. Engine-wise, the car was virtually the same as the J, with a 256 horsepower. A 320

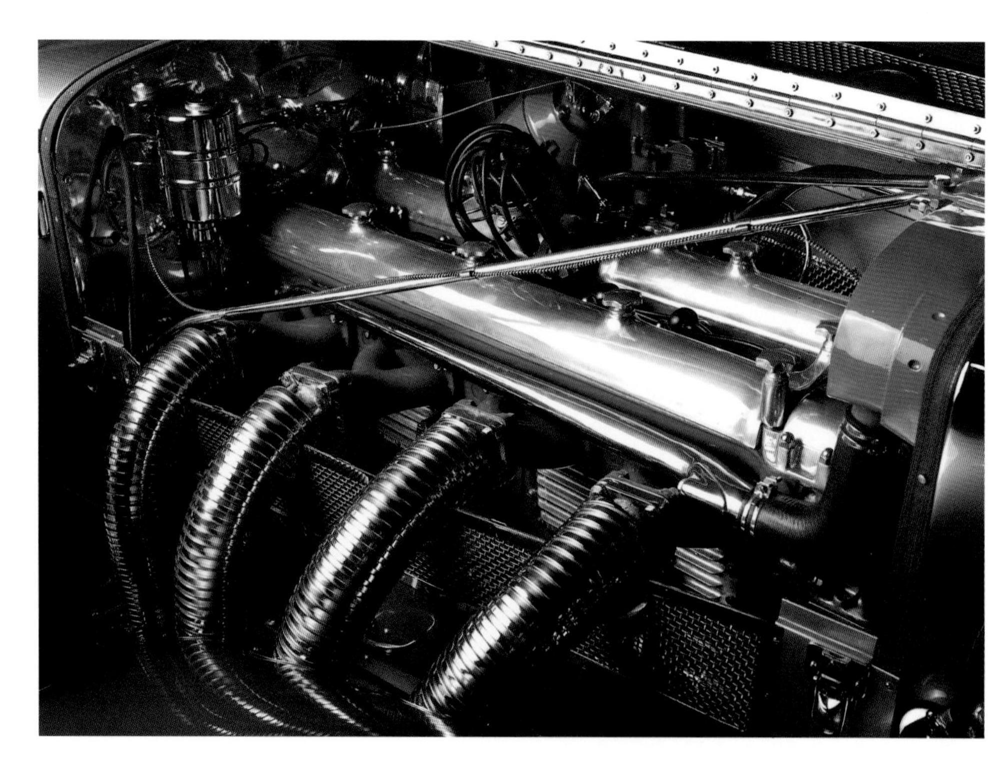

Right: The straight-eight model J motor was based on the company's successful racing engines of the 1920s and though designed by Duesenberg it was manufactured by Lycoming. In unsupercharged form, it produced 265 horsepower from dual overhead camshafts and four valves per cylinder. It was capable of propelling the car at a top speed of 119 mph and 94 mph in second gear. Other cars featured a bigger engine, but none of them surpassed its power.

horsepower version was also available. To control this engine power, the cars were fitted with Lockheed hydraulic brakes. The top speed was reported at 130 mph but for several models there were no records of how they could perform when pushed to the limit.

The Duesenberg was more than a status symbol – it was in a class all of its own – with owners ranging from robber barons and gangsters to an evangelist preacher. Duesenberg finally went out of production in 1937. Today, the company's cars are worth hundreds of thousands of dollars.

Above: A unique 1935 Duesenberg Convertible Coupe, one of only ten examples of the JN.

1935 Ford Coupe Roadster

Engine: Eight-cylinder
Displacement: 221 cid
Horsepower: 85 at 3800 rpm
Transmission: Sliding gear
Compression Ratio: 6.3:1
Fuel System: Stromberg EE-1 carburetor
Body Style: Coupe Roadster
Number of seats: 2
Wheelbase: 112 inches

Opposite page: A 1935 Ford Coupe Roaster with its distinctive large driving lights, curved mirrors and fully enclosed rear wheels.

Below: The interior is classically Ford –no nonsense and functional.

The 1935 Fords looked rounded and squat in shape, with smaller windows and a grille Vee that stood out more than in previous years. Few Ford enthusiasts would dispute Ford's claim of 'Greater Beauty, Greater Comfort, and Greater Safety' for this years models. The lively V-8 power plant was further enhanced by the addition of a new camshaft and better ventilation through the crankcase. Modifications were also made to the frame and the rear axle, both of which were strengthened.

The suspension was still extremely simple and effective, in fact it was a design retained from the old Model T, and had all the Ford benchmark qualities of durability and efficiency. It was a primitive set up of a solid axle bar on a transverse leaf spring in both the front of the car and the rear. Ford was also still using mechanical brakes, so the combination of

a crude suspension and out of date brake technology left the impression that Roadster was behind the times, design-wise.

Henry Ford was renowned for his inflexibility and desire to keep things simple, but this approach was not helping his company plan for the future. He did sanction one development: the 1935 cars now had 16-inch steel wheels to replace the old traditional wire designs. Styling for the year was credited to Phil Wright at Briggs Company and there was a vast range of 16 different Model 48s to consider.

The most expensive Ford of the year was the new convertible sedan at $750. It featured roll up windows if the weather turned, and when the collapsible top was lowered, the B-posts could be removed and stored in a bag underneath the trunk. A more practical car for warmer climates was the Deluxe four-door phaeton which cost $170 less than the convertible sedan. A special feature on this car was the side curtains that could be packed away beneath the rear cushion. The cabriolet convertible had rumble seats, which had the effect of keeping the passengers isolated from the driver when the top was up.

1935 Packard Model 1201 Victoria convertible

Engine: Straight eight-cylinder

Displacement: 320 cid

Horsepower: 130 at 3200 rpm

Transmission: Selective synchromesh

Compression Ratio: 6.5:1

Fuel System: Stromberg-Duplex

Body Style: Victoria Convertible

Number of seats: 5

Weight: 4835 lbs

Wheelbase: 134 inches

Base Price: $3200

Packard started an exciting year in 1935. They had a new management team in place. Max Gilman came in as the front man and George Christopher as the production king who had to be coaxed out of retirement. Gilman wound up the press and public while Christopher went through the plant modernizing it for high volume builds. The result of this activity and dynamism was the launch of the One Twenty, a car for the medium priced field.

But alongside this new model was the Twelve Series Eight that included the 1201 model. For this series, aluminum cylinder heads were standard and the compression ratio was 6.5:1, horsepower was up to 130 and the top speed was over 90 mph. On the outside of the car there was a completely new look: beginning up front, where there was a five degree slant to the radiator, carrying on to the pontoon fenders in the rear. Chrome was added to the side louvers as a feature.

For the 1201 model with the 134-inch wheelbase, ten body styles were available, which included the LeBaron Cabriolet and the Packard Victoria convertible. The eight cylinder power plant was fitted, the car had Bendix mechanical brakes, rode on 16 x 7 inch tires and had a fuel tank that could take 25 gallons. The two-door Victoria had the spare on the outside and a pair of prominent headlamps backed up with lower spots. Performance was smooth, with an easy ride and enough power for controlled speed.

1936 Cord 810

Errett Lobban Cord was the force behind some of the most magnificent cars ever built. In 1929 he built the Cord L-29 which proved less than he had hoped for, but it was the foundation for the 1936 Cord 810. The original L-29 was chiefly engineered by race car builder Harry Miller. It was a special design and according to one Cord authority it had over 70 unique parts. Fabulous though it looked, sales limped along and by 1935 one used car guide priced it at $145. But in 1936 the idea for the 810 was originated, using a V-8 rather than the original V-12, and incorporating front wheel drive, good weight distribution and better contraction than the L-29. Everything rode on a 125-inch wheelbase. There were many advanced features on the new 810. The front suspension, for example, was made up of independent trailing arms joined by a single transverse leaf sprint. The transmission had four forward ratios instead of the usual three and a Bendix Electric hand pre-selector. The power plant was a V-8, 288.6 cid unit made by Lycoming - another Cord company. The engine generated 125 horsepower, but with the aid of a Schwitzer-Cummins supercharger, it was boosted to 170 horsepower. This up-rated the engine to 190 horsepower. The basic 810 could top 90 mph and do 0-60 in 20 seconds. The supercharged car weighed in at 111 mph and hit 60 mph in 13 seconds.

Four models were launched in series 810. These were the Westchester and Beverley sedans (where the only real update was in the upholstery patterns) a two-seater sportsman and a four-passenger phaeton sedan convertible. 1,174 Series 810 cars were registered in 1936, all with prices around the $2000 mark. The coffin nosed car had first made its debut on November 2nd, 1935. These first cars were all hand built for auto shows around the country, where small development problems were ironed out (especially with overheating and an unreliable transmission).

Engine: Eight-cylinder

Displacement: 288.6 cid

Horsepower: 125 at 2500

Transmission: Selective synchromesh, Bendix electro-vacuum system

Compression Ratio: 6.5:1

Number of seats: 2

Weight: 3,715 lbs

Wheelbase: 125 inches

Base Price: 2000 lbs

Number Produced: 1,174

1936 Willys 77

Engine: Four-cylinder
Horsepower: 48
Body Style: Four-door Sedan
Wheelbase: 100 inches
Base Price: $540

Long before John North Willys's company built the Jeep, they were a visionary and inspired carmaker in the twenties and early thirties.

But this was not enough to stop the company being obliged to declare bankruptcy in 1934. It took a further three years before they emerged from receivership. In the meantime, Willys-Overland as the company was called, had to make a decision about its future and they decided to produce a low-priced small car. In fact, one could assume that this was all they had the funds to contemplate.

Harnessing all their ingenuity, they managed to develop and build the Willys 77 for very little money. It was given its first public viewing in 1932 and at that time they had a 100-inch wheelbase and a four-cylinder engine capable of producing 48 horsepower from only 134.2 cubic inches of displacement. The car survived and remained the only model sold by the company up until 1936. The 1933 reorganization of the company had bought in a new chairman, the bespectacled Ward Canaday, a Canadian with a strong sense of civic duty and a strong empathy with the workforce. It was he who wanted to get things moving again but his first real chance did not arrive until 1937 when he decided that enough had been achieved by the sturdy 77 and ordered a full restyle. The result was less than ideal. The cars had a lumpy appearance and an oddly bulging front end. The new sheet metal construction increased the overall length to 175.5 inches.

But the cars were moderately priced, at just below $600, and met with some success despite their odd appearance. But the general recovery of the car industry was also a turning point for the company. Although Willys produced twice as many cars in 1936, so did the rest of the industry. Willys only managed to improve their industry standing by one place, to 14th.

1937 Buick Century Sedan

Engine: Straight eight-cylinder
Displacement: 320.2 cid
Horsepower: 120 at 3200 rpm
Transmission: Sliding gear
Compression Ratio: N/A
Fuel System: Two-barrel Stromberg downdraft EE22 carburetor
Number of seats: 5
Wheelbase: 122 inches
Base Price: $1090
Number produced: 17,806

Sleek new styling was one of the many changes that occurred in 1936 as Buick adopted GM's all-steel turret top construction, which did away with the traditional fabric roof insert. Engines became more and more powerful and technological advances included aluminum pistons. The lexicon of car names began to take on a character that remained until the 1950's. 'Century sedan' was one of these, being the renamed series 60. Styling was a big factor in the resurgence of the brand. It was directed by Harley J. Earl, the founder and head of GM'S Art and Color Section, the first dedicated styling department opened by a mainstream carmaker. Streamlining drove Earl's styling in these years and Buick had it in full force for the 1937 models. The lines were rounded and set off by swept back features, fully integrated trunks and vertical bar grilles. 1936 was the first year that production returned to something like pre-Depression levels. The new Century was a fast car with a cruising capability of over 100 mph and a 10-60 acceleration of about 18 seconds. The cars quickly became known as 'factory hot rods' and were indeed about the fastest thing you could buy for around $1000. But was Buick impressed by this success? Hardly. They pressed on to introduce turret tops that had longer fenders and horizontal grille bars that became industry trendsetters. The 1937 models were more refined than any seen before with additional headlights in the same style as the fender parking lights. The line proved to be a real winner. 1937 was also the first year that Buick used plastic on some of the interior fittings. The door handles, dash knobs and steering wheel were all built in an ivory-yellow plastic. A modern horn was also designed into the steering wheel. Stabilizer bars front and rear gave the four-door sedan a better ride and windshield defrosters were added to make winter motoring more comfortable. The car was one of the most impressive of all the 1937 models. The Century's engine developed 120 horsepower from the 320.2 cid power plant.

1937 Cord 812 Cabriolet

Engine: eight cylinders	
Displacement: 288.6 cid	
Horsepower: 125 at 3,500 rpm	
Compression Ratio: 6.5:1	
Body style: Cabriolet	
Weight: 3,715 – 4,170 pounds	
Wheelbase: 125 inches	
Base Price: $2,445	
Number Produced: 1,146	

The styling of the 812 came from the synergy of the design teams. The team was made up of Cord regulars, Gordon Buehrig, assisted by Dale Cosper, Dick Robertson and Paul Laurenzen. Between them they came up with many innovations, and pushed the Cords graceful lines to the limit. Particularly striking design elements were the smoothly formed 'coffin' nose hood, the low silhouette, the unusual wraparound horizontal louvers instead of a standard radiator, the minimal trim and pontoon fenders. On the supercharged models, the racing style exposed exhaust pipes. The concealed headlamps on the car were an industry first.

These flipped up when needed by cranking a handle. Equally eye-catching design features were the full wheel covers and the concealed gas cap. The style impact then moved to the insides where turned-metal was used on the dash in needle gauges and a ceiling-mounted radio speaker. What became apparent was that Beuhrig and his team had 'borrowed' many of the appearance items from existing proprietary bits including leftovers from Auburn.

The engine was the V-8, 288.6-cid developing 125 horsepower power plant. The supercharged version could develop 190 horsepower. The 812 came in six

Below: The Cord 812 was also produced in Coupe form as seem here.

models: Westchester sedan four-door, Beverley sedan four-door, cabriolet, phaeton, Custom Beverley sedan four-door and the Custom Berline all with or without a supercharger. The custom cars came on an extended 132-inch wheelbase. But the Cord registered only 1,146 sales which just wasn't enough to keep the company out of trouble. As a result, the Cord Corporation had to work on a shoestring budget, and cut costs. This didn't work either. The company's problems moved to the production line, where haste became waste. Although build quality problems were ultimately resolved, the damage had been done, and unhelpful rumors circulated in the industry. Eventually, the E. L. Cord empire collapsed in 1937 and the Cord models followed the business into oblivion.

Above: A mean machine of outstanding beauty, the 812 Cord cabriolet with exposed exhausts and supercharger.

1937 Ford Five-Window Coupe Model 78-770

Engine: V-8 cylinder	
Displacement: 221 cid	
Horsepower: 85 at 3800 rpm	
Transmission: Sliding gear	
Compression Ratio: 6.3:1	
Fuel System: Stromberg downdraft carburetor	
Body Style: Coupe	
Weight: 2506 lbs	
Wheelbase: 112 inches	
Base Price: $660	
Number produced: 26,738	

Ford got the fashion message right in 1937. In a year of questionable styling throughout the industry, the Model 78 cleaned up. Ford produced one of the best looking cars of the decade. Even President Franklin D. Roosevelt bought a convertible sedan for use at his Georgia retreat.

The familiar 221 flathead was also given a new lease of life: becoming known as the V-8/85 after the development of the V-8/60 for the European market. A configuration that didn't work – fewer horsepower and better economy and a cheaper price wasn't what people wanted – power was king. The '37 engine had improved cooling due to the redesign and relocation of the water pump and new cast-alloy pistons.

Ford also made a radical change to the construction of the body. In response to GM's Turret Top bodies Ford made an all-steel body for closed models, dumping the old fabric roof inserts at long last. Headlamps were sunk into the new bodies. These were now buried in the fenders. A prominent grille, composed of fine horizontal bars which swept backwards as it approached the

top, was also fitted. The styling reflected the strong influence of the Lincoln-Zephyr. The sedans came in two body shapes, one with a trunk back and the other with the slant back.

The new coupe also came with a rear seat. All the models came with a rear-hinged hood, they were described as opening 'alligator-style'. With a larger V-8 under the hood, the model 78 Sedan proved to be formidable when equipped for Police duty. The mechanical breaks now had cable linkage and were advertised and promoted as being 'all steel from pedal to wheel'.

Above: An eye-catching red 1937 Ford Model 78 coupe, which by all contemporary accounts was the style champion for the year with its bold but smooth lines demonstrating that streamlining wasn't the only option.

Left: A surprisingly roomy interior for a two-door coupe

1937 Studebaker Coupe Express Pickup

Engine: Inline six-cylinder
Displacement: 217.8 cid
Horsepower: 90 at 3400 rpm
Fuel System: Stromberg carburetor
Number of seats: 2
Wheelbase: 116 inches

Right: The American motorized workhorse: the pickup truck. This bold yellow 1937 Studebaker Coupe Express Pickup was based on the Dictator chassis.

Below: The pickup used the power unit from the Studebaker Dictator a six cylinder flathead design that proved reliable, if not exciting.

The pickup truck is an essential part of the American auto experience. It was a practical workhorse designed to take the place of the original four-legged transport. By 1930, all of the Big Three manufacturers had active truck divisions producing pick-ups, factory complete all-steel, half-ton trucks. Studebaker had also dabbled in big trucks prior to 1930, but the company made its first entry into the world of the light-duty truck in 1937. The Depression had played a large role in the pickup's growing prominence, as the market for big trucks had contracted, so Studebaker felt that the time was right for them to enter this market. Their car-derived pickup was based around the 116-inch wheelbase Dictator, and had a handsome car-like shape and good lines for a working vehicle. The models were actually constructed by speciality body building firms. But from the doors forward the 1937 Coupe Express was a Dictator passenger car. The pickup was an attractive hybrid that reminded many witnesses of Hudson's Terraplane pickup model that had been introduced three years earlier. Its good lines were hardly surprising considering that the original design had come from the pen of Raymond Loewy.

This was a period when Studebaker's car production was going up and down. In 1937 some 98,000 cars were produced. This was up by a considerable 40,000 from the year before, and profits for the year were $812,000. An optional fully automatic overdrive followed in 1937, a further refinement of the semi-automatic that had been developed two years previously. The new hypoid rear axle had now become standard.

More innovations followed over the next few years and they too were incorporated into the pickup. 1938 saw windshield washers added, for example.

The Coupe Express 'half-car, half-truck' stayed in production for three years before this hauler was retired in favor of an 'all-pickup' pickup in 1940.

1939 Chevrolet Master Deluxe Sedan

Engine: six-cylinder	
Displacement: 216.5 cid	
Horsepower: 85 at 3200 rpm	
Transmission: Manual	
Compression Ratio: 6.25:1	
Fuel System: Carter IV model W1 carburetor	
Body style: Four-door Sedan	
Number of seats: 5	
Weight: 2875 lbs	
Wheelbase: 112 inches	
Base Price: $745	

Opposite: A Chevrolet 1939 Master Deluxe Sedan 4-door with gleaming white walled tires.

Below: The handsome Chevrolet Master Business Coupe.

Chevy's Master Deluxe model was a fancy version of the Master with its longer 1939 hood and fenders. The wheels and running boards were lowered for a more streamlined, longer appearance. The body shell was retained from the '38 model, but was restyled to look more modern. The grille extended back along the fender line at the top and then tapered to about four inches at the bottom. The car had a rounded look with horizontal grille moldings and a horizontal bar effect on the splash on aprons. The Deluxe Sedan had bumper guards as standard equipment as well as twin taillights. The four-passenger coupe was an all new body style, with folding opera seats replacing the rumble seat. The 'knee action' coil spring front suspension was another feature. Despite all this cosmetic buffing, the Master Deluxe was reported to be the fastest accelerating American passenger car of the year. It went from 10-60 in high gear.

Even with this speed advantage no cabriolets were built in 1939 but the 15-millionth Chevrolet was completed. As you would expect there were a hatful of options for the deluxe; radio, heater, clock, cigar lighter, radio antenna, seat covers, external sun shade, spotlights, fog lights, wheel trim rings, and a slip in coupe pick-up box. The power came from the Chevy six-cylinder engine with a 216.5 cid displacement. This developed 85 horsepower at 3200 rpm. The especially rare models from this year included the Master Deluxe Coach and the four-door, slant-backed Sedan of which only 68 were produced. The Master Deluxe Sports Coupe cost $715 and an identical body was used for the Business Coupe, which sold for $684.

UK918

1940 Nash Model 4081

Nash Motors was twenty years old in 1936 and Charlie Nash, now 72 years of age, was feeling his age. The company continued to produce fine cars and to hold their spot in the top fifteen auto manufacturers. Nash invited George W. Mason, Vice President of Kelvinator Fridges, to take over as Nash President. Mason agreed, provided that Nash took over Kelvinator. The sale duly took place with jokesters predicting Fridges with four-wheel brakes and cars with ice cube trays! The joint companies were a good fit and Mason was a good President.

Nash sold 86,000 cars in 1937, making a $3.5 million profit. Nash's cars always kept up with the times, and innovations of the late thirties included both overdrive and the unique Weather Eye ventilation system. The 1940 model range for the Ambassador series had improvements such as Sealed Beam Headlamps, a redesigned grille with closer spaced bars and Nash scripts on hood and trunk. There were 18 different model options including a one-off special cabriolet designed by Alexis de Sakhnoffsky with distinctive cutaway doors.

Although 1940 was a disappointing sales year with only 52,855 registrations, Nash remained profitable.

Engine: Straight-eight OHV

Displacement: 260.8-cid

Horsepower: 115 at 3400 rpm

Transmission: Three-speed selective sliding gears

Fuel System: Carter 485S carburetor

Body Style: Two-door Cabriolet

Number of Seats: 3/5

Weight: 3640 lbs

Wheelbase: 125 inches

Base price: $1,295

Number Produced: (Ambassador models) 63,617

1941 Lincoln Continental

Edsel Ford, a man of style and discerning taste is responsible for perhaps the most enduring of the Lincoln Models. The Lincoln line was running out of steam in the late 30s and Cadillac were challenging its position at the luxury end of the market. Edsel returned from a trip to Europe in 1938 and asked designer Bob Gregorie to build him a personal custom special that would look 'strictly Continental'. Using a Lincoln Zephyr, 125-inch wheelbase chassis, Gregorie created a fabulous car that caused a flurry of admiration wherever it went. So it was decided to put the design into production. The car was unveiled at the Ford Rotunda in Dearborn on October 2nd 1939.

The first 'Continental' was a cabriolet, this was followed by a coupe in May 1940.

Twelve-cylinder power provided by the 292 cubic inch L-Head unit amounted to 120 horsepower. The aluminum alloy heads were polished and had nice touches like the chrome plated acorn head nuts. The car's body was three inches lower than the Zephyr and the hood was seven inches longer, creating a sinuous, sleek look. The distinctive rear spare wheel cover was also present. The new model also had press button door operation.

The 1941 models were the first to be officially designated as Continentals and had their own tooling. 1,900 Continentals were manufactured produced up until production was suspended on the outbreak of war in 1942. Model production was resumed after the War, but very sadly, its inspiration, Edsel Ford had died in May 1943.

Engine: 75-degree V-12 L-Head.

Displacement: 292-cid

Horsepower: 120 at 3500 rpm

Body style: Two-door Coupe/ Cabriolet

Number of Seats: 5

Wheelbase: 125 inches

Weight: 3635 lbs

Base price: $2840

Number Produced: 404

1941 Oldsmobile Series 98

'The Car Ahead' was the cry from Oldsmobile's publicity department in 1941. 'The car that has everything modern' was another sales line, and indeed the range did have a lot going for it. New fastback styling was evident, as were redesigned grilles and simulated wood finishes. Oldsmobile also first offered a four-speed automatic gearbox option called Hydramatic. Over fifty percent of customers opted for this for an extra $100.

The company passed a couple of personal milestones in the year 1941. The two-millionth car left the production line and overall sales of 270,040 cars for the year was the best in the company's history.

The series 98 represented the top end of the Olds range. Powered by the straight-eight engine, it generated 110 brake horsepower at 3400 revs.

The relatively light two-door fastback coupe version was heading in the direction of an early muscle car. Other models on offer were the two-door convertible, four-door sedan, and the most expensive car in the Olds range, the four-door Phaeton (at $1,575).

These were amongst the last truly pre-war cars. The line was to be closed down in favor of ammunition production for the duration of World War II.

Engine: Eight-cylinder inline

Displacement: 257-cid

Horsepower: 110 at 3400 rpm

Transmission: Three-speed manual/ Hydramatic

Fuel System: Dual downdraft carburetor

Body Style: Two-door Fastback Coupe

Number of Seats: 2/4

Wheelbase: 125 inches

Weight: 3430 lbs

Base Price: $1059

Number Produced: 23,463

1941 Packard Clipper Touring Sedan

Packard recognized that something drastic needed to be done about their styling as they introduced a completely different shape for mid-year 1941. Whilst the old 120 series harked back to the 30s, the new Clipper was making a bid for the 50s! As a result this styling would see the company through the immediate post war sales period.

The Clipper was designed by staff stylist Howard 'Dutch' Darrin and modified by Werner Gubitz. The design was lower and much wider with faired in fenders, running boards and horizontal grille styling giving the car a mean low-slung look.

Although it shared chassis dimensions with the earlier 120 series, having an identical 127-inch wheelbase, the new chassis was of a completely new design. Under the hood, but not so apparent with the new styling, was the well-proven straight-eight power unit, adjusted to provide an extra five brake horsepower, courtesy of a higher compression ratio and an upgraded carbureter. A futuristic column-mounted gearshift for the three-speed manual transmission marked a new trend for the forties.

An optional Aero Drive overdrive was offered to take advantage of the straight-eight power on a long drive. The new car was strategically priced to fit between the 120 and 160 series cars, and was priced at $1420.

1941 Pontiac Torpedo

Engine: L-Head six- or eight-cylinder.	
Displacement: 239.2 /248.9-cid	
Horsepower: 90/103	
Transmission: Three-speed synchromesh	
Fuel System: Carter one-barrel / two-barrel	
Body style: Four-door Sedan	
Number of Seats: 5	
Wheelbase: 119 inches	
Weight: 3235/3285 lbs	
Base Price: $921/ $946	
Number Produced: 117,976 / 37,823	

The whiz behind Chevrolet's success came to Pontiac in 1934 as General Manager. Harry Klingler was a super salesman. Pontiac had had some tough years but Klingler quickly restored their fortunes, taking them from an all time low of 42,000 cars in 1932 to a production of 175,000 by 1935. He introduced a six-cylinder engine option to reduce costs and give the company a crack at the lower-priced market.

In 1937 he beefed up the engines, the six now had 223 cubic inches and 85 horsepower. The eight had 249 cid, which was good for 100 horsepower. All of Klinglers 'gut reaction' changes to the business seemed to bear fruit. The company had its best sales year so far with 235,000 cars. But there was still something missing. Klingler asked his design team (Bob Lauer and Joe Schemansky) to dream up a new streamlined style for the 40s. This became known as the Torpedo. The customers seemed to like this model because sales continued to rise, to 249,000 in 1940 and 282,000 in 1941 when the whole range went Torpedo!

Twelve different body options were offered with the new styling. The cars were built with either six- or eight-cylinder engines. Only a subtle hood badge distinguished between them. The new styling included faired-in fenders,

running boards and a wide horizontal grille that accentuated the width of the cars. Both cars were priced under $1000, which made them widely affordable.

Above This Torpedo shows off Frank Hershey's Silver Streak side detailing.

Left The affordable Pontiac Torpedo was drafted into wartime service.

1942 Hudson Super Six Station Wagon

Engine: L-Head Six-cylinder	
Displacement: 212 cid	
Horsepower: 102	
Transmission: Three-speed manual with Drive Master auto as an option	
Fuel System: Carter Duplex Downdraft carburetor	
Body Style: Four-door Station Wagon	
Wheelbase: 121 inches	
Weight: 3315 lbs	
Base Price: $1486	
Number Produced: 40,661	

In 1942, the Hudson line got a substantial styling uplift. The grille was restyled to look wider with chunkier, spaced horizontal bars, restyled fenders with chrome trim and two body-length chrome strips (one at belt level one at the bottom of the doors running onto front and rear fenders). Quite unique to Hudson were the two illuminating triangular badges on either side of the hood, towards the front. This was the final year for such frivolities - the War saw to that. In future 'Blackout Specials' (manufactured from January 1st 1942) would be limited to chrome bumpers only. Everything else would have to be coated in body-colored plastic. Running Boards were gone for good, just leaving a vestigial flare on the door bottoms. This year also saw the final model-year of Station Wagon production at Hudson.

Since people had become more mobile in the thirties, many people had been obliged to migrate to get work during the Depression, the automobile had become a vital part of the American way of life and culture. As the recession faded, people still had the desire to travel, and this was transmuted into vacation breaks using the automobile. The need grew for rugged powerful vehicles, which could transport entire families and their possessions over vast tracts of the USA. Not since the Mormon trek had there been so much mobility. The Station Wagon was part of this 'road' culture.

The wood bodies were made by the Cantrell Company and grafted onto the 121-inch Super Six wheelbase. The wagon seated eight passengers. The rear and center seats could be removed for even greater carrying capacity. Drive Master automatic transmission was available as an option.

Opposite top: A superb Cantrell wood-bodied Super Six with black paintwork, ready for any load.

Opposite bottom: Chrome niceties such as this Super Six script and hood rubbing strip became no-nos as the country approached WWII.

Right: The sedan version of the Hudson Super Six.

SUPER SIX

1946 Mercury Sedan Coupe

Engine: V8, L-head	
Displacement: 239.4-cid	
Horsepower: 100	
Transmission: Three-speed manual	
Fuel System: Holley 94 two-barrel	
Body Style: Two-door coupe	
Number of seats: 4	
Wheelbase: 116 inches	
Weight: 3190 lbs	
Base Price: $1495	
Number Produced: 24,163	

When production resumed on the Mercury line on November 1st 1945, it was without the guidance of its creator, Edsel Ford. Edsel had died of stomach cancer in 1943 (some say of a broken heart because of his poor relationship with his father) and Henry had taken back the reins of both Lincoln and Mercury at 80 years of age.

In 1946 Mercury became a separate division of the Ford Motor Company. But the cars took a few more years to cast off their image as 'glorified' Fords. At this time, the company was still suffering from 'war shortages' and new model production was held up. The Mercurys on offer were cosmetically restyled 1942 models, with new hood pressings and a new grille with thin vertical bars.

The model range included a Sportsman's Convertible with fancy wood paneling but priced at $2209, it was the most expensive Mercury to date. The model was not a success and was not repeated in the following year. The one model that did extremely well, with sales of over 24,000 cars, was our featured car, the very attractive notchback two-door sedan coupe. This was by far the

most tasteful and understated design of the model year, and if Edsel had been looking down from on high, he would have approved of it. At $1495, it also represented great value, particularly when compared to the over-the-top Sportsman's convertible. The V-8 was still an L-head, but this was eminently tunable and many of these cars saw service as hotrods in the '60s.

The overall sales total for the entire model range was 86,603 cars. This made Mercury America's twelfth largest manufacturer.

Above: The sleek and understated black two-door Mercury Sedan Coupe

Opposite: This postwar Mercury still relied on the Ford Flathead V8 for its power. Many of these engines ended up in Hot Rods as the unit was easily tuned up.

1947 Ford Super Deluxe Tudor Sedan

Engine: L-Head Six-cylinder or V-8	
Displacement: 226/239-cid	
Horsepower: 90/100	
Transmission: Three-speed manual	
Fuel System: Holley Model 847F single barrel/Mod.94 Two-barrel	
Body Style: Two-door Sedan	
Number of Seats: 6	
Wheelbase: 114 inches	
Weight: 3183 /3216 lbs	
Base Price: $1309/$1382	
Number Produced (Series total): 385,109	

Below: $73 bought you the Flat-head V8 option as shown here.

Whilst nothing very momentous happened to Ford's model lineup for 1947, and the pre-war body shape was still very much in evidence, the death of its 83-year-old founder, Henry Ford, made front page news around the world. Credited by many as the inventor of the automobile and of mass production itself, he was certainly the man who gave every American the chance of owning a car. Despite his egocentric style of management and his occasional failure to keep up with the times, he was the Citizen Cane of the auto world. The company was temporarily disorientated by his loss, and for a while, no 1947 models were announced at all.

It's possible that the management was floundering, with no Henry Ford to tell them what to do. But finally, a range did appear. Sales of 429,674 for the year put the company well behind Chevrolet. Our car, the Super Deluxe, offered the top range trim level and came with a new hood emblem, chrome body strip, chrome moldings on all the window frames, two sun visors, a chrome horn ring, a dash mounted electric clock, armrests on all doors and passenger assist straps.

Other small revisions made to the styling consisted of a red tracer removed from the grille and new round rear parking lights. There was a choice of six or eight-cylinder power across the range. Most people opted for the latter for an extra $73.

Above: Despite the prewar styling, the bright red finish shows a bit of postwar cheer.

Left: Stylish speedometer with chrome bezel and luminous gold numbers.

1947 Plymouth Special Deluxe Business Coupe

Engine: Six-cylinder L-head inline.

Displacement: 217.8 cid

Horsepower: 95 at 3600 rpm

Transmission: Three-speed manual

Fuel System: Carter Type BB one-barrel carburetor

Body Style: Two-door Coupe

Number of Seats: 3

Wheelbase: 117 inches

Weight: 2982 lbs

Base Price: $1209

Number Produced (Series): 350,327

With the return to the civilian economy in the late 40s, there was an increased demand for salesmen to promote new products in a rapidly expanding consumer goods market. The concept of a traveling salesman became a phenomenon. The Plymouth two-door coupe was ideal for this purpose, as it was inexpensive to buy at $1200, and had a large trunk and good carrying capacity. It also had a lively 95 horsepower six-cylinder engine.

Our model is identifiable as a Special Deluxe by the lettering at the rear of the hood, below the belt line trim, by the chrome windshield surround. Larger 10-inch hubcaps were fitted to this model and it also had a lot of useful extras like dual wipers and glovebox locks. The car was unusual in that it had proper rear quarter windows as many business coupes had blind rear quarters.

The rear fenders were re-flowed from earlier models to be more streamlined, and had smaller wheel apertures. The bumpers were wrap around and the grille had horizontal chrome bars. Options included an external sun visor and a light up hood ornament. Although the overall design was clearly pre-war in origin, sales boomed, and over 350,000 cars were produced in the model year maintaining Plymouth's third place in the auto industry.

1947 Pontiac Streamliner Station Wagon

Full production resumed at Pontiac in June 1946, and 131,000 cars were produced in the first postwar calendar year. Pontiac's General Manager was Harry Klingler who had joined the company back in 1934. In those days, he had a reputation as a super-salesman and had restored the company's fortunes with a series of smart marketing moves. But if his energy may have been burned out by this time, as his model range was looking positively elderly.

Another Pontiac old timer, Ben Anibal, who had designed the company's cars back in the late '20s, now retired. He was replaced by designer George Delaney. One can only imagine how hard it would have been for Delaney to make substantial changes in a conservative company like Pontiac.

Despite their stylish names, such as Streamliner and Torpedo, the Pontiac models were deeply pedestrian and predictable. However, they were also reliable.

The 'woodie' wagon is an American classic however. Ash framing over dark maple paneling compliments the dark cherry body color. The vehicle could carry eight passengers in comfort, although the third row of seats was removable to allow give the model a greater luggage capacity. The eight-cylinder power option would gave the car both rugged reliability and long-legged pulling power on a cross-country journey. Aluminum fender panels were available for the 1942-48 model years and many owners fitted these to reduce weight.

Engine: L-head, Eight-cylinder inline

Displacement: 248.9-cid

Horsepower: 103 at 3500 rpm

Transmission: Three-speed Manual

Fuel System: Carter WCD two-barrel carburetor

Number of Seats: 8

Wheelbase: 122 inches

Weight: 3845 lbs

Base Price: $2037

Number Produced: 86,324

1948 Cadillac Series 62

Engine: L-Head V 8	
Displacement: 346-cid	
Horsepower: 150 at 3400 rpm	
Body Style: Two-door Convertible	
Wheelbase: 126 inches	
Weight: 4449 lbs	
Base Price: $3442	

Nervous Cadillac dealers entertained doubts about the radical new design for the 1948

'60s' series. Although the inspiration for the styling came from a wartime theme , that of a P38 Lightning Fighter, it was beautifully translated into a great line of the postwar luxury car market. As the Cadillac ads said, the new models were 'The New Standard of the World'. The Harley Earl design team had done General Motors proud and dealer anxiety soon gave way to pleasure at sharply rising sales figures. Rising from around 30,000 cars in 1942, to 52,000 sales in 1948. Earl's new 'Futuramic Styling' had certainly done the trick.

Fins inspired by the 'twin boom' fighter's distinctive appearance made their first appearance with this design. They were to rise to a crescendo of extravagance in the following decade. Windows and windshield were curved, as was the dashboard, complete with a rainbow shaped instrument panel. The

wheelbase decreased by 3 inches to 126, and the body had handsome stone guards, which began behind the rear wheel and ran back along the rocker panel with a matching trim at the base of the rear fender. The rear tailfin on the driver's side flipped up to reveal the gas-filler. The front of the car had a redesigned grille, and torpedo shaped over-riders on the bumper. Everything about the car made it stand out from the opposition, making Cadillac the style innovator in the luxury market.

Above: The handsome black Series 62 Convertible showing off the side stone shield trim to its beat effect.

Opposite page: The series 62 also came as a four-door sedan as shown here.

1948 De Soto Custom Club Coupe

- **Engine:** L-head Six-cylinder
- **Displacement:** 236.60-cid
- **Horsepower:** 109 at 3600 rpm
- **Transmission:** Three-speed manual
- **Fuel System:** BB EV1 carburetor
- **Body Style:** Two-door Coupe
- **Number of Seats:** 6
- **Wheelbase:** 121 1⁄2 inches
- **Weight:** 3599 lbs
- **Base Price:** $1874
- **Number Produced:** 98,850

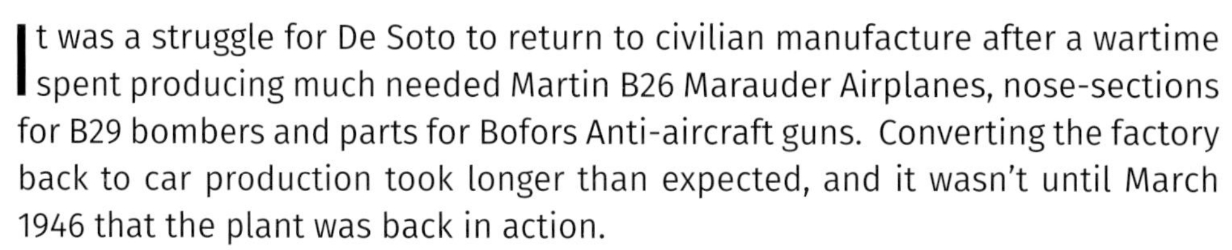

It was a struggle for De Soto to return to civilian manufacture after a wartime spent producing much needed Martin B26 Marauder Airplanes, nose-sections for B29 bombers and parts for Bofors Anti-aircraft guns. Converting the factory back to car production took longer than expected, and it wasn't until March 1946 that the plant was back in action.

Like most manufacturers the 'new cars' on offer after the war were really revamped 1942 models. In fact in De Soto's case, they were actually somewhat less exciting than their forebears. Gone was the neat 'Hidden Headlight' feature, conventional exposed lights were fitted instead. The cars also had rather predictable restyled grille and body trim. Moves toward some form of automatic transmission were catered for by Fluid Drive and the Vacumatic options. Fluid Drive was not a fully automated system, the driver had to 'Tip-Toe' to shift and spend an extra $121 for the privilege.

Low-pressure tires were adopted in 1948, as were stainless steel wheel 'beauty rings'.

The company's brief flirtation with plastic whitewall tires was discontinued at the same time. Body weights came down this year, just as prices went up. The postwar car boom was gathering pace and an average of $300 was slapped on the price of most models, according to the principal of supply versus demand. Nobody seemed to mind because sales were over 98,000 in the year. This gave DeSoto a twelfth place in the sales charts. Standard wheelbase (121.5 inch) models were a four-door sedan, a two-door sedan, our featured car, the two-door Club Coupe and a two-door Convertible Coupe.

Opposite: This Cherry-red Custom Coupe with Fluid Drive badge on front panel would have been good value at $1874.

Right: Side access to the coupe looks a little restricted.

Below: Dashboard styling echoes the vertical chrome slats of the front grille.

DeSoto
1948
Desoto

1948 Plymouth Convertible Special Deluxe

Engine: L-head Six-cylinder	
Displacement: 217.8 cid	
Horsepower: 95 at 3600 rpm	
Transmission: Three-speed manual	
Body Style: Two-door Convertible	
Number of Seats: 5	
Wheelbase: 117 inches	
Weight: 3225 lbs	
Base Price: $1857	
Number Produced: 15,295	

Plymouth was the Chrysler Company's budget division, and things were looking good for the company in the late '40s. Model-year production of 412,000 cars lifted them to second place in the auto producers' charts for 1948. Although there were no major styling or mechanical alterations made this year, and despite the fact that the car was essentially a pre-war design, sales kept up remarkably well.

In fact, nearly all US manufacturers were still peddling models designed at the end of the 30s, so Plymouth was in good company. Prices did go up this year by an average of $200 but in the optimistic glow of the immediate post war years this didn't seem to matter either. Customers seemed actually more inclined to pay more for the upper level model choices, such as the featured Special Deluxe. Plymouth offered six different body styles in the Deluxe range: a two and fourdoor sedan, a two-door Club Coupe, two-door Coupe, two-door convertible and a four-door station wagon. The station wagon was the most expensive model at $2,068, the least expensive was the two-door coupe at $1440. The former was a splendid creation with white ash framing contrasting to dark maple paneling. These 'woodie' bodies were largely made by hand, hence the extra cost. Next down the list was our featured car, the convertible Special Deluxe, priced at $1857. Powered by the L-head inline six, it was good for 95 horsepower. The car had all the extras associated with the series: a bright metal surround on the windshield and front quarter light windows, dual windshield wipers and sun visors, glovebox locks, and bumper fender guards. The convertible also had leather seats, a body colored painted dashboard and the pre-war style 'blind' top in white. Upholstery for the rest of the range was done in pencil-stripe broadcloth. Special Deluxe emblems on the hood identified the models in the range.

Below: Plymouth's inline six power unit was essentially a pre-war design.

Above: The handsome profile of the convertible Special Deluxe in Chevron Blue, one of ten colors available for 1948.

Left: Plymouth Convertible looks good with top up or down.

1949 Lincoln Series 9EL

Engine: L-head V-8	
Displacement: 336.7-cid	
Horsepower: 152	
Transmission: Three-speed manual	
Fuel System: Holley Two-barrel carburetor	
Body Style: Two-door Convertible	
Number of Seats: 6	
Wheelbase: 121 inches	
Weight: 4224 lbs	
Base Price: $3116	
Number Produced: 38,384	

Below: A wide horizontal grille laden with chrome detail was part of the Lincoln late forties look.

The Continental was dead! The car that The Museum of Modern Art was to judge as one of eight great automotive works of art had been cast aside. Edsel Ford would not have been pleased. In its place was the Cosmopolitan, which critics said looked too much like the basic Lincoln series 9EL, which in turn was judged to looked too much like its cheaper brother, the Mercury.

As Lincoln headed toward the '50s it had at least one good card in its hand. The re-designed body shape was a huge improvement on the early '40s series 876H styling. The body was sleeker, and slab-sided with a wide horizontal grille and loads of chrome. The Cosmopolitans had a single-piece windshield, but the series model retained the two-piece.

In keeping with the times, the base model had its wheelbase cut to 121 inches, whilst the Cosmo stayed at 135 inches. Weight was still high at over 4000 pounds, so the car's performance required every ounce of strength that the 337 cubic inch L-head V8 could muster.

The Lincoln range comprised of seven different models, three Series 9EL cars and four Cosmopolitans.

Our featured car is the base model convertible, recognizable by its two-piece windshield. At $3116, this car represented better value for money than the comparative Cosmo model, priced at $3948. Eighteen different body colors were available, including seven metallic options.

Right: A Lincoln Convertible in Teal Blue.

Left: Lincoln script in chrome-plated Mazak.

1950 DeSoto Two-door Hardtop Sportsman

Engine: L-head six-cylinder, cast iron block

Displacement: 236.7 cid

Horsepower: 112 at 3600 rpm

Transmission: Tip-Toe Hydraulic shift with Gyrol Fluid Drive

Compression Ratio: 7.0:1

Fuel System: B-B E7L3 or B-B E7L4

Body Styles: Eight Custom styles in all, including a sedan, coupe, hardtop, convertible, two station wagons, longwheelbase sedan and suburban

Number of Seats: 6/8

Weight: (Two-door Hardtop) 3735 lbs

Wheelbase: 125.5 inches

Base Price: $2489

Number Produced: 4,600

The DeSoto division of the Chrysler Group had been very active in war production, assembling aircraft and cannon components for the USAF. But a return to civilian production in the post-war years was fraught with labor problems, shortages of raw materials and difficulties re-converting plant to auto production.

Their first post-war model, the 1946 Deluxe arrived in 1946, but it wasn't until 1949 that the company launched a fresh, practical look – 'The Car Designed with With You in Mind' ran the advertising slogan. This new look was to remain in production until 1954.

By 1950, the company was twelfth in the ranks of auto producers, with an output of 136,203 vehicles in the year. 39.6 million passenger cars were now on the American road, with 60% of families owning at least one.

The new top-of-the-range Custom Series was launched in 1950, and the cars reflected the same minor styling changes as the Deluxes to the original 1949 design. There were six six-seated Customs models, and two eight-seat examples that had a longer wheelbase and extra seats.

The Two-door hardtop version was the only body style with the 'Sportsman' logo. In fact, the car marked DeSoto's sole return to the hardtop market for this model year. It was an all-new pillarless coupe, with whitewall tires and full wheel covers fitted as standard.

1950 Ford Custom Deluxe Two-door Convertible

Like so many car producers in the post-war period, Ford production relied heavily on slightly re-styled 1942 models. The cars sold very well into a car-starved market, and were only slightly tweaked until the newly designed 1949 cars. These were the first totally new cars that Ford had produced since the end of the War. They were built on a wishbone chassis, and had smoother and lower body shells than those of the early to mid-1940s. Slab-sided, the cars had now lost the bulge over the rear wheels. Horizontal chrome moldings, and a heavy chrome grille were simple and attractive adornments.

The 1950 cars were essentially unchanged, though minor trim changes were made. The ad slogan for the year was '50 improvements for '50' but these consisted of tiny adjustments in the styling rather than anything substantial. Things like a redesigned hood ornaments and push-button handles. Three model series were marketed in the year, the Deluxe, Custom Deluxe and Custom Deluxe V-8. Between them, they accounted for sales of over 1.2 million cars. This was partly accounted for by the general upsurge in demand for automobiles, a hitherto unheard of number of 6.7 million cars were sold in the year. The fear that entry into the Korean War would result in the same consumer shortages as WWII had done fuelled demand for autos, as had the general return to civilian life.

The convertible Custom Deluxe V-8 in our photograph was one of six models in this range. It also included a four-door sedan, four-door station wagon, two-door sedan, two-door Crestliner, two-door club coupe and two-door station wagon. The Club Coupe has a short top, and the Crestliner was a special model with vinyl roof covering, extra chrome trim, a special steering wheel, special paint and full wheel covers. All came with V-8s. With model year updates, this body style lasted until 1955, when the entire Ford range was substantially restyled.

Engine: L-head, V-8 cast iron block

Displacement: 239 cid

Horsepower: 100 at 3600 rpm

Compression Ratio: 6.8:1

Fuel System: Holley two-barrel Model AA-1

Body Style: Two-door Convertible

Weight: 3263 lbs

Wheelbase: 114 inches

Base Price: $1948

Number Produced: 50,299

1950 Ford Custom Deluxe V-8 Sedan

Engine: L-head six-cylinder, cast iron block
Displacement: 226 cid
Horsepower: 95 at 3300 rpm
Compression Ratio: 6.8:1
Fuel System: Holley one-barrel Model 847F5
Body Style: Two-door Sedan
Number of Seats: 6
Weight: 3031 lbs
Base Price: $1590
Number Produced: 396,060

Opposite: This two-door sedan version of the Custom Deluxe series was available with both sixcylinder and V-8 engines.

Below left: Neat dash layout with Ford crest in horn button and modern look speedo.

Below right: The flat head six-cylinder engine option.

Despite President Truman's order to US troops to enter the Korean 'conflict' in 1950 and the pressure to conserve raw materials exerted by the National Production authority, 1950 nevertheless saw a huge increase in car production and purchasing. Despite a slip from first to second place in the league of auto producers for the year, and trailing Chevrolet by nearly 300,000 units, Ford were still riding high from their restyle of 1949. As a company, Ford sold slightly more cars than in the previous year, but the increase in Chevy sales was enormous, nearly 50% in fact, as the company improved their output by nearly half a million cars. Traditionally however, Fords appealed to customers whose priority was performance.

Part of Ford's problem was that there were still gaps in their model line-up – they had no pillar-less hardtop, for example. Their Crestliner model (a twodoor sedan in the Custom Deluxe V-8 series, with wild two-toning and deluxe interior) was a slightly half-hearted attempt to resolve this particular omission. This gap was finally filled in the following year by the Custom Deluxe Victoria, which sold a staggering 110,286 units in its first year of production.

The two-door sedan Custom Deluxe was available with both of the 1950 engine options, the L-head six-cylinder and L-head V-8. Like the other (five) models in the line-up, the body was only slightly modified from the redesigned cars of the previous year, with minor styling changes. The Custom Deluxe models had the top trim for this model year.

The two-door sedan was by far and away the best-selling model in the Custom Deluxe range, (which had total sales of 818,371 units). In fact, sales of this option were nearly 50% more than the next best-selling model, the fourdoor sedan ($47 more expensive). On numbers alone, the highly specified Crestliner was the least successful, with a total number of 8,703 sales. Both the Crestliner and convertible models were only available with V-8s.

5 8 1
23 367 G

1950 Oldsmobile 'Rocket' Deluxe Holiday 88

Engine: Overhead valve vee-block,cast iron block

Displacement: 303.7 cid

Horsepower: 135 at 3600 rpm

Compression Ratio: 7.25:1

Number of Seats: 5

Weight: 3535lbs

Wheelbase: 119-1/2 inches

Base Price: $2035

Number Produced: 12,682

Oldsmobile has the longest heritage of any U. S. auto manufacturer, its handbuilt, wooden carriage-style vehicles having arrived on America's roads in 1897. This Lansing, Michigan based division of General Motors was successful right from the start, and its Curved Dash was the most popular car manufactured in the US in the early part of the twentieth century. In the '50s, Olds was benefiting from the 'Futuramic' restyling of 1949 and technological advances of the '40s. The latter included Hydra-Matic transmission (the first version of which was introduced back in 1940) and the new highcompression overhead valve Rocket V-8 under the hood (first offered in '49).

This engine continued to roar around the racetracks in 1950, winning 10 out of 19 major NASCAR stock car races, handing the championship to driver Bill Rexford. Fitted to a Rocket 88, the engine broke a class speed record at Daytona, averaging 100.28 miles per hour, developing 263 lbs of torque per foot at 1800 rpm.

These assets contributed to Oldsmobile's sixth position in the league of motor manufacturing output for the year, of 408,060 cars. This was achieved under the management of General Manager S. E. Skinner.

The 88 series was Oldsmobile's mid-range line-up, with a total of seven body styles available. The 88 Holiday was introduced as a new hardtop body style to the junior series, and accounted for part of Oldsmobile's 8% share of the US hardtop market. Deluxe and standard trim versions were available.

The series was now in its second year, and continued to be one of the hottest performing ranges available. One-piece windscreens were added to the range in the middle of the year, as exemplified in our photographed car. The less expensive 76 series would be discontinued in 1951, and the 88 would become the standard Olds. The top-of-the-range 98 series was Olds first slab-sided car, and was larger and more luxurious, but less performance orientated than the slightly smaller 88s. Both series had the 'Rocket' V-8 engine, but the 98 had nearly 250 lbs more to haul, and a seven-inch longer chassis to move.

1950 Plymouth Deluxe Sedan

Horsepower: 97 at 3600 rpm

Transmission: Manual with optional overdrive

Weight: 2946 lbs

Base Price: $1492

Number Produced: 67,584 (Two-door Sedan)

All-new Deluxe and Special Deluxe Plymouths had been introduced in 1949, so 1950 was a quiet model year for the company as it was for many other American motor manufacturers. K. T. Keller, who had succeeded Walter Chrysler in 1940, had introduced 'box styling' as the philosophy behind Plymouth products. This meant that comfort and practicality were valued over beauty.

His dictum was that the Plymouth autos should be practical transportation pieces in which one could sit bolt upright, wearing a hat. Although many buyers appreciated this common sense approach, and Plymouth's reputation for reliability, it did mean that the Chrysler styling fell behind that of its competitors until the mid-fifties. The cars were efficient and roomy, and modestly priced, true people's cars. These simple merits were emphasised in their 1950 advertising campaign, which offered the buyer 'spacious travel comfort'. Keller was promoted to being Chairman of the Board in this year, and Tex Colbert succeeded him as the President of Chrysler Corporation.

The '49 models were still pretty fresh in 1950, so the styling up-date for the year was limited, of the 'bolt on' type. Despite this, the cars were promoted as the 'Beautifully New Plymouth'. They were still using the L-head six-cylinder engine of the 1940s, however, and this contributed to a slow down in sales growth, combined with the general slackening of consumer demand. The Deluxes were available in both short and long wheelbase versions. The latter was seven and a half inches longer. The evolutionary changes to the Deluxes consisted of trim updates, and the rear fender was now attached with bolts so that it could be replaced if subject to collision damage. The fender design was also simpler than that of the previous year, with plainer moldings. The grille was modernised, and the rear windows were enlarged and widened. The windshields were unchanged. So were the dashboards, but these are a design classic that many people consider to be the most attractive ever used on a low priced car.

1950 Studebaker Champion

Engine: Champion Six-cylinder L-head, cast iron block	
Displacement: 169.6 cid	
Horsepower: 85 at 4000 rpm	
Compression Ratio: 7.0:1	
Fuel System: Carter Model WE-715S one-barrel	
Body Style: Four-door sedan	
Number of Seats: 6	
Weight: 2730 lbs	
Wheelbase: 113 inches	
Base Price: $1519	
Number Produced: 16,000	

Studebaker entered the 50s in great shape, in eighth position in the auto manufacturers production league table. Having gone bust in the Great Depression – mostly through serious mismanagement at a corporate level – the company management was taken over by the successful team of Paul G. Hoffman and Harold S. Vance. Between them, these two able men guided the company through the Depression, and introduced the Champion in 1939. They were still running the company when car production resumed in 1945, and were responsible for Studebaker becoming the first major US auto manufacturer to introduce new styling in the post-war period. The fact that it was a 'seller's market' in the late '40s meant that their striking new car could really fulfil its potential, and the company made healthy profits in the final years of the decade.

There was a superficial re-style of the successful models in Studebaker's line-up to welcome the new decade. The famous 'bullet-nose' grille at the front end and unusual vertical taillights were new, and looked very futuristic. Studebaker coined the phrase 'next look' to describe this styling and their advertising pictured the car parked next to a USAF F-86 Sabre jet fighter to emphasise its aerodynamic sleekness, the blurb describing their model as a 'beautifully flight- streamed car'. This was promoted as the gas-saving virtue of 'thrift'. Two model lines were now on offer, the base model Champions were available in a total of thirteen different body shapes in three trim levels (Custom Line, Deluxe Line and Regal Deluxe Line), mostly sedans and coupes with a single convertible model. The cars were economically priced at around fifteen hundred dollars. The company was taken over by Packard in the following year.

Right: The convertible version of the Champion.

Below: The bullet nose was very distinctive.

Above: The Champion may look quite bulky to modern eyes, but in 1950, it lines were extolled as having the 'goingsomewhere distinction of a jet-plane'.

Left: The L-head straight six engine.

1951 Buick Special (Two-door Coupe)

Engine: Straight-eight cylinder
Displacement: 263.3-cid
Horsepower: 120 at 3,600 rpm
Compression: 6.1:1
Number produced (two-door coupe): 2,700

The '51 Special was launched as one of four members of an 'All-Star (Fireball powered!) Line-Up for '51'. The advertising copy hailed the car as 'the newest car in the world – new in structure, new in power, new in dimensions, new in thrift – and potent in price appeal.' '51 versions of the Roadmaster and Super models were launched simultaneously.

The 1950-'52 General Motors Buicks were solidly built and heavily trimmed, so solid in fact that they are affectionately know as 'lead sleds' to this day. The heavy chroming, especially the cheesy grimace of the 'toothy' vertical-pillar grille was followed through in all the models of that year, plus a row of three chrome dimples on each front wing of the car that render the Special instantly recognisable as a Buick. The Specials were introduced with two- and four-door versions, all with Ned Nickles's 'hardtop convertible' styling, and all with sideopening hoods. A full convertible two-door model (the Special Deluxe) was also available for a further $515. Standard and Deluxe trim versions were offered (the photograph shows a Deluxe model). Interiors on Standard and Deluxe cars were cloth, but of a plusher grade in the latter.

Specials had a unique instrument panel, with speedometer and gauges housed in two large round units flanking the steering column notch. Controls were centered vertically, flanking the radio speaker grille. Standard Specials were the only 1951 Buicks to retain a split-windshield, the Deluxe models had the new one-piece version.

The cars often appeared in classic pastel two-tone combinations, pale blue and cream being especially popular, with white wall tire options. Wheels came with either solid chrome hubs or wires, which look particularly sporty.

In fact, this post war period was a great era for Buick styling, engineering and sales. Sales in 1950 reached 550,000 cars, fuelled to some degree by the

Below: The straight eight power unit was one of the few overhead valve engines around in the early 1950s.

Below right: Pale blue and cream was favorite color scheme that year and it is reflected in the interior décor.

fear that the impending Korean War would result in a shortage of consumer goods, as the Second World War had done.

With their straight 8-cylinder 4.3 liter engine block and dynaflow automatic transmission (first introduced on the 1948 Roadmaster, and the first torque converter automatic transmission), the cars still managed to impress original owners with their gas mileage. Yet, they were somehow a bit lacking in excitement and, even today, are not ranked among the most popular Buick models of this decade. The straight-eight era ended for the larger cars in 1952, when a fourbarrel carburetted, 170-hp Roadmaster version made a fitting farewell.

Above: This 1951 Special is in the classic pale blue and white wall tire combination, the chrome grille more massive than ever in the post war era.

1952 Lincoln Capri Convertible

Like virtually every other US car manufacturer, Lincoln entered the post-war years by re-offering their pre-war models. However, they were still fantastically luxurious up-market cars and even in 1946, the top-of-the-range Continental model was selling for a staggering $4474.

But in 1952, Ford was the only one of the 'big three' car manufacturers to instigate a complete re-design of its entire model range, and Lincoln was a beneficiary of the 'All new for '52' look. For the first time at FoMoCo, both Lincoln and Mercury offered true pillarless hardtop models in their respective ranges. At Lincoln, this was the Capri two-door hardtop model. In fact, the entire '52 range shared the Ford corporate styling, and had a lean racy look.

The bumper and grille were integrated in the car bodies instead of being recessed. All the new Lincolns were fitted with Ford's first modern overhead valve, 5.2 litre V-8s. Hydra-Matic auto transmission was fitted as standard, and Ford is reputed to have fitted the first ball-joint front suspension to these Capri models. Their reputation for performance was reflected in the racing successes of the marque. Lincolns took first, second, third and fourth positions in the 1952 Pan American 2,000 mile Road Race in Mexico (although '53 cars were used). A 'maximum duty kit' was available as an optional extra for owners who wished to race their own cars. The watchword of the Lincoln range was that they were fine cars with outstanding performance. But although Lincoln was to retain its reputation for luxury and comfort, the Chrysler 300 letter cars were about to steal its racing crown.

The 'Capri' badge was first used in 1950 as a limited edition version of a Continental, but became a product line in its own right as the new top Lincoln series for '52. The cars were well appointed, featuring fabric and leather upholstery. The gas tank filler was hidden behind the rear licence plate. The line remained virtually unchanged for a further three models years, until the heavy re-style of 1956. Cosmopolitan was the base line model for '52, though this being Lincoln, the cars were very well equipped and sold for well over $3,000.

Engine: Overhead valve V-8, cast iron block

Displacement: 317.5 cid

Horsepower: 160 at 3900 rpm

Transmission: Hydra-Matic automatic

Compression Ratio: 7.5:1

Number of Seats: 6

Weight: 4350 lbs

Base Price: $3665

Number Produced: 1,191

1953 Buick Roadmaster Convertible

Whatever excitement had been missing from Buick in the 1950-'52 models was back in 1953. The big news was under the hood – a 322-cid overhead valve V-8 for Super and Roadmaster models. For the first time ever (except for two sedans), all Buicks shared a common wheelbase. Celebrating its 50th, anniversary (and sharing it with the Ford Motor Company), the Buick Division of General Motors was locked into aggressive price war with Ford – its most direct competitor. GM retained a 7.9% market share, with a model year total of 488,755 cars built. The seven-millionth Buick came off the production line in the June of that year.

The company owed its success in part to the upsurge in consumerism that resulted from the end of the Korean War. Detroit accurately reflected the national desire for good times by launching an endless stream of dazzling new models onto the market. The average family now aspired to owning two cars, and as the middle classes became increasingly better off, this became a reality for many.

The 1953 Roadmaster (series 70) was the top of the Buick's range of cars. Their finest model had a fore-shortened nose to emphasise the compact power of the newly introduced 'Fireball' V-8 engine under the hood.

Roadmasters sported chromed rear fender gravel shields and the same sweepspear side detail that appeared on the slightly less expensive Super models. Four chromed ventiports adorned each wing. The interior options were nylon, broadcloth or leather with Roxpoint nylon carpeting as standard.

Color co-ordinated instrument panels were incorporated into the dash, which followed the glittery jukebox look favored in the American cars of the period.

Engine: V-8 overhead valve. Cast iron block

Displacement: 322 cid

Horsepower: 188 at 4,000 rpm

Transmission: Dynaflow fitted as standard

Compression: 8.5:1

Price (two-door convertible): $3506

Number produced (2-door convertible): 3,318

1953 Buick Skylark Anniversary Convertible

Engine: V-8 overhead valve. Cast iron block	
Displacement: 322 cid	
Horsepower: 188 at 4,000 rpm	
Transmission: Dynaflow	
Compression: 8.5:1	
Fuel System: Stromberg 4AUV267 carburetors, or Carter 996 or 2082S four-barrel	
Body Styles: Convertible, two-door	
Number of Seats: 6	
Weight: 4315 lbs	
Price: $5,000	
Number Produced: 1,690	

Designed as a model to celebrate Buick's 50th Anniversary, the Skylark convertible (style 53-4767SX) was based on the top-of-the range Roadmaster chassis, but had its own fenders with open wheelhouses painted white or red. It also did without the classic Buick ventiports, and looked all the more streamlined for that.

Designers were the true kings of Detroit in the '50s, supplying the increasing market for cars with a new selection of models every year. This hugely increased the American car market, as the average driver now did in excess of 10,000 miles a year. Motels now outnumbered hotels two to one, catering for the needs of an increasingly mobile population. Special editions like the Skylark, which weighed in at a fairly punchy $5,000 stimulated the market, and pushed design and specifications further on. Even at this price, the model proved immensely popular, and the cars are still very collectible today. A total of 1,690 cars were produced.

The Skylark sported a shapely lowered topline to its body, 40-spoke Kelsey-Hayes wire wheels and a full complement of luxury accessories were included. The interior of the car was leather, and the car was trimmed with slim-cast sweepspear moldings and special bodyside emblems on the rear quarters. The seats, windows and convertible tops were all controlled hydraulically.

The Skylark was an essentially '50s car, adorned with the glitter, go and gadgets that the American car buyer of the time insisted upon. With gas at a quarter a gallon, fuel economy and safely features were fairly tertiary considerations when car buying, glamour being the most important at this end of the market.

Below: An open topped view of the chromed dash and steering wheel.

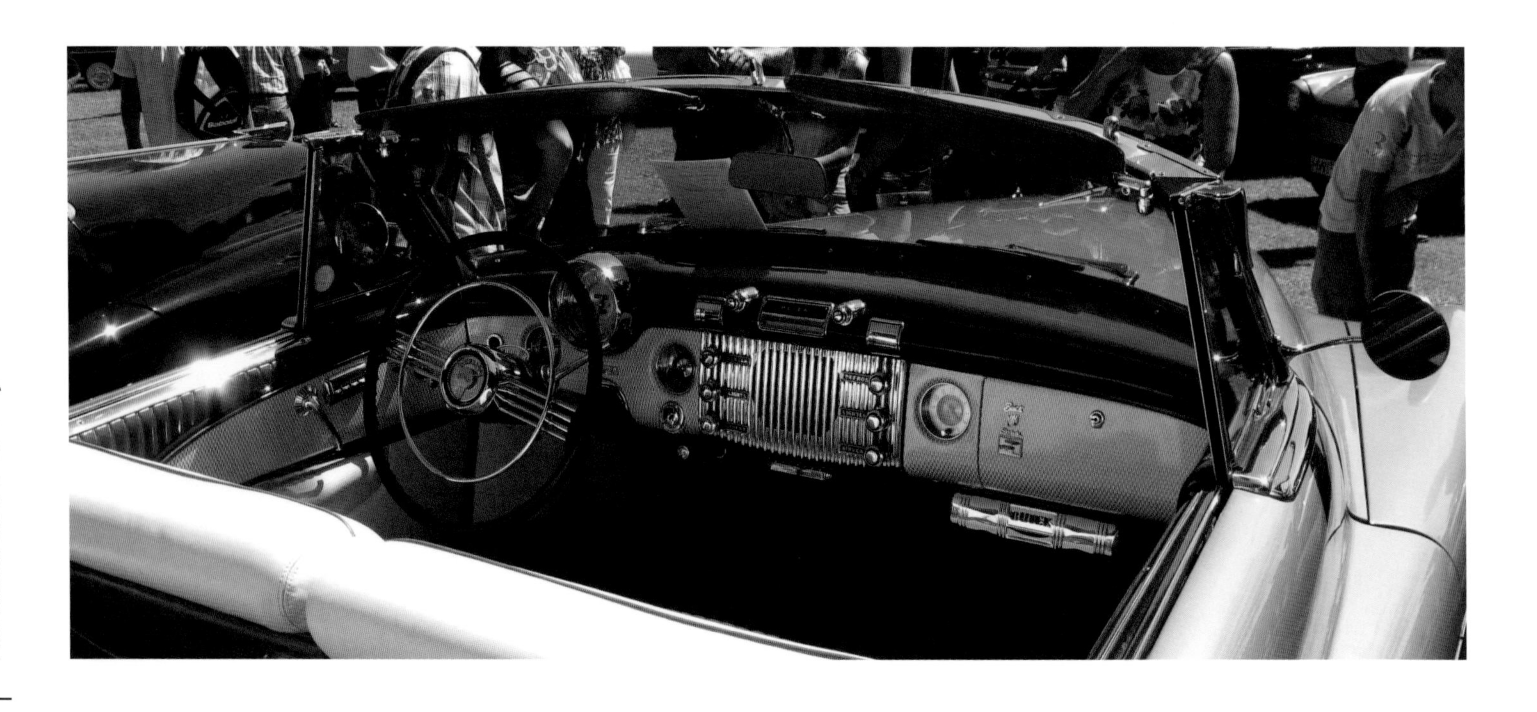

Above: This elegant cream and cherry Skylark demonstrates the elegant flowing lines of this anniversary model.

Left: A handsome car from all angles.

1953 Chevrolet Corvette Roadster

Engine: Inline six, overhead valve.
Displacement: 235.5 cid
Horsepower: 150 at 4200 rpm
Transmission: Powerglide automatic
Compression: 8.0:1
Body style: Convertible
Number of seats: 2
Weight: 2705 lbs
Wheelbase length: 102 inches
Price: $3498
Number Produced: (1953) 300

The Corvette was launched in 1953 and was a huge departure from the norm for the American car industry. It is said that GIs returned from the War were the inspiration behind the car, as they looked for American models that reflected their enthusiasm for British and Italian sports cars. It is ironic, therefore, that the Chevrolet division of General Motors– famous for its stable of family cars should become the manufacturer of such a unique and sporty car. Two men were the inspiration behind the car – chief stylist Harley Earl and chief engineer Ed Cole anxious to prove that GM could build a car to emulate the Jaguars and Ferraris they revered.

The original concept car behind the Corvette brand, the EX-122, had toured the States in the GM Motorama and generated huge interest, having also appeared in the GM promotional film 'Halls of Wonder'. The car was consequently put into production with few alterations. A tiny run of 300 cars was built in 1953. But cost saving in the manufacturing process somewhat diluted the original concept, and the engine in particular was disappointing compared to the European cars that had inspired it. Ironically, the convertible top for the first Corvette was very much an afterthought and it looked awkward and inelegant. It was also tedious to put up and down – the sidescreens had to be stored in the trunk, and the hood reduced the driver's visibility considerably.

The first cars were roundly criticized by the high-profile customers who were lucky enough to receive them, and at nearly $1,000 more than an imported MG the price tag meant that the car was well out of reach of the young drivers it was aimed at. The 'Blue Flame Six' engine was particularly derided. But Cole and Earl refused to accept that the car was a failure, and they gradually developed the model line into the American classic it remains to this day. Their first attempts are now regarded as Fifties classics, and the forerunners of a breed of true sports cars.

Right: A view of the Corvette cockpit. Note how extreme the wrap around nature of the windshield is.

Above: '53 'Vette in Polo white with a stunning background of Montana scenery.

Left: The Blue flame six was not a performer even with triple side draught induction.

1953 Dodge Coronet Convertible

Like every other auto manufacturer, Dodge entered the post-war auto market with face-lifted pre-war models. Like the other Chrysler lines, the cars changed very little until 1949, when the 'Second Series' cars, with their rather boxy exteriors, were introduced. Venerable model names like Deluxe and Custom were replaced by the newer names of Coronet, Meadowbrook and Wayfarer. In fact, they were some of the first Chrysler models to be designed by Virgil Exner (who had formerly worked for Studebaker).

Introduced in 1949, the Coronets were the most expensive cars Dodge produced in the year, and received the top level of trim. The basic body shell for the model was identical to the Meadowbrook, but was more richly appointed inside and out. Six models were introduced, ranging from a two door convertible to an eight-seat four-door station wagon.

President Eisenhower lifted production controls on domestic cars in February 1953, opening the gates to hugely increased production levels from all the major manufacturers. Dodge retained its position of seventh in the league table of car producers for the year.

Apart from a general Dodge price cut in March 1953, the big news for the Coronets was the introduction of a hemi head overhead valve V-8 under the 'Red Ram' hood ornament. Despite the extra sizzle introduced by the V-8, the six-cylinder engine was still much in evidence. In fact, the V-8 was fitted to only two out of the five Coronets, the four- and two-door sedans. The convertible was only ever available with the six-cylinder option.

The Coronet remained the top trim level Dodge model in 1953; the new six-cylinder models were launched in March. The cars were assembled in three Dodge plants, Detroit, San Leandro and Los Angeles.

Dodge outperformed all other American eight-cylinder models in the 1953 Mobilgas Economy Run (achieving 23.4 mpg), whilst Dodges also set 196 AAA stock car speed records at the Bonneville Salt Flats, as well as clocking up six NASCAR victories in the year.

Engine: Inline, L-head, six-cylinder

Displacement: 230 cid

Horsepower: 103 at 3600 rpm

Compression Ratio: 7.1:1

Fuel System: Carter one-barrel Model D6H2

Number of Seats: 6

Weight (Convertible): 3480 lbs

Base Price (Convertible): $2503

Number Produced: 4,100

1953 Ford Customline Two-door Sedan

The year 1953 was the 50th anniversary of Ford production. A special Ford crest appeared in the center of the steering wheel hub of cars manufactured in the year, containing the words '50th Anniversary 1903-53'. The company celebrated its birthday by manufacturing a greatly increased number of cars, and closing the gap between itself Chevy, to less than 100,000 cars. This despite a Chevy restyle. One major innovation for the year was the introduction of Master Glide power steering. 1952 had seen the introduction of the first completely new Ford body shape since 1949.

The new shape was introduced to make the company's cars more effective competitors to their Chevrolet and Plymouth rivals. The new models featured a one-piece curved windscreen, fullwidth rear window, protruding parking lights and new body trim. Inside, there was a completely redesigned control panel and suspended clutch and brake pedals. Under the hood, the six-cylinder Ford engine was overhauled, but the flat-head V-8 remained. Customline had been introduced as part of the general re-style as the mid-price range model in Ford's new line-up. This continued to be the case in the following year.

Mainlines were the base Ford model, and the Crestline had the top trim option. In the second year of the new shape Fords, the body panels were used, but the grille was redesigned, and elongated to wrap around the front of the car, whilst parking lights became rectangular instead of round. The Customline series hugely outsold the other Ford series, accounting for 761,662 units of the total years production of 1,244,540.

Mainlines accounted for 302,714 cars, and Crestline 180,164. The fourdoor sedan was the most popular Customline model with sales of 374,487 units. Customlines were available with both six-cylinder and V-8 engines in three of the four body styles offered.

Engine: Six-cylinder overhead valve or L-head V-8. Both had cast iron blocks

Displacement: 215 cid/ 239 cid

Horsepower: 101 at 3500 rpm/ 110 at 3800 rpm

Compression Ratio: 7.0:1/ 7.2:1

Number of Seats: 6

Weight: 3067 lbs/ 3133 lbs

Base Price: $1724/ $1809

Number Produced (division between V-8 and 6-cylinder sales not shown in Ford sales figures): 305,433

1953 Henry J

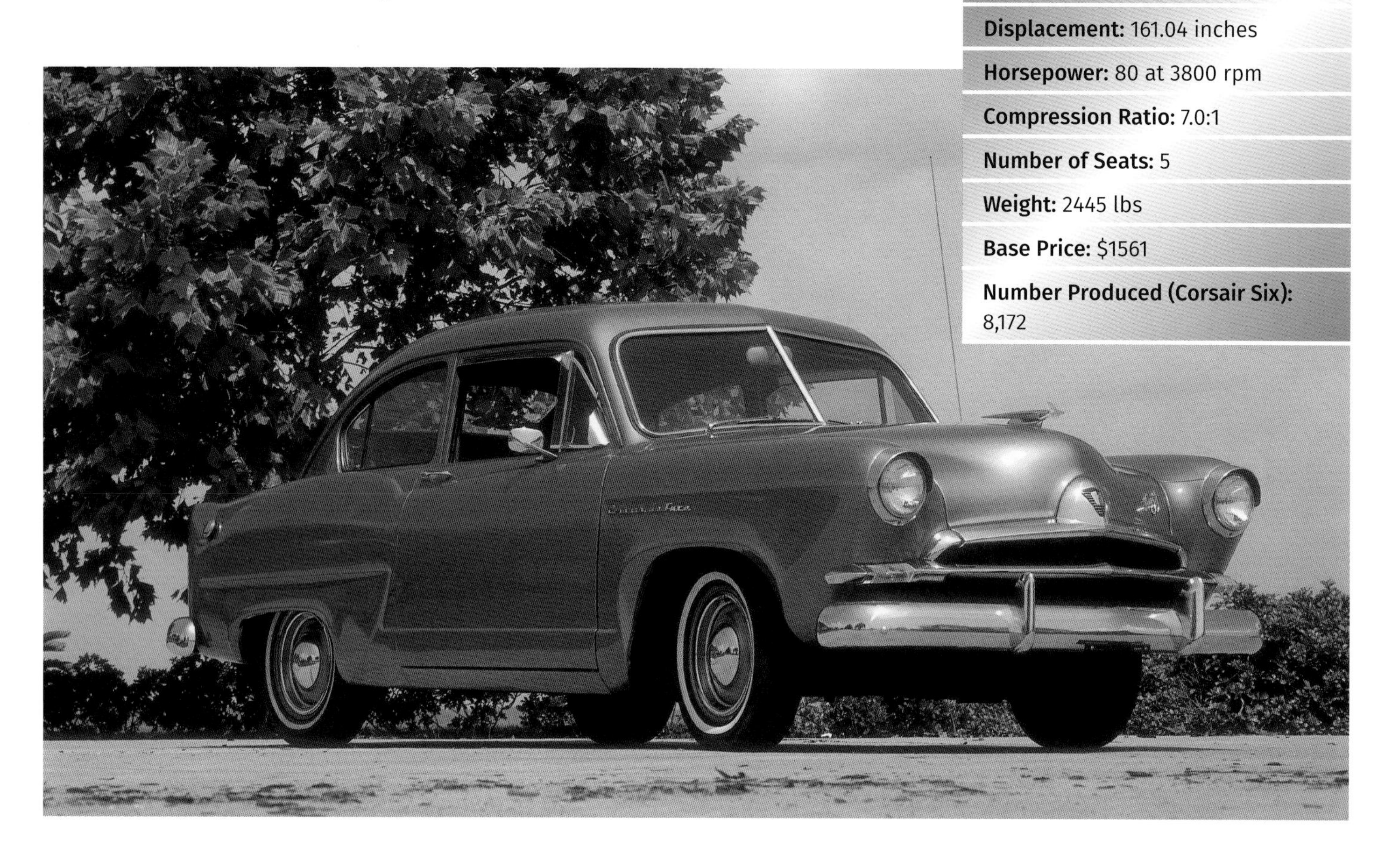

Engine: Inline six-cylinder, valves in block
Displacement: 161.04 inches
Horsepower: 80 at 3800 rpm
Compression Ratio: 7.0:1
Number of Seats: 5
Weight: 2445 lbs
Base Price: $1561
Number Produced (Corsair Six): 8,172

The Henry J was a low-priced, compact economy car introduced by Kaiser-Frazer in 1951. Henry Kaiser had been very successful in his manufacturing career, and believed that there was an untapped mass market for a car that could be sold on price alone – just as the Model T Ford had been. 'Dutch' Darrin had styled other Kaiser models, and wanted the commission to design the Henry J prototype along the lines of these earlier models. But the temperamental Kaiser employed American Metal Products of Detroit instead, with a brief to evolve something completely different. Something that he hoped would change the fortunes of the company. AMP named the new model line after their patron, and the 'Henry J' was born. The car was perched on a short hundred-inch wheelbase and the car had optional four- or six-cylinder power options.

The rather sweet little car was fundamentally a two-door fastback sedan, and had miniature bench seats inside. The New York Fashion Academy named the car 'Fashion Car of the Year' and this initial model sold 81,942 units of the Standard and Deluxe models. But only the top of the range car that had an opening trunk lid, which gives an idea of exactly how basic a car this was.

The 1951 model was marketed as a Vagabond in the following year, and had some modest styling updates. A second Henry J, the Corsair, replaced it later in '52. The cars were manufactured at Willow Run, Michigan. But sales figures were following the pattern for the rest of Kaiser's automotive production – downwards - to around 22,000 units. Sales remained similarly depressed in 1953, with 17,505 units produced, despite a general 'tweaking' of the car's styling. This consisted of a new hood ornament, new chrome wheel discs and a curved rear bumper. Kaiser had managed to get a deal on with Sears to market the car in their stores, using their Allstate brand name. But the associate failed miserably, and Sears pulled out of the deal in '53 after two unsuccessful sales seasons.

In May that year, a Henry J won the Mobilgas Economy Run, but this was about the only unqualified success enjoyed by the car. The Henry J experiment collapsed in 1954, and is estimated to have cost Kaiser in excess of a hundred million dollars.

1953 Mercury Monterey Sedan

Ford had formed the Lincoln-Mercury division in 1945, but it wasn't until the 1949 models were introduced that Mercury shed its image of being a line of glorified Fords. Ford had intended to introduce an updated overhead valve V-8 to replace the old flathead in 1952, but it wasn't ready in time, so the old engine was retained until '54. That was only one of the innovations in the 1954 models, but in 1953 this was all still to come. The entire line was to be restyled in 1955.

The Monterey model name was introduced in 1952, as an up-scale version of the base model Custom. This was the first time Mercury offered two distinct model lines in the same model year. Standard features included a two-tone paint job, and a leather and vinyl interior. Suspended pedals were also standard in both series. Mercury introduced three body styles in the original series, a hardtop coupe, sedan and convertible. The following year, the car was almost identical, and a station wagon was also added to the Monterey range. The car had a smoother cleaner look than the earlier Mercury series, being designed with 'unified' styling. Essentially, this meant a one-piece rear window, one-piece windshield and down-sweep hood for better forward visibility. The bumper and grille were now combined in a single unit, as were the rear lights and bumper. The Monterey interior was roomy and comfortable, and simply designed. The truck was a capacious thirty cubic feet. The model was available in fifteen solid and metallic paint colours.

'53 was also Ford's 50th anniversary year, and a Monterey convertible was used as one of the promotional vehicles for the celebrations. It was supposedly the 40 millionth car to roll off a FoMoCo production line. Ford brothers Henry II, Benson and William Clay were photographed driving the car.

Mercury sold a total of 305,863 cars in the year, and came eighth in the league of motor manufacturers.

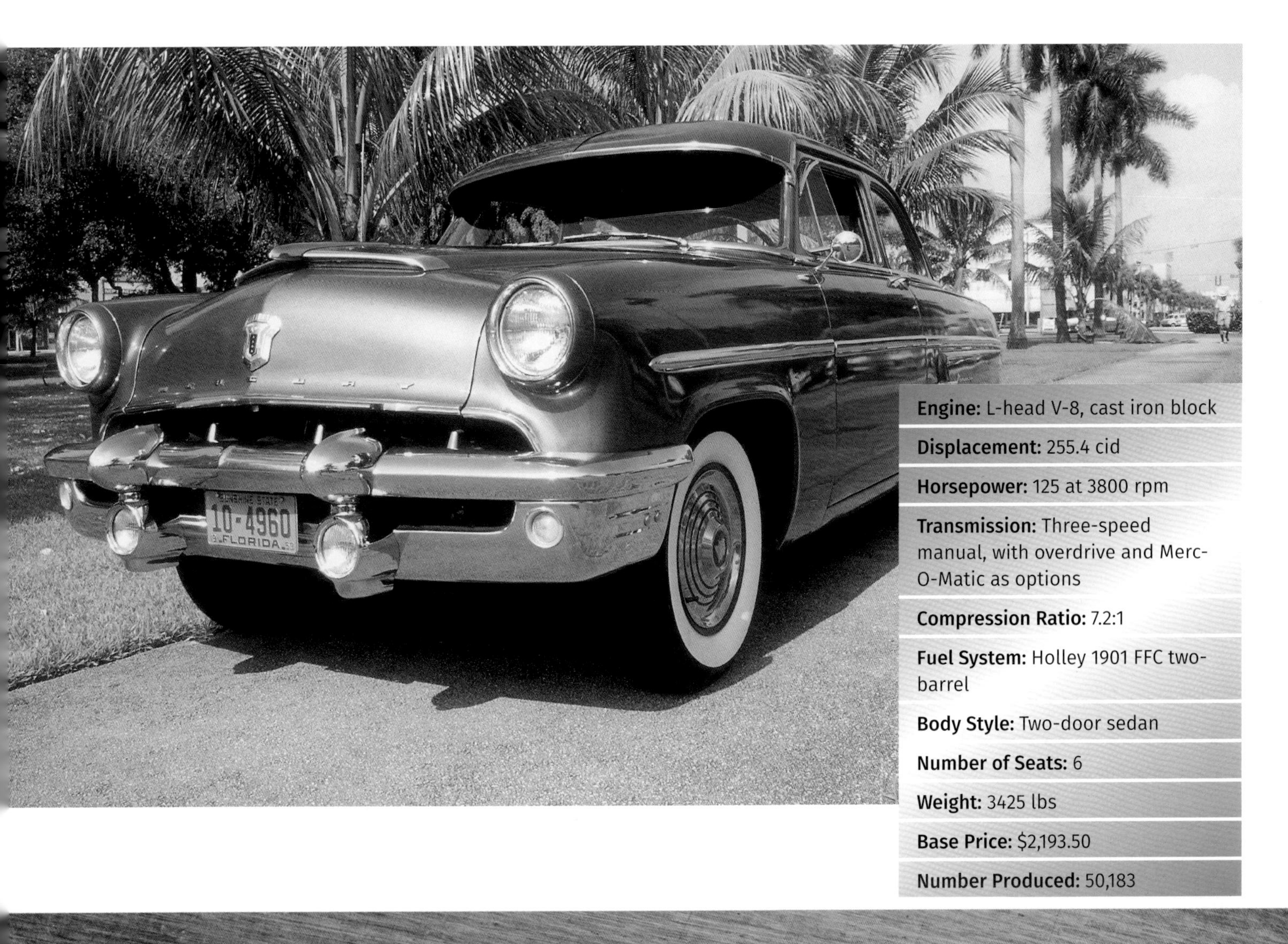

Engine: L-head V-8, cast iron block

Displacement: 255.4 cid

Horsepower: 125 at 3800 rpm

Transmission: Three-speed manual, with overdrive and Merc-O-Matic as options

Compression Ratio: 7.2:1

Fuel System: Holley 1901 FFC two-barrel

Body Style: Two-door sedan

Number of Seats: 6

Weight: 3425 lbs

Base Price: $2,193.50

Number Produced: 50,183

1953 Muntz Jet

Engine: Lincoln L-head V-8	
Displacement: 336.7 cid	
Horsepower: 154 at 3600 rpm (218 at 4200 rpm option available)	
Transmission: GM four-speed Hydra-Matic	
Compression Ratio: 7.5:1	
Body Style: Two-door Convertible	
Number of Seats: 4	
Weight: 3560 lbs	
Base Price: $5114	
Number Produced: 136 (includes sales for '54)	

Muntz began life in 1951, when the television magnate Earl 'Madman' Muntz bought the California-based Kurtis-Kraft business. He modestly gave the car his own name, and made some relatively superficial changes. He moved production of the car to Evanston Illinois.

Frank Kurtis had begun the original business in 1949, having built many winning cars for the Indy races. He used his performance know-how to go into general production cars, aiming to offer a high-performance sports car to the public. He launched a two-door roadster that was practically custom built.

Although there was a standard Kurtis bodyshell, any kind of technical kit could be fitted. When the customer didn't specify an engine, Kurtis fitted a modified Ford Flathead V-8. In this option, the cars were retailed at $4,700. He produced 36 cars in this incarnation of the business.

Muntz launched the Jet two-door hardtop coupe in 1951. The styling looked virtually identical to the Kurtis, but the wheelbase was stretched by three inches to make it a four-passenger car. He also changed the standard engine to a Lincoln V-8, which gave the 1951 Muntz a top speed of over 108mph. The car changed again in 1952 to a roadster-convertible with a removable hardtop, retaining the Lincoln engine. The final version of the car, launched in 1953 remained as the '52 model in every respect, except that the bodies were now constructed from steel rather than aluminum, and were consequently 260 pounds heavier, as well as being $178 more expensive than in the previous year. Muntz ceased car production in '53, but leftover stock was designated a '54 model. Kurtis returned to car production in the same year, basing his new model on the victorious Indy 500 car he had constructed for Bill Vukovitch. The new car was equipped with a Cadillac V-8 in a McCulloch Motors-supplied body. He put 20 of these 500M two-door roadsters together, which were capable of 135 mph, before discontinuing sales to the general public in '55.

Right: A hopped-up version of the Ford (Lincoln) flathead V8.

Opposite top right: This 1953 Muntz Jet had simple and elegant lines. The hardtop was removable to make it a truly versatile convertible.

Opposite bottom: The interior of the Muntz was sports car orientated.

1953 Oldsmobile Fiesta Convertible

Engine: Overhead valve V-8 with cast iron block
Displacement: 303 cid
Horsepower: 170 at 4000rpm
Compression Ratio: 8.3:1
Base Price: $5715
Number Produced: 458

The year 1952 had been great for Oldsmobile. They had risen to fourth in the league of automotive producers, which was a new high for the company.

During the year, they also continued to be the top-selling automatic range in America. Oldsmobile also continued to be prominent in the NASCAR stock races, winning the Southern 500 at Darlington, South Carolina on Labor Day 1950, which was one of three NASCAR victories for the year.

Things weren't quite so rosy in 1953. Despite an increase in cars produced, the company fell back to seventh in the production league. There was also a terrible fire at the General Motors Hydra-matic plant that completely destroyed the factory. This meant that many automatic Oldsmobiles produced in the year were actually fitted with sister GM company Buick's Dyna-flow transmissions.

On the positive side, the sporty fibreglass-bodied Starfire concept car (complete with aircraft-inspired oval grille) toured with the year's GM Motorama car show for six months, winning acclaim from 1.7 million visitors. In fact, the Starfire name was destined to become the inheritor of the Fiesta legacy. At NASCAR events, the '53 Olds, complete with heavy-duty factory parts, were victorious in nine races, whilst Bob Pronger set a new speed record of 113.38 mph in the Flying-Mile at Daytona in an 88 two-door sedan. Bill Blair drove another Olds to victory in the 1953 Daytona stock car race (run over 160 miles).

Oldsmobile launched a trio of flashy limited-edition convertibles in 1953, the

Below: The Ninety Eight series also came as a four-door sedan.

Fiesta appearing mid-year. Classed as part of the Ninety-Eight series, it joined a range of already luxury cars, but was even plusher than the other models with 1954 styling advances such as the 'panoramic' wraparound windshield, sloping windows and virtually every other Oldsmobile option as standard (with the exclusion of slightly redundant air conditioning). If required, Oldsmobile/Frigidaire air conditioning was available at $550. The fitted options included Hydra-matic transmission, a padded dashboard, windshield washers, and a deluxe steering wheel with a horn ring. The upholstery in the photographed model is white leather. The Fiesta V-8 engine was slightly more powerful than that fitted to other ninety-eight models, developing 170 horsepower at 4000 rpm rather than 165 at 3600. But considering that the car was priced at nearly double that of the regular ninetyeight convertibles, it is hardly surprising that it only achieved 6% of the sales volume of the cheaper model.

Above: This 1953 Fiesta became a milestone convertible car.

1953 Packard Caribbean Custom Convertible

Engine: Cavalier/Mayfair Eight. Inline eightcylinder L-head engine, cast iron block
Displacement: 327 cid
Horsepower: 180 at 4000 rpm
Compression Ratio: 8.0:1
Base Price: $5210
Number Produced: 750

The Packard Motor Car Company had had an up-and-down time of it in the '40s, and it took some while to get the company back to full level peacetime production. President George T. Christopher steered the company through the worst of these problems, and PMCC launched its first post-war model, the Twenty First Series, in 1946 (although production was still limited by the availability of raw materials). PMCC celebrated its Golden Anniversary in 1949, and was poised for the '50s as the oldest fine-car company in America.

During its illustrious history, Packard had been responsible for many automotive innovations, including the 'H' shaped gear slot, hydraulic shocks and pressurized cooling. The first car to beat the mile a minute record was a Packard, and a Packard Thirty was the first automobile to drive coast to coast in 1908.

Unfortunately, the resolution of the war-imposed production problems could not disguise the fact that the automobile market had undergone a radical change. Packard, 'a gentleman's car built by gentlemen' did not revise their styling or models on an annual basis, relying instead on traditional values of elegance, efficiency, distinction and prestige. The company believed that their customers wanted durable cars of quality, and that they were uninterested in the vagaries of fashion. But by 1949, Packard realised that this was no longer the case, and that they had to keep abreast of new trends. Ironically, this realisation was the root of the company's ultimate demise, as the expense involved was unsustainable for an independent auto manufacturer.

The Cavalier/Packard Eight-cylinder series was introduced in 1953, and its sedan model was directly comparable to the Packard 300 of the previous model year. However, the series also included some more sporty Packards,

Below: Packard's L-Head straight eight was old hat by 1953.

Below right: Dash and upholstery match the external paint color.

including the new semi-custom Caribbean convertible inspired by the Pan American show car of 1952. This ragtop was introduced in January '53 as a midyear model, as the most expensive car in the series. The Caribbean was powered by an eightcylinder engine, which pointed to another of Packard's problems, the lack of a modern overhead valve V-8. Its body was designed by the Mitchell-Bentley Corporation of Michigan, and the car listed a host of standard luxury features including a full-leather interior, chrome-plated wire wheels, a de-chromed body style, integrated fishtail rear fender, continental tire kit and custom paint finish.

Right: Produced in just four custom paint finishes, this lovely Caribbean is resplendent in Gulf Green Metallic with white leather interior.

1954 Buick Century Estate Wagon

Engine: V-8 overhead valve, cast iron block
Displacement: 322 cid
Horsepower: 195 at 4100 rpm
Compression Ratio: 7.2:1
Number of Seats: 6
Weight (Station Wagon): 3975 lbs
Wheelbase: 122 inches (206.3 overall length)
Base Price (Station Wagon): $3470
Number Produced (Station Wagon): 1,563 (out of total of 81,982 Century models manufactured in 1954)

Buick launched its hot new Century series in 1954, shoehorning the powerful Roadmaster engine into the lightweight Special body for extra performance. The company was joining the national horsepower race, fifteen out of eighteen makes announced improved performance and higher engine ratings in the year. GM were also locked in a damaging price war with Ford, and were anxious to compete as aggressively as possible by offering new models to a newly-prosperous American public.

The Century name was revived for the first time since 1942 for a new range of high performance cars. The line was launched with four model options, a four-door sedan, a two-door hardtop coupe, a two-door convertible coupe and a four-door station wagon. GM show car styling was visible on all four cars. Some of the distinctive Ned Nickles's Buick styling from the early '50s was retained however - the 'toothy' vertical grilles and ventiports for example. In fact, the Specials and Centurys were virtually identical in their exterior trim, with the exception of the different model scripts. But the actual dimensions of the Special/Century chassis was larger than in the previous year, having grown lower and wider. A new Panoramic windshield was also fitted, with slanting side pillars. The Century/Special models also shared the new dual bullet taillamps. The interiors were also very similar, with identical instrument panels (gauges set into twin round housings), and upholstery cloth.

Some of these design modifications were to accommodate the new overhead valve V-8 powered engine block, which was fitted to the Specials, and shared by the Centurys. But the Centurys were far more powerful beasts than their stable mates. Even in the station wagon, the new powerhouse developed a respectable 195 horsepower at 4100 rpm (200 with Dynaflow transmission), compared to the 143 at 4200 rpm achieved by the Specials.

1954 Chevrolet Bel Air Convertible

Like almost all American manufacturers, Chevrolet had entered the postwar years with a warmed over version of a 1942 model. However, Chevrolet quickly assumed its position as the best-selling American marque, earning their reputation as value leaders in the low-price field. Chevrolet manufactured 1,143,561 units in 1954 and outsold Ford (even though Ford actually produced more cars). This was the last year for the exclusive use of six-cylinder power plants in all Chevrolet models. The only two engine options were Inline Six Syncromesh and Inline Six Powerglide.

Styling improvements for 1954 Chevrolets' although limited, gave the impression of wider, more modern cars. The full-width wraparound grilles with their five vertical 'teeth', newly designed front bumpers headlight rims and hood mascot were all introduced in this model year. The taillight housings were also brought up to date with contemporary looking tailfin styling.

The new 1954 Bel Air convertible (in fact, all the new models were introduced in December 1953) had the traditional assortment of extra equipment and features such as full genuine carpeting, newly designed full wheel discs, two-tone vinyl trimmed interiors, snap-on boot cover and an electric clock. Instead of the inconvenient dashboard-mounted rearview mirror, this was now mounted at the top of the bright metal windshield frame.

Full-length chrome sweepspear moldings identified all the Bel Air models, and double moldings on the rear fenders enclosed the Chevy crest and Bel Air name badge.

Engine: Inline Six, overhead valve

Displacement: 235.5 cid

Horsepower: 115 at 3700 rpm

Transmission: Syncromesh or Powerglide

Compression: 7.5:1

Fuel System: Rochester one-barrel 'B' type Model 7007181 or Carter one-barrel 2102S

Body Styles: Two- and Four-door sedan, twodoor Sport Coupe, Two-door Convertible, four-door Station Wagon

Number of Seats: (Convertible) 5

Weight: 3445 lbs

Wheelbase Length: 115 inches

Price: $2165

Number Produced: 19,383

1954 Ford Ranch Wagon

The 1954 Fords were introduced to the public on January 6. They utilized the 1952 bodyshell, with a few cosmetic trim changes. The grille was redesigned with a central spinner and horizontal chrome bar, with round parking lights at either side. Several new convenience options were also introduced, including power windows, power seats and power brakes. Ford was also the first to update their suspension with ball-joints.

But the main change to the 1954 Fords was under the hood. The old flathead V-8 was replaced across the range by the more modern Y-block engine with overhead valves. This engine was nearly 25% more efficient than the previous power source, and followed the national trend towards more powerful autos.

For the first time in years, Ford was number one in the auto production charts, beating Chevy into second place by a slim margin of 22,381 units and keeping their fifteen manufacturing plants busy. The new V-8 got the credit for this reverse in fortune; economy and safety were still rather peripheral ideals in the auto industry.

The main news for the Ford Division itself was that the following year would see the introduction of the completely new Thunderbird concept in the 1955 Ford model line-up.

The Ranch Wagon was one of three new models introduced to the line-up. It was a two-door station wagon in the base trim Mainline V-8 series, but with Customline appearance features. These included the body-long chrome accent strip, and a chrome stone shield at the bottom of the back quarter panels. The 1954 hood ornament was also rather more streamlined and 'aeronautic' in appearance. The station wagon was still a staple of the American car market, and the vehicle of choice for the average suburban family in need of practical, flexible transport. There were three other options in the Mainline V-8 series, a four-door sedan, a two-door sedan and a two-door business coupe. A total of 233,680 units were sold.

1954 Hudson Hornet

Engine: Inline six-cylinder L-head, chrome alloy block

Displacement: 308 cid

Horsepower: 114 at 4000 rpm

Transmission: Three-speed manual with optional Overdrive and Hydra-Matic automatic

Compression Ratio: 7.5:1

Weight: 3288 lbs

Wheelbase: 124 inches

Base Price: $3800

Number Produced: 24,833 for the Hornet/Hornet Special series in this model year

Hudson began production in 1909, taking its name from J.L. Hudson, the financier and backer of the fledgling company. Hudson quickly became identified with modestly priced closed cars. During the war, the company had made a small profit in airplane production, returning to car manufacture in 1946. Like virtually every other US car produced, their first post-war offerings were slightly face-lifted pre-war models.

When production really got going in '48, Hornet introduced one the great automotive ideas of this era, the so-called 'Step Down' models. Being low and sleek, the car handled very well indeed, and stayed in production until '54. Despite their lack of a V-8 power plant, Hornet introduced several new models in the early '50s, including the Pacemaker (1950) and four body styles of Hornet (1951). The first Hornet line-up included a convertible. In the years '51-'54, possibly the most successful aspect of the Hudson Motor Company was their 262-cid Super Six engine, which was the king of the stock car racing circuit in this period. A Hornet won the NASCAR championship in 1952, 12 out of 13 AAA stock car races in 1953 and 65 NASCAR victories in 1954. However, declining sales, and a couple of lemon models (the Jet and Jet Liner) resulted in the company merging with Nash-Kelvinator in 1954 to form AMC.

Production ceased at the Detroit Hudson factory in October that year.

The 1954 Hornet models were divided into two versions, the Hornets and Hornet Specials. These were cheaper and had a more subdued exterior trim, sharing the interior trim of the Hudson Super Wasps. The Convertible Brougham came only as a full-up Hornet and was the most expensive in the series. The cars were fitted with the Hornet 'Big' Six engine that had achieved 112mph for Marshall Teague in the NASCAR races.

1954 Kaiser Darrin

Engine: Inline six-cylinder with F-head

Displacement: 161 cid

Horsepower: 90 at 4200 rpm

Compression Ratio: 7.6:1

Fuel System: Carter type YF one-barrel

Number of Seats: 2

Weight: 2175 lbs

Base Price: $3655

Number Produced: 435

Henry Kaiser had parted company with his original business partner, Joseph Frazer by 1954 and the company was in some difficulty by 1954. Sales had slumped, and the models were struggling price and innovation-wise against the offering of the big established automobile manufacturers.

In an attempt to stave off failure, Kaiser followed the path of other independent carmakers, and bought Willys-Overland of Toledo, Ohio. But Kaiser had always been a very innovative company, and they certainly followed this inclination with the Kaiser-Darrin sports car, introduced to the range in 1954.

The car was Howard 'Dutch' Darrin's brainchild, and had first been announced in 1952. It was based on the chassis of the Henry J (q.v.). Initial prototypes were shown in the following year, and a limited number (between 12 and 62) experimental cars were produced for testing and exhibition purposes at Kaiser's Jackson, Michigan plant. The merger with Willys resulted in a production move to Toledo, and this somewhat delayed the manufacture of the new roadster. Darrins were built in fibreglass, and were tremendously sculptural in appearance; long, low two-seaters with strange angles that just couldn't have come out of sheet metal. The car incorporated a whole raft of innovative ideas, including forward-sliding doors, and a three-way folding 'landau' soft-top. Introduction of the Darrin pre-dated that of the Thunderbird, but was a year later than that of the Corvette. Ironically, even the feeble sales of the 'Vette were nearly twice as good as those of the Kaiser- Darrin. This production model was fitted with a Willys F-head six-cylinder engine, so performance was even more compromised that that of the 'Vette. Darrin manufacture was discontinued in mid-'54. Darrin himself bought the final 100 cars in the range, and fitted them with Cadillac V-8 powerplants that enabled the car to achieve 140 mph. He then retailed them through his own Los Angeles dealership for a whopping $4,350.

Kaiser ceased manufacture in 1955, although many of the company's innovations are now standard in modern cars.

1954 Nash Ambassador Country Club

In Nash history, 1954 was a big year. The long-planned merger with Hudson was effected in April of that year, and Nash President George Mason died prematurely in October. His position was taken over by George Romney.

Several new models were also introduced, including a two-door version of the Rambler American, a four-door sedan and a four-door station wagon.

Nash had three lines for '54, the Statesman, Ambassador and Rambler. The Rambler series was very extensive, having every kind of model from a two-door convertible to a two-door utility wagon. Eleven different models were offered, selling between 56 and nearly 10,000 units each. They were the most competitively priced Nash products.

There were two divisions of both the Statesman and Ambassador classes, the Super and Custom lines. Our photographed car is the most expensive Nash option for this model year, the two-door hardtop coupe. Whilst sharing the same basic bodyshell introduced in the '52-'53 Golden Anniversary model year, both classes underwent the same facelift for the '54. This included a new front concave grille, new chrome headlight bezels, revised instrument panel and redesigned interior. The custom models for both lines had continental rear tire carriers fitted as standard. The Ambassador models had a seven-inch longer wheelbase, but all the Nash vehicles were fitted with a version of the company's inline six-cylinder engine. Low-priced airconditioning was introduced as an option by Nash in '54, the first to be offered on mass-market vehicles. Their system remains the basis for all modern systems to this day.

Nash was the country's thirteenth largest American automobile manufacturer in 1954, with a model year production of 62,911 cars. The Ambassador model was to receive its final restyle in 1955, but with dramatically falling sales, AMC dropped both the Nash and Ambassador names in 1957. The only recognizable model name to survive, briefly, was Rambler.

Engine: (Ambassador Six) L-head Inline six-cylinder, cast iron block. LeMans 'Dual-Jetfire' Six available as an option

Displacement: 252.6 cid

Horsepower: 130 at 3700 rpm

Compression Ratio: 7.6:1

Weight: 3120 lbs

Base Price: $2730

Number Produced: 3,581

1954 Packard Panama Hardtop

Packard took Studebaker over in 1954. The new company was known as Studebaker-Packard Corporation. This was a common phenomenon at the time, as small independent carmakers were finding survival so very difficult.

The main problem was generating high enough profits to fund annual restyling and re-tooling. Being unable to do so made competing with larger manufacturers impossible. James Nance (President since 1952), had been forced to seek a financial partner, and had formulated a plan to merger four independents, Packard, Studebaker, Nash and Hudson, so that the resulting company would be able to compete directly with the 'big three' car companies.

Ironically, Studebaker turned out to be a disastrous acquisition, effectively dragging its parent company into oblivion. This was largely because the appalling financial position at the company was not fully disclosed. One of the Packard board members likened the merger to the captain of a sinking ship trying to save his craft by tying it to another sinking ship.

By 1954, Packard was down at sixteenth position in the league of auto producers, on a production of 31,291 units. The company had hoped to be able to afford a complete styling revision for this year, but were obliged to carry over the '53 models. There were name changes – Pacific replaced Mayfair as the top hardtop line for '54, with a larger straight eight engine (212 rather than 180 horsepower) to distinguish the car from the earlier version. The Caribbean returned as the glamour leader for PMCC, but with a price increase to a staggering $6100, the company was only able to sell 400 of these.

However, there were positive aspects to this shaky year. Packard had established a tradition of introducing a show car for each new model year. The supercharged Panther-Daytona (Grey Wolf II) experimental car, the first fibreglass-bodied car, was driven by Dick Rathmann at the Daytona Speed Weeks, and clocked an unofficial time of 131 mph. This was the fastest time ever recorded by a car in this class. Two '54 models, the Caribbean and Patrician, were honored as Milestone Cars.

Engine: Super Clipper Eight-cylinder, Inline, L-head, cast iron block

Displacement: 288 cid

Horsepower: 150 at 4000 rpm

Compression Ratio: 8.0:1

1954 Willys Aero Lark Sedan

Willys-Overland of Toledo, Ohio had a long history of manufacturing small cars before the war, but on the general return to civilian production in 1945, they decided to try and capitalize on the success of their wartime production military jeeps, by developing civilian versions of these. This meant that there was far less retooling necessary at the Willys plants than for most auto manufacturers in the post-war period, and the company sold their one-millionth Jeep in 1952. A wide choice of jeep-based vehicles was developed. Although they were marketed as commercial vehicles, many customers picked up on their utility, and ran them as passenger cars. Willys introduced the Jeepster in 1948 as a slightly 'softer' Jeep option. But this model was discontinued in 1950, and the company re-entered the 'true' passenger car market in 1952 with the Aero sedan.

The original Aero model was the Lark, a rather basic two-door sedan, but it had pleasing looks, having slab sides, high fenders and a low hood. Kaiser took Willys-Overland over in the company's 50th anniversary year of 1953, and expanded the line by adding a four-door sedan model. The line continued to expand, and twenty compact cars, including nine Aeros were offered for the 1954 season. At a first glance, the 1954 Aero models looked almost identical to the 1953 output, but it was now equipped with a Kaiser engine. The cars handled better than before with the introduction of a new front suspension, including a cross member to eliminate torque, and this contributed to a reputation for excellence in this area.

As the advertising blurb went, 'Aero Willys leads them all – with more horsepower per pound of car than the low-priced cars produced by the Big 3'. Sadly, this did not seem to affect sales positively, and only 7,867 Aeros were sold in this year.

The Lark trim was the base trim level for the Aero model for '54, having rubber moldings around the windshield and rear window and standard headlight trim. All Aeros were given new bumper guards, aluminum scuff plates, taillight assemblies and chrome wheel covers. Power came from the six-cylinder L-head engine.

One last attempt for the Willys Aero design was made in 1955, when two models were offered in four body styles. But only around 6,500 cars were sold. Kaiser's CEO, Edgar F. Kaiser decided to abandon the American passenger car market, maintaining vehicle production only in its Jeep line. The company could sell as many of these as it could manufacture. However, Aero production continued in Brazil through to 1962.

1955 Cadillac Coupe De Ville

Engine: V-8 Overhead valve. Cast iron block
Displacement: 331 cid
Horsepower: 250 at 4600 rpm
Compression: 9.0:1
Transmission: Hydra-Matic
Fuel System: Two Carter WCFB Four barrel carburetors
Body style: Two-door hardtop coupe
Number of seats: 6
Weight: 4428lbs
Wheelbase: 129in
Length: 223.2in
Price: $4305
Number produced: 33,300

Cadillac total production in 1955 reached a record time high with 140,777 units produced. Not bad for a luxury brand.

The 1955 Series 62 Coupe De Ville held its own, with nearly a quarter of total output.

The new model included a lower sleeker body, a new egg-crate grille insert, and inverted gull wing front bumpers with tapered dagmar style front bumper guards. The parking lights were repositioned directly below the headlights. The extended side rub rail molding formed a right angle with the vertical trim of the rear fender. Round jet style dual exhaust outlets were incorporated into the vertical bumper extensions as part of a complete redesign of the rear bumper. An Eldorado style wraparound windshield was seen on all models. For coupes a smoothly curved wraparound rear screen was referred to as the "Florentine" style. Coupe de Ville gold script was located on rear corner pillars of the luxury hardtop, which also had wider sill moldings. Tubeless tires on Saber silver alloy wheels were standard on the Coupe De Ville. Popular Mechanics rated to 0-60mph time as 17.3 seconds.

Coupé deVille

Above: The Saber alloy wheels were of split rim construction with an inner steel half.

Right: Cadillac crest above a gold plate V.

Opposite page:
Top left: One of two induction motors in the trunk for the air conditioning

Top right: Tailpipe built into the rear fender.

Bottom: Overhead valve V8 with various power take offs.

Above: Dash with wide angle speedometer.

Right: Air cond scoops on the rear panel.

Left: Dagmar type front fenders.

Below: Egg-crate style front grille.

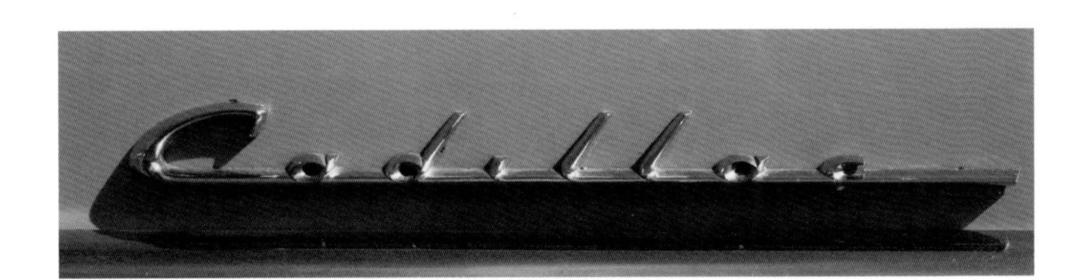

1955 Chevrolet Bel Air Two Door Sedan

Engine: Six cylinder overhead valve, cast iron block	
Displacement: 235.5 cid	
Horsepower: 123 at 3800 rpm	
Transmission: Three speed manual with column mounted gearshift	
Compression: 7.5:1	
Fuels System: Rochester one-barrel carburetor	
Number of seats: 6	
Wheelbase: 115 inches	
Price: $1825 (with two-tone paint option)	
Number produced: 249,105	

The six-cylinder car featured here is regarded as a separate series from the V8 powered cars. Nonetheless at 136 brake horsepower, as opposed to 162, it wasn't far behind in the power stakes. In 1955 Chevrolet had drastically changed its body design from the previous year. Ford had already adopted what became known as the shoebox body design in 1950 and Chevrolet styling was gradually evolving along the same lines through the early 1950s. By 1953, the Chevy had advanced to smoothed out straight side panels for a flatter side look, but retained the somewhat outdated faired in rear wheels , the 1940s-style humpy hood and flat glass windshield. In 1955 Chevy finally caught up with Ford by leaping to a real shoebox look. The new car had smooth straight panels on the sides and hood, wrap-around glass on the windshield and rear screen, and triangular tail lights that jutted outward. And this new look, combined with new power and engineering, made the 1955 car an instant hit and a critical success. This model also offered a wide array of colors. Standard for the Bel Air was one solid color...or the option of nineteen different two-tone color combinations like the car shown here. Other new options for 1955 included air conditioning, power windows, and even power seats; power steering and power brakes, as were automatic light dimmers, door handle protectors, bumper protectors and "wonder-bar" radios. This array and goodies led to the moniker "Chevy's little Cadillac" because never before had so many options been offered for a car in the medium price field.

So successful was this classic design that Chevrolet retained the same body and chassis for 1955, 56, and 57. These model years are extremely sought after by collectors and enthusiasts, and are often referred to by the given nickname of "tri-fives" because the series success spanned 3 crucial years in the 1950s. The test of any car design is whether it dates and the 1955 Bel Air looks as good today to our eyes as it ever did.

Above: Two-tone paintwork became very popular around this time.

Left: Chevy badge.

Left: Jet plane chrome hood mascot.

Opposite page
Top left: White wall tires were another 1950s craze.

Top right: Chevy badge embossed into the tail pipe.

Right: Overhead valve six putting out a credible 123 horsepower.

CHEVROLET

Left: The Bel Air motif is repeated throughout the interior.

Below: Exterior paint color is strongly repeated in the interior décor and trim.

Right: Rear taillights are finlike but restrained.

Bottom: Bullet-style sports mirrors are a nice touch!

1955 Chevrolet Bel Air Convertible

Engine: V-8 overhead valve, cast iron block

Displacement: 265 cid

Horsepower: 162 at 4400 rpm

Transmission: three-speed manual with column mounted gearshift, overdrive available as an option. Powerglide two-speed automatic transmission optional at $178 extra, optional V-8 'powerpack' also available with dual exhaust

Compression: 8.0:1

Fuel System: Rochester two-barrel carburetor

Wheelbase Length: 115 inches

Price: $2305

Number Produced: (V-8 and 6-cylinder Convertibles) 41,292

In economic terms, 1955 was a boom year. It saw the opening of the first McDonald's drive-in restaurant and Disneyland California. Rock and Roll was now in its infancy, and Bill Haley and the Comets topped the charts with 'Rock Around the Clock'. The automotive industry experienced a record production year - 7,920,186 units being churned out by Detroit. Chevrolet was still outperforming Ford by over a quarter of a million cars, selling over 1.7 million with the advertising slogan 'The Hot One!' The average car now cost $2,300, with average pay at $3851 per annum, so instalment buying is becoming more and more common for the average family.

The Chevy cars of '55 are some of the most popular restoration projects in the world – the combination of great fifties styling and the superb new V- 8 engine. They were difficult to improve upon.

The Bel Air was the company's top series, and standard equipment was generous and interiors comfortable and stylish. The exterior styling was also sharpened up, and the chrome accents were far more subtle. But the most important change by far was the introduction of the V-8 engine to the series. Although all the models in the Bel Air range were still available in the six-cylinder engine type, the V-8 was so fundamental a change that It was classed as a separate series rather than an option. In fact, this remained the case for at least some of the model shapes until the Bel Airs were eventually phased out.

1955 Chevrolet Corvette Roadster

Car sales for the decade peaked in 1955 with record-breaking sales of seven-million plus. The frantic GM/Ford price war of 1953-54 had resulted in the mortal wounding of several surviving independent manufacturers, and the market was polarising in the hands of the three great companies – GM, Ford and Chevrolet. Throughout these years, Detroit was in a 'horse-power race' to rival the technological Cold War between the East and West.

At Corvette, the big news for 1955 was the introduction of the V-8 engine in nearly all of the cars produced in that year. An enlarged golden 'V' embellishing the Chevrolet logo on the front fender was a quick way to identify cars fitted with the new engine, and a 12-volt electrical system. The earlier cars had had a six-cylinder engine and six-volt electrical system. In this transitional period, Chevrolet used three engine number suffixes to denote the three types of car available - YG for the six-cylinder engines, FG for V-8 and automatic transmission models and GR for V-8 cars with manual transmissions. Ironically, although the six-cylinder engine was offered as standard, with the V-8 as an extra (costing $135), only six non V-8s were sold. The V-8 'Vettes could go from 0-100 miles per hour in 24.7 seconds.

The car retained its original styling for '55, and the distinctive Corvette grille was now established as an identification trademark of the car's design – the thirteen chrome 'teeth' were originally designed to carry the Chevrolet family identity. The very first '53 models had had only three, which had led to concerns that the fibreglass bodywork was unprotected against parking knocks.

Various (rather basic) options were available for the car at extra cost – signal seeking AM radio at $145.15, windshield washer at $11.85, heater at $91.40 and parking brake alarm at $5.65. However, the continuing low volume of sales – just 700 for 1955 meant that, yet again, Chevrolet came close to dropping the model from the range. So ironic, in view of its continuing success to this day, when so many better selling contemporary cars are remembered only in the archives.

Engine: (Most popular for this model year) V-8 Overhead valve. Cast iron block

Displacement: 265 cid

Horsepower: 195 at 6000 rpm

Compression: 8.0:1

Body Style: Two-door convertible roadster

Weight: 2705 lbs

Price: $2934

Number Produced: 700

1955 Chevrolet Nomad Station Wagon

Engine: V-8 overhead valve, cast iron block (also available with six-cylinder engine)

Displacement: 235.5 cid

Horsepower: 162 at 4400 rpm

Compression: 8.0:1

Fuel System: Rochester two-barrel Model 7008006

Number of Seats: 6

Weight: 3270 lbs

Wheelbase Length: 115 inches (overall length 197.1 inches)

Price: $2571

Number Produced: 6,103

Right: This two-tone Nomad looks very at home in front of suburban white clapboard and manicured lawns.

As the newly affluent middle classes increasingly moved to the suburbs, the station wagon became a more and more popular form of transport for the typical American family – shopping, dropping the kids off at school – everything now involved driving.

Chevrolet introduced the Nomad to the Bel Air range as a midyear model addition, pioneering the introduction of the 'hardtop' wagon to the US market. Although the car was really groundbreaking, the model attracted only 6,103 buyers in 1955. The four-door Beauville station wagon was also available as part of the Chevrolet Bel Air series, and outsold the Nomad by about 4 cars to 1 in that model year.

The front-end styling was classic Bel Air, but although the designers carried through the tailfin design of the other range models at the back of the car, it seems somewhat anachronistic with the tailgate arrangement.

Interestingly, 2004 saw the launch of a brand new Chevrolet Nomad concept car. In a wave of retro nostalgia that has seen many automakers dusting off their family jewels, Chevrolet have chosen to reinterpret the very original Nomad concept (a lightweight sporty two-door wagon) for a twentyfirst century generation of customers. Similar in style and function to the 1955 model, the styling of the new car is fresh, whilst paying direct homage to the original in many details – the toothy (typically Chevy) front grille, and the seven chrome ribs that adorn the tailgate, for example. But the engine is completely modern in conception – four-cylinder and turbocharged. If the new design goes into production, it will be almost fifty years to the day since Nomads first graced the suburban streets of America. That really would be a case of back to the future...

Right: Chevrolet 235.5-cid V8 optioned Nomad.

1955 Chrysler C-300 Coupe

Engine: Firepower 'Hemi' V-8

Displacement: 331.1 cid

Horsepower: 300 at 5200 rpm

Compression: 8.5:1

Base Price: $4109

Number Produced: 1725

Chrysler had a difficult start in the post-war return to manufacturing consumer goods rather than military hardware and offered only warmed over pre-war models until managing to get some truly new cars off the blocks in the mid '50s. However, Chrysler had introduced three major mechanical improvements in the first half of the decade, the Firepower V-8 engine (with hemispherical combustion chambers), power steering and fully automated PowerFlite transmission. But the company failed to be counted amongst the top ten automotive producers during these years, and remained out of this select coterie until design supremo Virgil Exner launched the company's 'Hundred Million Dollar Look' in 1955. A major new model was launched in this year, the 300 Letter Car. It was to become known as Chrysler's 'beautiful brute', with a brilliant synthesis of performance and luxury styling that would be successful for more than a decade. Exner's innovations took the company to second place in the high-priced sales field for this year, and ninth overall in the sales war.

The C-300 was designed to be a 'sports touring car designed to bring Chrysler the benefit of a high-performance reputation.' Although it was a big, heavy car based on the existing New Yorker body shell, the original 1955 model was still almost as fast as Chevy's much lighter newly introduced Corvette. In fact, the C-300 was named after its stunning performance statistics – it was the very first American production model to deliver 300 horsepower as standard.

The 'C' is reputedly a reference to the racing driver Briggs Cunningham who had used Chrysler's V-8 hemispherical engine in the early '50s.

The statistics are truly impressive, the C-300 could go from 0-60 mph in 9.5 seconds and cover a quarter-mile in 17.6 seconds at 82 mph. The slightly revised 1956 version of the car, the 300B, broke the world's passenger speed record in competition at Daytona Beach, Florida, with an average speed of nearly 140mph. The production model was fitted with a 150-mph speedometer to underline the latent power available under the hood.

All the 1955 C-300s were two-door hardtop coupes fitted with Powerflite transmission, power brakes, and optional power steering. Only three model colors were introduced, Chrysler Platinum White, black and the colour of our car – Tango Red. The styling of the model was comparatively modest for 1955 with discreet tailfins and only a moderate amount of chrome ornamentation - although flashy wire wheels were optional. The interior was truly luxurious, with full leather trim.

Opposite top: Successful as both a race and stock car, the C-300 has been described as America's first muscle car.

Oposite bottom: Styling details on the 1955 letter car refer strongly to its performance appeal-300 horsepower equals a checkered flag.

Below: Tail-end view of the C-300 coupe. The car's European-influenced styling looks at home in this country house setting.

300

1955 DeSoto Fireflite Two-door Convertible

Engine: V-8 overhead valve, cast iron block

Displacement: 291 cid

Horsepower: 200 at 4400rpm

Compression Ratio: 7.5:1

Number of Seats: 6

Weight: 4090 lbs

Wheelbase: 126 inches

Base Price: $3151

Number Produced: 775

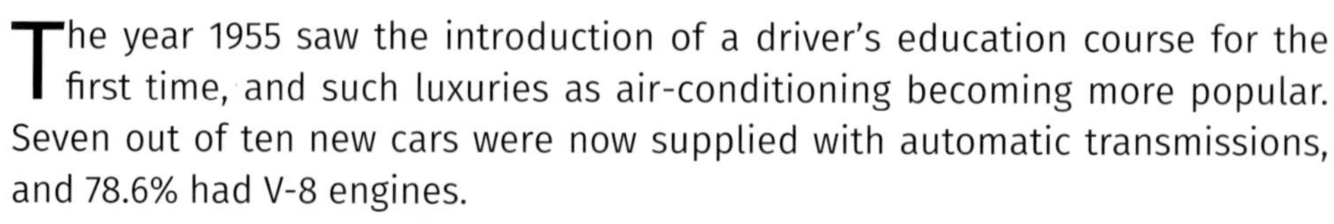

The year 1955 saw the introduction of a driver's education course for the first time, and such luxuries as air-conditioning becoming more popular. Seven out of ten new cars were now supplied with automatic transmissions, and 78.6% had V-8 engines.

At DeSoto, as in the rest of Chrysler, Virgil Exner's aircraft-inspired 'New Look' was introduced, with streamlined body styling and wraparound windscreens (though the rather toothy front grille remained). The re-style was applied in its most extensive form to the top-of-the-range Fireflite series. Under the hood, the Firedome V-8 became the standard power train, with two hundred horsepower, a four-barrel carburetor and hemispherical segment combustion chambers.

Identification features included Fireflight front fender script, chrome fender top ornaments running back from headlamps; and rocker panel beauty trim. Special side beauty panels were The year 1955 saw the introduction of a driver's education course for the first time, and such luxuries as air-conditioning becoming more popular. Seven out of ten new cars were now supplied with automatic transmissions, and 78.6% had V-8 engines.

At DeSoto, as in the rest of Chrysler, Virgil Exner's aircraft-inspired 'New Look' was introduced, with streamlined body styling and wraparound windscreens (though the rather toothy front grille remained). The re-style was applied in its most extensive form to the top-of-the-range Fireflite series. Under the hood, the Firedome V-8 became the standard power train, with two hundred horsepower, a four-barrel carburetor and hemispherical segment combustion chambers.

Identification features included Fireflight front fender script, chrome fender top ornaments running back from headlamps; and rocker panel beauty trim. Special side beauty panels were standard on Fireflite convertibles. Interior styling was of 'cockpit' inspiration, with standard upholstery of silky nylon with nylon carpeting.

This was also the year when DeSoto pioneered the use of triple-tone paint jobs, introducing this trim on the Fireflite Coronado. Distinctive 'Flight Control' gearshift levers protruded from the dash in the automatic models from the range.

Below: Top up or down Virgil Exner's "New Look"worked well in 1955.

Above This DeSoto Fireflite has elegant aeronautic lines and sweeping body trim.

Left: The Firedome V8 option was ordered by 78.6% of the customers.

1955 Dodge Royal Lancer Hardtop

Engine: Red Ram V-8 overhead valve, polyshpere combustion chambers, cast iron block	
Displacement: 270 cid	
Horsepower: 175 at 4400 rpm	
Transmission: three-speed manual fitted as standard, overdrive and two-speed PowerFlite automatic available as optional extras	
Compression Ratio: 7.6:1	
Fuel System: Stromberg two-barrel Model WW3-131	
Body Styles (Lancer): Two-door Hardtop	
Weight: 3425 lbs	
Wheelbase: 120 inches	
Base Price: $2370	
Number Produced: 25,831	

Dodge slipped to eighth place in the production stakes in 1955, despite the introduction of several new models and a 160.3 percent jump in sales. In fact, even more cars could have been sold, but for the fact that Dodge shared their V-8 plant with Plymouth. The 1955 models were grouped into three series and all had longer, lower, wider bodies; new hardtops, threetone paint and increasingly plush trim variations. Some of the economy Dodge models were discontinued. The Royal series had been originally introduced as the top trim Dodge series for 1954, and was constructed at plants in Detroit and San Leandro. A total of four Royal models were introduced: a four-door Sedan and three assorted two-door models. This was reversed in the following year, with one two-door model (the Lancer) and three four-door cars, a sedan plus two versions of the Sierra Wagon.

By 1955, the Royal was the mid-range trim model, with the Custom Royal Series now Dodge's top range. Even so, the Royals were at least 50% more expensive than the average new car.

The 1955 Royals had all the Coronet features, plus hooded, chrome headlights and the Royal name in script on the front fenders. The series was now fitted with V-8 'Red Ram' engines with overhead valves and polysphere combustion chambers.

The two-door, hardtop Royals were designated as Lancers, and had a special trim a standard. The cars were ornamented with narrow chrome strips trailed back from the hood scoop, dipped at the C-pillar, then continued high on the rear fenders to the taillight housings.

1955 Ford Victoria

The year 1955 was important in Ford history. Not only was their range completed restyled inside and out, but some truly beautiful new lines were also introduced. The Fairlanes (named after Henry Ford's Dearborn, Michigan mansion) were instantly popular, and sold better than any other model introduced since the end of the War. They replaced the Crestline as the top trim series. The line had particularly attractive chrome and tone-tone trim, and included the popular Sunliner convertible. Wagons were now designated as a separate Ford series. The Country Sedan remained the most popular station wagon, selling 106,284 into the market, but five other wagons were also available with six- to eight-seat options.

Of course, the Thunderbird was also introduced in this model year – a twoseater for the young market. Despite all this positive change, Chrysler's 'Hot One' – the Virgil Exnerstyled C-300 complete with small block V-8 engine – was the biggest automotive news of the year. Chevy also got 'hot', aiming the new range at the young buyer for the first time. Their 1955 'renaissance' thrust the company back into their habitual first position in the league of auto manufacturers, with Ford in second. Despite this, 1955 was Ford's second production best year, so far.

Over at Ford, the bodies were longer, lower and wider and the instrument panel and interior trim were substantially new. The powerblocks were even more powerful in this model year, with the introduction of a larger overhead V- 8, and the majority of the 1955 cars were sold with this option.

The Victoria model was a Fairlane two-door pillarless hardtop. A Crown Victoria model was also available, with 'tiara' foor trim. The Crown Victoria Skyliner is a similar model with a forward roof panel constructed from transparent green plexiglass.

Engine: Overhead valve V-8, cast iron block

Displacement: 272 cid

Horsepower: 162 at 4400 rpm

Transmission: Manual with overdrive and Ford-O-Matic optional

Compression Ratio: 7.6:1

Body Style: Two-door Victoria Hardtop

Weight: 3318 lbs

Wheelbase: 115.5 inches

Base Price: $2195

Number Produced: 113,372 (including the 6- cylinder engine version)

1955 Hudson Custom Hornet Sedan

By 1955, Hudson had lost its independence, and was a part of the American Motors Corporation. As production had ceased at the Hudson factory in Detroit, and been transferred to the Nash plant at Kenosha, Wisconsin, the models were really little more than AMC model lines. All but Nashes in disguise, the cars became known as 'Hashes'.

The 1955 Hornets were based on Nash bodies, but retained the Hudson model names – Wasp and Hornet, whilst the Nash Ramblers were re-badged as Hudson Ramblers.

The '55 Hornets has the same styling as the Wasps, but were built on a longer wheelbase. High quality interior trim and continental spare tire carriers fitted as standard distinguished the Custom level cars. A new V-8 engine option was now available, bought in from Packard. George Mason, the President of Nash, had intended to continue his automotive acquisitions by adding Packard and Studebaker to the AMC group, but his early death in 1954 meant that only the engine made it. In cars where the V-8 was fitted, Packard Twin Ultramatic transmission was mandatory.

Model sales for the year actually increased, partly due to various dealer and customer incentives, including free trips to Disneyland. Dealer incentives were also widely used, with promotions such as the 'Dealer Volume Investment Fund' and 'Sun Valley Sweepstakes'. Of course, Nash models were also now included in the sales figures. Although these tactics did have something of a ring of desperation about them, the company climbed 7% in the automotive sales league. In 1956 AMC introduced its own V-8 engine. But only modest styling and engineering updates to the range, in an era of huge revision, conspired to undermine sales dramatically. Wasps and Hornets were the models most severely affected. AMC decided to discontinue these models in 1957, and concentrate on Rambler production. This proved to be a very good move for the company, and AMC continued in production for nearly three more decades, until the company was finally taken over by Chrysler in 1987. Sadly, the company name has now ceased to exist.

Engine: Hornet 'Big' Six: Inline six. L-head, chrome alloy block

Displacement: 308 cid

Horsepower: 160 at 3800 rpm

Compression Ratio: 7.5:1

Body Style: Four-door Sedan

Wheelbase: 121.5 inches

Base Price: $2760

Number Produced: 5,357

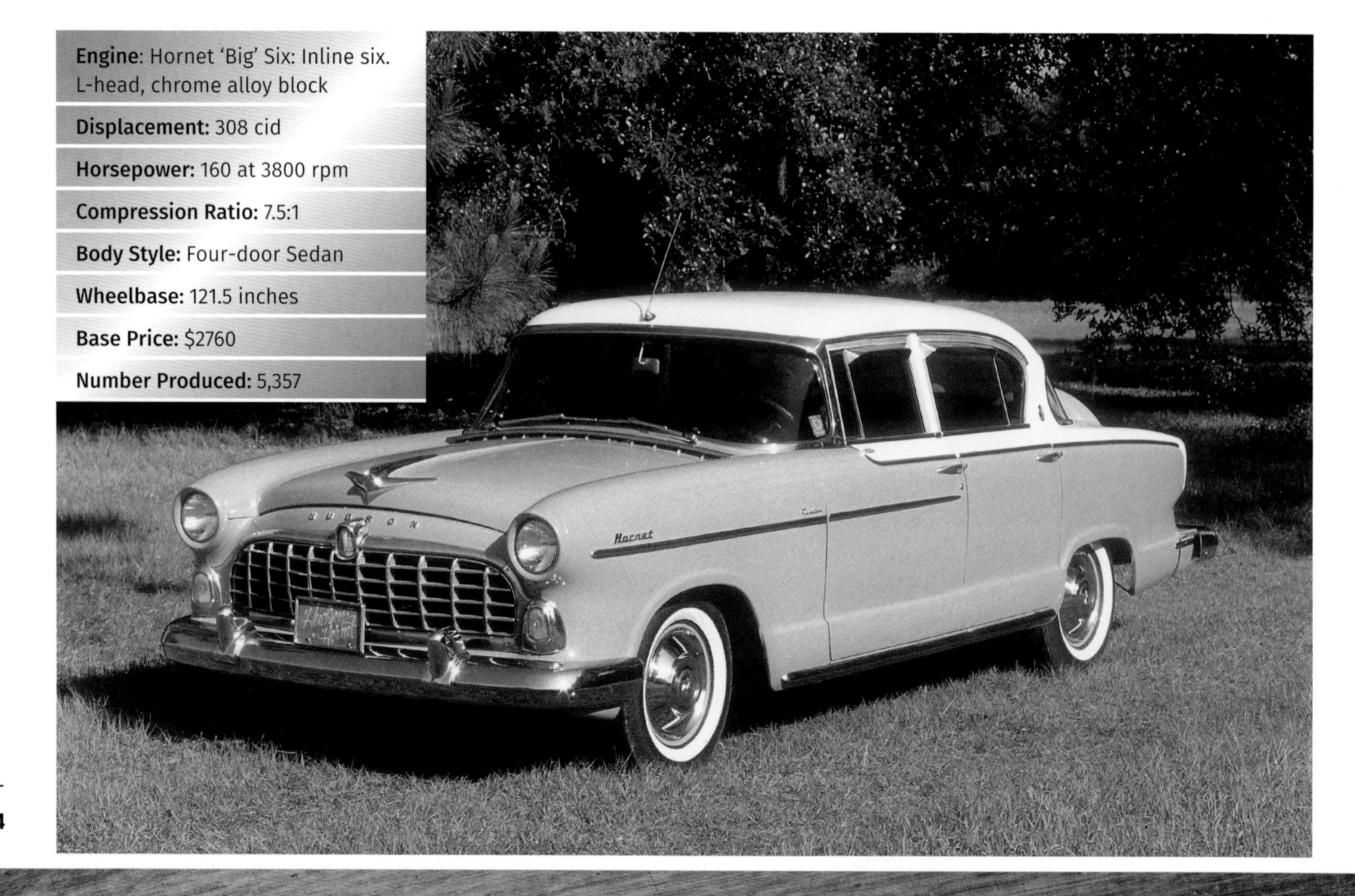

1955 Mercury Montclair Two-door Sun Valley Hardtop

Mercury and Lincoln became separately managed companies in spring 1955, as befitted their differing aspirations. The entire Mercury output was restyled in 1955, their Ford-based body shell was completely overhauled, and the Montclair range was introduced as the top line model. Generally, the cars were bolder, brighter, longer, lower and wider. But they were still recognisable from the '54 models. Montclairs were built on the same chassis as the rest of the lineup, and were similar in body style, though slightly lower than the other Mercurys.

Having first appeared in the Monterey series, the Sun Valley was now placed in the Montclair line. This was a true all steel pillarless hardtop, unlike its Ford rival the Fairlane Crown Victoria. Like the other Montclairs, (which included a four-door hardtop model), the Sun Valley cars were trimmed with an additional narrow band of chrome under the side windows, but also had a tinted plexiglass section in the front of the roof. This led to them being nicknamed the 'bubbletop'. This unique styling conceit got a mixed reception.

Some customers complained that it made the cars uncomfortably hot, a serious problem with no air-conditioning available on Mercury models. Custom remained the base model for the year, whilst Monterey now became the mid-range series.

Chevrolet and Mercury shared the 'Car of the Year' title awarded by Motor Trend magazine. This was substantially due to the mechanical improvements (begun in the previous year) – the enlarging of the updated overhead valve V- 8 engine to 292 cid, and the upgrading of the front suspension to ball-andsocket joint. Dual exhausts were standard on Montclairs. The cars managed a reasonable fuel efficiency of 16 miles per gallon. Sadly, the Sun Valley was discontinued in this, only its second year, with a production total of fewer than 12,000 units. Mercury was an impressive seventh in the table of 1955 car production, with sales of 329,808 cars.

Engine: Overhead valve V-8, cast iron block

Displacement: 292 cid

Horsepower: 198 at 4400 rpm

Compression Ratio: 8.5:1

Body Style: Two-door hardtop

Weight: 3560 lbs

Wheelbase: 119 inches

Base Price: $2712

Number Produced: 1,787

1955 Packard Caribbean

Engine: Packard Line V-8, overhead valve,cast iron block	
Displacement: 352 cid	
Horsepower: 275 at 4800 rpm	
Transmission: Twin Ultramatic	
Compression Ratio: 8.5:1	
Fuel System: Two Rochester Type 4GC fourarrel carburettors as follows: (front) Model 476010; (rear) Model 476011	
Body Style: Two-door Convertible	
Number of Seats: 6	
Weight: 4755 lbs	
Wheelbase: 127 inches	
Base Price: $5932	
Number Produced: 500	

Things looked as though they were picking up in 1955, Packard climbed to fourteenth in the production league, output climbed to 55,247 units. Packard Styling Director, Dick Teague, conjured a remarkable facelift for the year that almost made the '51 body look like new. The new models were introduced in January 1955. Packard continued to introduce technical innovations in this model year. The most important development was the replacement of the eight-cylinder engines with modern V-8s. This massively improved the performance of the Packard range, and generated a lot more buyer interest in the cars. Packard's innovative 'Torsion Level' suspension (designed by William D. Allison, formerly of Hudson) was also introduced in '55. This consisted of full-length motor-controlled torsion bars (instead of springs) with electric ride-height control. Allison had touted this excellent invention around the motor industry, but Packard was the first manufacturer to realize its potential.

Exterior running lamps were added to the senior Packard models, including the 'Four Hundred' hardtop, named after the 'four-hundred' members of the social elite of a bygone era. The two-door custom convertible Caribbean topped the Packard line again. The car also topped the industry power league with twin carburettors and other tuning adjustments enabling the car to achieve 275 horsepower. The Packard line models had the same shaped grille opening as the Clippers.

New features included wraparound parking lamps and sweep-around windshields. Teague also introduced brand new cathedral shaped taillights and a flat-rounded tailfin. The two chrome sidespears allowed for a three-tone paint finish, with one color on the roof, one between the panel moldings, and one below the lower bar. The car also had a twin scoop hood with no ornamentation, except on the front edge of the scoops. Except for air conditioning, almost every cost-plus option was offered as standard on the car, including dual outside mirrors and rear antennae and the new Twin Ultramatic Drive.

Other highlights for the year at Packard included a 25,000-mile run for a Packard Patrician at the company's Packard Proving Ground in Utica, Michigan. The car averaged 104.737 mph. More negatively, Pan American Airlines started a suit against

Below: The dash and the interior reflected the body colors.

Above: This Caribbean's three-tone paint job is in White Jade, Fire Opal and Onyx.

Left: Overhead valve V8 achieved 275 horsepower.

the company for its use of the names 'Caribbean', 'Clipper', 'Pan American', 'Constellation' and 'Panama'. The suit lingered on until 1958 when, sadly, the names had lost any relevance to Packard.

The Packard Division made a profit this model year, but the Studebaker-Packard Group as a whole made an operating loss of nearly $30,000,000. The writing was on the wall.

1955 Packard Clipper Custom

Everything seemed new for the Packard base-level Clipper line this year, and the introduction of the new Packard V-8 was particularly beneficial to sales. At the top end of the model line, the Custom Clippers were fitted with larger engines awarded to the senior Packard line. Although the engine was quite similar in size to the old eight-cylinder, it was much more efficient, developing 245 horsepower at 4600 rpm. In fact, the Packard V-8 represented the best in technical automotive design of the time. Torque was very high at 355 pounds per foot, which required a strengthened transmission. This led to the introduction of Twin Ultramatic. Packard's new V-8s also had a more generally positive impact on the finances of the company, as these were supplied to Nash and Hudson (together with Twin Ultramatic transmissions) to be fitted to their 'Eight' models. In fact, the V-8 was so good for Packard, that the company seriously considered developing a V-12, but the huge development costs this would have entailed put a stop to this.

Styling for the Clipper range was heavily revised from the straight-eight models. Richard Teague fitted the cars with massive new bumpers with bulletshaped guards, complete with full-width grilles with delicate vertical fins, bowed upper bars and 'ship's wheel' centre medallions. The front fenders were bent over to hood the headlamps, and sweep-around windshields also appeared. However, the rear of the car was substantially the same as in the previous model year – this was just a facelift, after all.

The Clipper four-door models were available in either Deluxe, Super or Custom series trim levels. The Custom was available with the same leather interior as the Packard senior Patrician range, and was generally luxurious. The cars had a special integrated two-tone paint job. Everything above and below the chrome side moldings was painted in one color, the space between them in another.

The Clipper Custom series also contained the two-door Constellation Custom hardtop sports coupe. Both models sold well, and in similar numbers.

Engine: Clipper Custom V-8, overhead valve Displacement: 352 cid

Horsepower: 245 at 4600 rpm

Compression Ratio: 8.5:1

Fuel System: Carter Type WCFB four-barrel Models 2232S or 2284S

Base Price: $2926

Number Produced: 8,708

1955 Plymouth Belvedere Sport Coupe

The 1955 Plymouths were completely redesigned, with longer lower wider bodies. Every panel of the high-priced Belvedere line, including the hardtop was revised, from the side the car is one long gentle sweeping arc stretching uninterrupted from headlight brow to tail light tip. The highlights were hidden from side-view by forward-canted hoods. At the opposite end of the car, the taillights are also hooded with a complimentary rearward cant.

The whole design gives an impression of the car moving forward, and this design style is labelled 'Motion Design'. The competition at Ford and Chevrolet had also come out with great new styling and some excellent new models in the year, including the Thunderbird.

The revision came not a moment too soon. Plymouth had been dumped from their tradition third place in the car producers' league and had gone down two full places to fifth position. They had been caught in the crossfire of an extremely vicious price war between Ford and Chevrolet. Both companies had been overproducing to flood the market with their product, to top up their production figures, then dumping the overproduction on the market. By comparison to the early 1950s offerings from these two companies, the Plymouths looked clunky and old-fashioned. But the '55 re-style was expressly designed to make the cars appealing to the 'young at heart'. So determined were they to make the cars more youthful in appearance that they even gave up their hat-wearing premise by decreasing the height of the vehicle.

The wheelbase of the car was an inch longer than in the previous year, but the overall length of the car was a full ten inches greater. The tri-sectional grille was a continuation of the '53 theme, and would continue until the early 1960s. The hood ornament is a very abstract sailing ship motive indeed, and was due to disappear altogether in 1956. The windshield was now wraparound, 'The New Horizon' Plymouth called it, and the backward A-post contributed towards the 'Motion Design' effect.

The dashboard came in for some criticism with a rather bizarre layout of the instruments and dials. There was also a dash-mounted gearshift lever.

Engine: Overhead valve Hy-fire V-8

Displacement: 241 cid

Horsepower: 117 at 3600 rpm

Compression Ratio: 7.6:1

Weight: 3261 lbs

Base Price: $2192

Number Produced: 47,375

1955 Studebaker President Speedster

Engine: President V-8 overhead valve Displacement: 259.2 cid	
Horsepower: (later version) 185 at 4500 rpm	
Compression Ratio: 7.5:1	
Base Price: $3253	
Number Produced: 2215	

Opposite: Although the grille has been beefed up and chrome beauty bars added, the basic Raymond Loewy styling of '53 is clearly visible in this '55 model President Speedster.

By 1955, Studebaker's fortunes appeared to be in something of a decline. The company had been taken over by Packard in 1954 and became the Studebaker-Packard Corp. of America. They had slipped to eleventh place in the league of auto manufacturers, on a reduced production level of 116,333, but this was a huge improvement on the 68,708 cars produced in the takeover year.

After an absence of thirteen years, the top-line President name was revived. The original Studebaker President had been styled by industrial designer Raymond Loewy. His design partnership with the company stretched from his original partnership with company President Paul Hoffman in 1938 to the iconic 1962 Avanti. In fact, he started the eponymous 'Raymond Loewy Design Division' at Studebaker, working with bright young engineers such as Harold Churchill (who became company President in 1956). Loewy's design principle for Studebaker was 'simplification', effectively streamlining the cars to remove visual and literal weight and 'drag' that obliged manufacturers to fit more and more powerful powerplants to maintain performance in big heavy models.

Through his work with Studebaker, Loewy is credited with the introduction of many design features that seem obvious today – the down-slanted hood, the integrated bumper, and the replacement of the grille by an air scoop. He was passionate about the aerodynamic wedge-shaped superstructure, with uncluttered body styling, and seems to be one of the first car designers to consider designing safety into the structure of the car itself. Many design elements by which Studebaker's are instantly recognisable, like the 'bulletnose' came from Raymond Loewy's pen, and the classic 1953 model line was his, from the coved body styling down to the plain conical hubcaps.

Loewy was finally dropped by Studebaker after their merger with Packard.

To secure a release from their contract with him reputedly cost the company one million dollars.

The 1955 Studebakers were fundamentally restyled '53 models, but the President Speedster was probably the closest in appearance to the classic Starliner, retaining the overall grille style (although this was now more massive).

The tri-level painted hardtop model was introduced in January 1955, and featured the 'butter knife' side trim. The Speedster was the top-of-the-range President model and was loaded with every conceivable extra including an engine turned instrument panel and Automatic Drive or overdrive as standard.

ENJ91
MIKE X

1956 Cadillac Eldorado

Engine: V-8 Overhead valve, cast iron block	
Displacement: 365 cid	
Horsepower: 305 at 4700 rpm	
Transmission: Hydra-Matic	
Compression: 9.75:1	
Fuel System: Two Carter WCFB four-barrel carburetors , Model 2371	
Body style: Two-door hardtop coupe, twodoor convertible Biarritz model	
Number of Seats: 6	
Weight: 4880 lbs	
Wheelbase Length: 129 inches (overall length for Eldorados 222.2 inches)	
Price: $6501	
Number Produced: 2,150	

In 1956 the Federal government approved a massive building programfor 41,000 miles of Interstate Highway. But car production eased slightly to around 6.3 million units. Cadillacs accounted for 154,577 vehicles, moving from tenth to ninth position in the American sales race. Eighty per cent of all cars manufactured in America in 1956 were fitted with classic V-8 engines.

The postwar years were good ones for Cadillac, as they competed with Packard for the luxury American car market. The trademark aircraft inspired tailfin look debuted on selected 1948 models and from then, Cadillac became the style innovator at this high-priced end of the market. Sales hit 10,000 in 1950, outselling other prestige makes, and were at 265,000 by 1975.

Cadillac's Eldorado was introduced in 1953 (when 532 examples were produced), but – like all Cadillacs – the 'Eldo' was re-designed and re-blended almost every year. The 1956 Eldorado retained the distinctive shark fins, but benefited from the deft restyle given to its sister models, with a new grille and the repositioning of the parking lights in the bumper below the wing guards.

The convertible now carried the Biarritz name to distinguish it from the first Eldorado hardtop, a new two-door named Seville, which sold 3,900 copies to the ragtop's 2150. Both models were priced at a princely $6501. For the first time since 1949, all Cadillacs had a larger V-8, a bored-out 365-cubic-inch unit with 285 horsepower. With dual four-barrel carburetors, the figure rose to 305 bhp for Eldorados.

Various luxury options were also available including air conditioning, gold finished grille, and E-Z eye safety glass.

1956 DeSoto Fireflite Two-door Convertible

In a year when general manufacturing production was eased, DeSoto moved up to number 11 in the output league table, but the company's sales fell yet again to 109,442 cars. A refined Fireflite convertible was introduced to the market in 1956. To re-inforce the more widespread use of the V- 8, the new perforated mesh grille with a large central 'V' dominating the front aspect of the car.

The taillight clusters were redesigned, but color sweep two-tone remained a feature (although the shape of the panels was redesigned). The headlamps still had chrome plated headlamp hoods.

Mechanically, all Fireflite models were now fitted with PowerFlite automatic transmission as standard. This was now controlled by push-button gear selection.

On January 11 1956, DeSoto announced that a Fireflite convertible with heavy-duty underpinnings, but standard engine, would act as a pace car at the Indianapolis 500 and that a limited 'Pacesetter' convertible would be available to the public. These cars had the same special features and a heavy complement of power accessories but were not lettered like the authentic pace car shown in the photograph.

The gold and white trim of this limited edition car was commemorated in the slightly later 'Adventurer', an exclusive sub-series of the Fireflite range launched to compete with the Chrysler 300-B, Dodge D-500 and Plymouth Fury. The convertible pace car carried a premium of $110 over the regular convertible model in the Fireflite range, but no separate production total was released.

Engine: V-8 overhead valve, cast iron block

Displacement: 330.4 cid

Horsepower: 255 at 4400 rpm

Compression Ratio: 8.5:1

Weight: 4070 lbs

Wheelbase: 126 inches

Base Price: $3565

Nos. Produced: (All Fireflite Convertibles) 1485

1956 Dodge D-500 V-8 Two-door Custom Royal/Lancer Hardtop

Dodge began a major restyling project in 1956, with back fenders sprouting fins, and technologically, a push-button PowerFlite automatic transmission was introduced to both Dodge and Chrysler models. The D-500 model was launched as a flag-waving performance car for the Royal range.

But the re-style did nothing to improve sales, which actually tumbled by 22.4% to just over 200,000 units, and gave the company a market share of exactly 3.7%. M.C. Patterson became the President of the Dodge division in 1956, with a brief to improve its performance.

The 1956 Custom Royal was Dodge's top-of-the-range and most prestigious line, with the most intense trim options. They had everything the Royal models did and more – hooded and painted headlight housings, and extra grooved taillight assemblies. Several special models were introduced as part of the Custom Royal two-door hardtop Lancer line-up for the year, including the Golden Lancer and the rather bizarre Dodge La Femme. The first American car specifically introduced to appeal to women buyers, the 1956 La Femme appeared in a sickly livery of two-tone lavender, and gold-flecked interior. It also had a patronising array of matching feminine accoutrements such as a cap, a pair of boots, an umbrella, a shoulder bag and floral upholstery fabric... sales were reassuringly miniscule. The almost equally tasteful Golden Lancer has a Sapphire white body, with a Gallant Saddle Gold exterior and interior trim.

At the other end of the Custom Royal market, the D-500 (complete with hemi V-8) was introduced to demonstrate the performance capabilities of the marque. These were underlined by the Dodge performance at the year's NASCAR races, where they won eleven victories. A Dodge Custom Royal fourdoor sedan was driven 31,224 miles in 14 days, and set 306 speed records at the Bonneville Salt Flats in Utah.

Engine: D-500 V-8 overhead valve, hemispherical combustion chambers, cast iron block

Displacement: 315 cid

Horsepower: 260 at 4400 rpm

Compression Ratio: 9.25:1

Fuel System: Carter four-barrel Type WCFB

Weight (Lancer Hardtop): 3505 lbs

Wheelbase: 120 inches

Base Price: $2658

Number Produced: 40,100

1956 Ford Convertible Sunliner V-8

The Ford Motor Company became a publicly owned company on February 24 1956, and immediately gained about 350,000 additional stockholders.

This was part of Henry Ford II's vision for the future of the company, and was the platform for a great expansion of FoMoCo in the 1960s.

1956 saw a general easing of industrial production, but a stepping-up of the horsepower race. Like several other manufacturers, Ford offered a 'power pack' option consisting of a four-barrel carburettor, dual exhaust and slightly stronger camshaft to boost the power of the new V-8 as an option. Ford's version displaced 312 cid. FoMoCo manufactured in excess of 1.4 million units in this model year, but Chevrolet achieved an output of 1.56 million to take first position in the league of car producers for '56.

So far as model revises went, Ford re-used the '55 body shell with new grilles and trim, but the car interiors were completely new. Safety was now an important sales feature, and a padded dashboard, sun visor and less aggressively shaped steering wheel were all designed to lessen injuries in an accident. Seat belts were also offered for the first time. A twelve-volt electrical system and 18mm anti-fouling spark plus were also introduced.

The Fairlane series was the top trim for the 1956 model year, and included chrome window moldings, chrome 'A' pillar trim on the Sunliners, and simulated exhaust outlets. Chrome side sweep bars were similar to those used in the previous year, but subtly revised. All seven versions of the Fairlane series came with either six-cylinder or V-8 engine options, and sold a combined number of 645,306 units in the year. Apart from the Sunliner convertible, the 1956 Fairlane model range comprised the four-door Town sedan, a two-door Club Sedan, a four-door Town Victoria, a two-door Club Victoria, a two-door Crown Victoria and a two-door Crown Victoria Skyliner.

The Town Victoria was released in direct competition to Chevy's established range of hardtops, but with limited success. Mainline were the base models for 1956, Customline the intermediate. The Thunderbirds were considered separate to the main Ford output.

Engine: Overhead valve V-8, cast iron block

Displacement: 272 cid

Horsepower: 174 at 4400 rpm (176 at 4400 rpm with Ford-O-Matic)

Compression Ratio: 8.0:1

Wheelbase: 115.5 inches

Base Price: $2459

Number Produced: 58,147 (6-cylinder and V-8s combined)

1956 Packard Executive

'Ask The Man Who Owns One' was the Packard advertising slogan for decades, the most famous in the motor manufacturing world. Through the strength of this 'word of mouth' recommendation, Packard became the car of Rockerfellers, Vanderbilts, Astors, European royalty, the last Russian Czar, movie royalty and US Presidents. The Packard crest (featuring roses and a pelican) became a graceful symbol of success.

From its humble beginnings in Warren, Ohio, Packard moved to the forefront of the motor industry, opening the first gas-filling station (the 'Red and White' in Los Angeles), and was very active in the national 'good roads' movement. Even their showrooms were elegant and impressive.

The Packards built for 1956 may have been the company's best ever products, when measured by their advanced engineering and styling, an outstanding and beautiful car. Packard was intensely proud of its lineup for '56, but buyer interest was reduced in a generally depressed market. Packard had always had an excellent and close relationship with its dealer network, but the dealers now began to defect in droves, feeling that they might be pulled down if the company crashed. The company was now completely lacking in direction, and decided to commission a report from the management consultants Ernst & Ernst. Their shock recommendation was that Packard production should cease immediately, and the Studebaker- Packard Corporation should go into liquidation.

The Executive Line was introduced in 1956 as a replacement for the entire Clipper Custom lineup, and was designed to bridge the gap between the lower price Clipper and top line Packard ranges. The Clipper V-8 was used for power, and was available in two-tone color schemes.

The company name continued on some redressed Studebaker models until 1958, but 1956 was the final year for true Packard-built automobiles. Sadly, a car of enduring quality and style could not survive in this era of planned obsolescence.

1956 Plymouth Fury Hardtop

Plymouth was number four in the production league in 1956, very slightly behind Buick in the number three position, with an output of 571,634 vehicles. J. P. Mansfield was Chief Executive Officer of the Division.

Introduced in 1956 with a 240-horsepower 303 V-8, the Fury had sharply peaked tail fins, a Cadillac-like logo and special side trim. It was a great example of Plymouths 'All new Aerodynamic' styling for the year, which was instigated as part of Virgil Exner's 1955 Chrysler-wide 'Forward Look' re-style. The front and rear windshields were wraparounds, and the side windows were pillarless.

The Fury was introduced to heat up Plymouth's performance image, for '56 was a seminal, powerorientatated, year for the entire US automotive industry. Effectively, it was the Division's first muscle car. Plymouth chose to introduce the production version of the car with a 240-horsepower 303 V-8, a polyspheric head engine, rather than the 'hemi' used by the other Chrysler divisions.

Sister company Chrysler introduced its 300 'B' letter car in the same year, which was specifically designed to outperform every other US production car, including newly-introduced Corvette and Thunderbird rivals. But at the Daytona Beach February Speed Week, the Plymouth Fury FX, driven by Phil Walters beat even the 'B', with a 'resounding boom and a flash of gold', by achieving a record pace of over 149 miles per hour. This made it the fastest Plymouth every built. Evidently, this was not a source of unalloyed joy to the board at Chrysler, who had hoped the 'B' would be totally invincible by allcomers, including their own. As well as being truly quick, the Fury also handled well, with its body lowered an inch from the other Belvedere models.

The production version Fury was a sub-series of the top of the range Belvedere.

Engine: Overhead valve Fury V-8

Displacement: 303 cid

Horsepower: 240 at 4800 rpm

Transmission: Three-speed manual with optional overdrive and automatic

Compression Ratio: 9.25:1

Body Style: Two-door Sports Coupe

Number of Seats: 6

Weight: 3650 lbs

Wheelbase: 115 inches (overall length 204.8 inches)

Base Price: $2807

Number Produced: 4485

1956 Studebaker Golden Hawk

Engine: Golden Hawk V-8 Packard engine. Overhead valve, cast iron block

Displacement: 352 cid

Horsepower: 275 at 4600 rpm

Compression Ratio: 9.5:1

Opposite: The comparative staid lines of the Studebaker must have contributed to its slump in sales.

Engineer Harold Churchill was selected as Studebaker's President in 1956.He was a loyal company man, who had worked in the company since 1926,and had cooperated with Raymond Loewy in his Studebaker Design Divison. Hewas determined that the company should survive, even with its difficulties.These were pretty legion in this year.

Model production slumped to 69,593 units, despite a full-on restyle of every model range. Studebaker fell back to thirteenth position in the league of auto production. The California plant was mothballed, leaving only two factories in production, at South Bend, Indiana and in Canada. The company also diversified somewhat, manufacturing the Ambulet (based on the station wagons) was launched and the Custom four-door sedan (the cheapest Stude at $1717) was marketed as a police/taxi vehicle.

This model was also available in a Canadian 'export' edition. The Loewy 1953 design classic, the Starliner, was the chief inspiration behind the pillarless Hawk quartet - the Golden Hawk, Power Hawk, Sky Hawk and Flight Hawk hardtops. Loewy himself supervised their deft makeover into a line of family sportscars with an effective restyle. The cars in this vein ran from the six-cylinder Flight Hawk, priced at a reasonable $1986 to the Packard V-8 equipped Golden Hawk at $3061. The big block Packard engine displaced 352 cid, and there was a dealer option to install a dual four-barrel set up borrowed from the Packard Caribbean. The 160 mph speedometer and matching 6000 rpm rev-counter made it clear that this was one car at least that Studebaker did not consider an economy model. But true to his principles, Loewy also designed safely into the car, and with a padded dash and floor-anchored seatbelts, it excelled in this area.

The Golden Hawk was the top-of-the range Studebaker for 1956, and its styling and equipment reflected this. Exclusive fibreglass fins were fitted on top of the rear fenders, which did not appear on the other Hawk models. Like the President Classic the Golden Hawk had wide grooved horizontal moldings just above the rocker panel. The interior specifications were also high, with all vinyl interiors or vinyl with cloth inserts in a variety of colors and patterns, with co-ordinating carpets. The fiberglass dashboard was also attractive, with its engine turned facing, featured black dials with white numerals and indicators. Standard items fitted to the Golden Hawk included a cigar lighter, glove compartment, electric windshield wipers and a dual exhaust system. A long list of cost-plus options was also available, including an AM radio, power seats and windows and tinted glass.

Below: The Golden Hawk used the Packard overhead valve V8 by this time.

1957 Chevrolet Bel Air Sport Sedan

Jack Kerouac published 'On The Road'in 1957, a book that was to set themood for a whole generation, demonstrating how Road Culture was now a subliminal part of America's national psyche.

Chevrolet and Ford reversed their sales positions in 1957, Chevrolet slipped to second position manufacturing just over 1.5 million vehicles, but nearly 1.7million Fords came off the production line that year. In fact, by the end of the year, symptoms of the economic recession of 1958 were already beginning to appear. But so far as automotive history goes, the Chevrolet Bel Air goes downas the most important car of the year, examples are treasured to this day and the model may well be the most highly regarded car of the decade: a style and performance classic.

1957 was also the year when Chevy introduced optional fuel injection - 'Ramjet'- for the first time on its V-8 engines. They claimed that the innovation made them the first American car manufacturer to deliver one horsepower for each cubic inch of displacement. But at $500 the system was tremendously expensive, especially as a basic Chevy came in at just a tad more than $2,000. Rivals Pontiac also introduced fuel injection in '57. Turboglide transmission was also introduced as an optional update to Powerglide. It was based on Buick's successful Dynaflow system, but proved less reliable. Safety seat belts and shoulder harnesses were also offered as options for the first time that year. Just as well. Extra richness characterized the update styling and trim for the Bel Airs, which were now three years old. The marque seemed to be looking towards Cadillac for its influences, with more style and go fitted as standard for the new models. The massive front bumper became an integral part of the lower profile body. Chrome sidespears formed branched on the rear wings, and were filled with silver anodized aluminum beauty panels. Three gold chevrons accented the forward sides of the front fenders. The Chevrolet and Bel Air scripts were gilded for 1957, and so were the V-8 ornaments on models where this engine was fitted. Rocker sills, roof and windows outlines and the entire edge of the fins were all trimmed with rather glitzy metal moldings.

Engine: Six-cylinder overhead valve, cast ironblock (V-8 also available for this model)

Displacement: 235.5 cid

Horsepower: 140 at 4200 rpm

Compression: 8.0:1

Wheelbase Length: 115 inches

Price: $2238

Number Produced: 62,751

1957 Chevrolet Corvette

The Corvette styling for '56, '57 and '58 remained pretty well unchanged except for modest tweaking of the chrome between model years. But the new 283 V-8s under the hood offered one horsepower per cubic inch with Chevy's new 'Ramjet' fuel injection. As the adverts of the time proclaimed, 'For the first time in automotive history – one horsepower for every cubic inch.' Of course, this wasn't actually true, the Chrysler 300 series was the first American car in this field.

The spirit of the car also remained unchanged, it was designed as a fun and economical sports car for young people, and remains so to this day.

The original car was built in fiberglass to lower the initial production cost. Amongst the standard features fitted to the car in this model year were dual exhaust pipes, all-vinyl bucket seats, three-spoke competition style steering wheel and carpeting. Seven optional colors were available, Onyx black, Polo white, Aztec copper, Artic blue, Cascade green, silver or, the color of our photographed car – Venetian red. In two-tone cars, the 'cove' in the side panels could be picked out in white, silver or beige.

For a model that Chevy only retained to compete with Ford's Thunderbird, Corvette sales were virtually doubled from 1956, and continued to climb for the rest of the '50s.

Interesting option included a detachable hardtop ($215), Whitewall tires ($32), Hydraulic power top ($99), automatic transmission ($175), three versions of the fuel-injected engine were available (the most expensive of which could from 0-60 mph in 5.7 seconds, and achieve a top speed of 132 mph) and heavy-duty racing suspension ($725). In fact, only 1040 of the 1957 'Vettes were manufactured with fuel injection.

Engine: v-8 overhead valve, cast iron block. Fuel injection available as an option.

Displacement: **283 cid**

Horsepower: 220 at 4800 rpm

Transmission: Three-speed manual floor shift, or four-speed manual floor shift or Powerglide as optional extras

Compression Ratio: 9.50:1

Fuel System: Carter four-barrel Model 3744925

Weight: 2730 lbs

Wheelbase: 102 inches (overall length 168 inches)

Base Price: $3,465

Number Produced: 6,339

1957 Chrysler 300-C Two-door Convertible

Engine: V-8, overhead valve with hemispherical combustion chambers	
Displacement: 392 cid	
Horsepower: 375 at 5200 rpm	
Transmission: TorqueFlite	
Compression ratio: 9.25:1	
Number of Seats: 6	
Weight: 4390 lbs	
Wheelbase Length: 126 inches (overall length 219.2 inches)	
Base Price: $5359	
Number Produced: 484 Convertibles	

The 300-C was the third (following on from the C-300 and 300-B), and first convertible generation of the Chrysler 300 series. The model continued to fly the flag as the fastest production car in America. A completely new direction in styling, 'The Forward Look' had been applied to the 1957 300s by Exner and the Chrysler design team. The main features of the new body style were the elegant tailfins and masculine grille. The cars were restrained compared to the flamboyance of several contemporary models (the Chevy Bel Air Convertible, and Plymouth Fury are just two examples), but won the company several design awards. The cars were also built with a new torsion bar front suspension, replacing coil springs, that was to be used by Chrysler well into the 1980s.

Motor Trend Magazine awarded the model 'Car of the Year', mainly for its excellent handling and engineering. However, in the annual league of car producers, Chrysler was back at eleventh place.

The Firepower hemi was again fitted to the car, but this would be its penultimate year in the model, the Golden Lion V-8 would be used from 1958.

Although considered part of the New Yorker series by Chrysler, the 300s were really a world away from even their production stable mates for power and handling. They were also much more simply trimmed, with a just a single sidespear running along the rear the rear quarter panels on both sides of the car. The wheels, at 14 inches, were smaller than the previous model, and typical of the wheel size of contemporary US cars. The smallness of the wheels necessitated more effective brake cooling, which was achieved by adding rectangular openings under the headlamps, from where the in-coming air was ducted onto the brake drums. The 300s were now produced exclusively in monotone colors.

1957 Dodge D-500 Convertible

Engine: D-500 V-8 overhead valve, cast iron block, hemispherical heads

Displacement: 354 cid

Horsepower: 340 at 5200 rpm

Compression Ratio: 10.0:1

Fuel System: Carter four-barrel Type WCFB

Body Style: Two-door Convertible

Number of Seats: 6

Weight: 3975 lbs

Wheelbase: 122 inches

Base Price: $3635

Number Produced: 6,960 Two-door Convertibles across all ranges

The year 1957 saw the introduction of Virgil Exner's second generation 'Forward Look' applied to all the Chrysler products, including Dodge. The Dodge brand of the styling was denoted as 'Swept-Wing Styling'. The Dodge tailfins made the cars look as though they could take off, and with the D-500 super package, they virtually could. This supercharged version of their Hemi V-8 developed up to 340 horsepower, which put the car into some very good Chrysler company.

Body-wise, the cars were longer, lower and wider than any previous Dodge, and the wheelbase was longer than anything the company had built since 1933. The change over to fourteen-inch wheels helped to lower the silhouette of the car, and the torsion bar suspension (introduced over at Chrysler) was also new. The grille featured a gull-wing shaped bar that surrounded a huge Dodge crest. All the 1957 models had a single chrome bar along the body, and chrome trim under the rear fins. But the new styling could not disguise some quality control problems that were to dog the company and hinder sales for a couple of years.

In fact, the D-500 was actually designed as a high-performance engine that could be used in all the 1957 series, but the specifications given are for the engine fitted to two versions of the Custom Royal series (the two-door sedan and the two-door convertible) – but the model shown in the photograph is a Coronet (two-door Lancer convertible) fitted with the D-500. The Coronet was now the base Dodge model, and sales of this series made up over 160,000 of the sales figures for the year.

As a company, Dodge was performing much better, climbing to seventh in the league of sales.

1957 Ford Thunderbird

Engine: Standard V-8. Overhead valve

Displacement: 272 cid

Horsepower: 190 at 4500 rpm

Compression Ratio: 8.6:1

Wheelbase: 102 inches (181.4 inches overall)

Base Price: $3408

Number Produced: 21,380

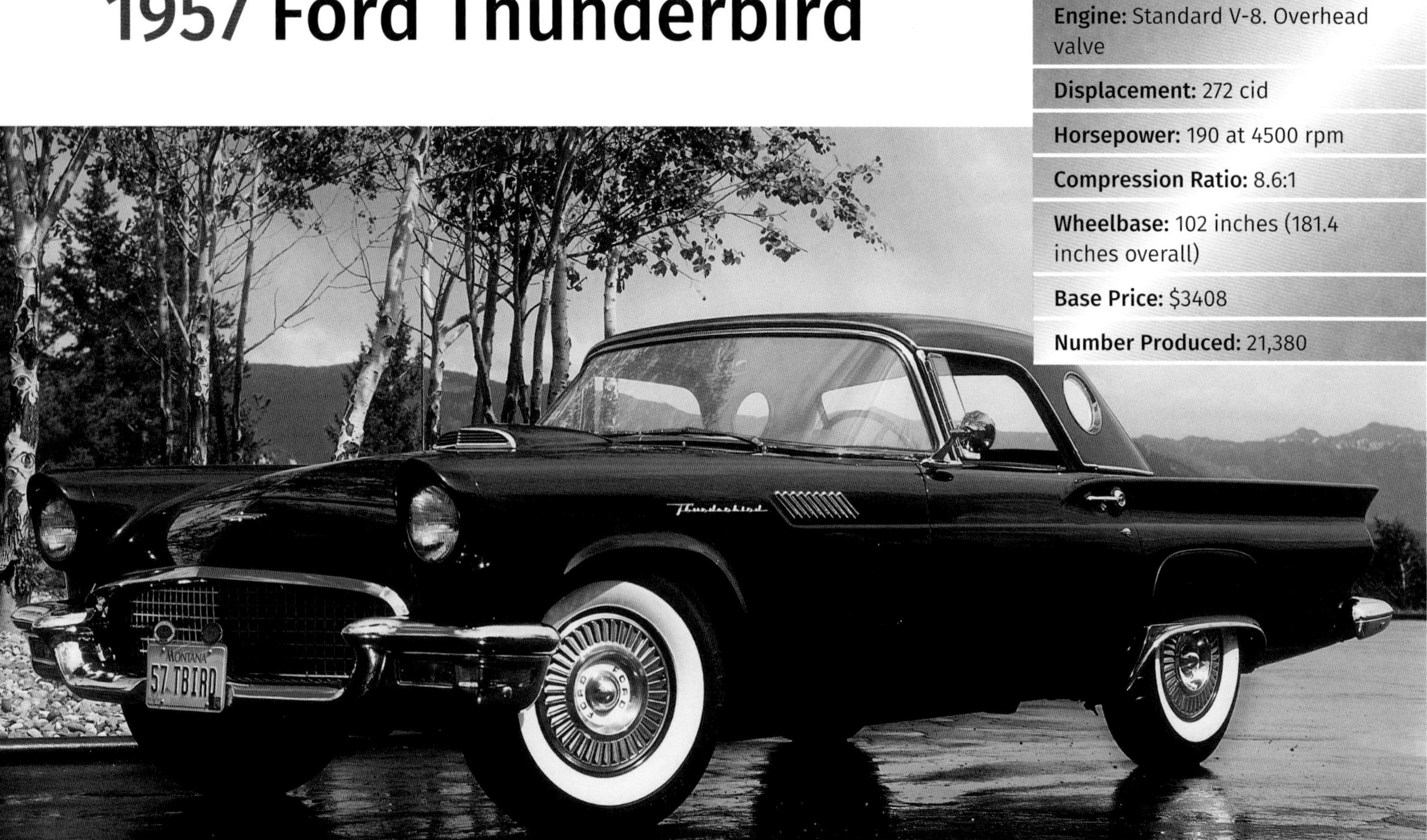

Ford had first introduced the Thunderbird two-seater 'personal' car in 1955 to counter Chevrolet's 1953-introduced Corvette. At almost exactly the same base price as its more established rival, the T-bird hugely outsold the 'Vette in its first year. Buyers loved the sleek good looks of the car, and the practical lift-off hardtop. 16,155 examples were sold in the car's first model year, as opposed to a slightly pathetic 700 Corvette roadsters.

By 1957, the Thunderbird was still well ahead in the sales race, (21,380 to 9,168) although the Corvette was considerably more expensive by this model year, and followed the general trend as Fords outsold Chevys for the first time in a good while. The Thunderbird was also available as the more powerful option by the time the new 1957 model was introduced in October 1956. Whilst Chevy offered 283 horsepower via its optional Ramjet fuel injector (one horsepower per cubic inch) Ford was offering a far more massive power option. The supercharged Thunderbird Special Supercharge V-8 belted out 340 horsepower at 5300 rpm in the NASCAR version of the car. Both Thunderbirds and Corvettes were available with a single body option in 1957, the standard two-door convertible.

Thunderbirds had been heavily re-styled for the 1957 model year, as were all the cars in this decisive year in Ford's history. The redesign included a revised body shape, including huge tailfins (or 'high-canted fender blades') and the abolition of the externally mounted spare wheel. Interior changes included a new instrument panel and 'Lifeguard Design' safety features (including a padded dashboard and dished steering wheel). Six engine options were available for the T-bird, a six-cylinder version (serial number code 'A') and five increasingly powerful V-8s (serial number codes 'B', 'C', 'D', 'E' and 'F'). All kinds of optional refinements were available at extra cost – everything from power windows to white sidewall tires.

Despite its success, 1957 turned out to be the final year for the two-seater version of the Thunderbird. 1958 saw the introduction of the luxury fourpassenger 'Square Birds'. A ragtop version was also launched. Effectively, this took Thunderbirds out of the American sports car market.

1957 Mercury Turnpike Cruiser

The 1957 Mercurys were completely restyled again, and for the first time, the marque had bodyshells unique to it, not shared with either Ford or Lincoln.

The new restyle made the cars looks as though they had just landed from outer space with extreme body styling, including bizarre fins and V-shaped taillights sculpted into the rear deck. A gold anodized insert in the concave section of the upper rear fender section led to the taillights. The front grille consisted of concave vertical bars. A large chrome 'M' between the grille and fender proudly claimed the heritage of the cars. Later in the year, the cars went from two to four headlights in a chrome eyebrow assembly.

Three model lines were introduced for the year. Monterey was the base option, the Montclair now became the mid-range product, and the Turnpike Cruiser series became Mercury's top-of-the-range output. The Mercury station wagons also became a separate series this year, with the Colony Park as the top model in the range of six vehicles.

The Turnpike Cruiser had the most modern and extreme styling that was controversial at the time, and resulted in the designers being derided as 'hacks and chrome merchants'. The Turnpike Cruiser convertible is perhaps the most gaudy and extreme looking member of the range, its length and massively overhanging rear deck are emphasised by its low lines. The ragtop is completely invisible once down, thanks to the 'Breezeaway' roof with a retractable back window. Unique to Mercury, the roof gained a reputation of its own. The Turnpike Cruisers were possibly the most gadget-laden cars every built in Detroit, the veritable home of gadgetry. All power items – Seat-O-Matic, steering and brakes - were fitted as standard in the range. The hardtop model had a startling roof overhang front and rear – the front to accommodate air ducts mounted on the windshield, complete with fake aerials – the rear window was retractable.

Inside, the innovations didn't stop. The power seats had a memory dial, the instrument bezels were made of rubber and there was a special starter button.

The series included a fastback hardtop to improve Mercury's chances at winning stock car races. The convertible pictured was used as a pace car at the 1957 Indianapolis 500. The third model in the range was a four-door hardtop sedan. All three cars shared the Whacky Races range styling. The Turnpike Cruiser models weren't a roaring sales success, however, and were re-absorbed into the Montclair series for '58.

Engine: Overhead valve V-8, cast iron block

Displacement: 368 cid

Horsepower: 290 at 4600 rpm

Compression Ratio: 9.7:1

Base Price: $4103

Number Produced: 1265

1957 Oldsmobile Starfire Convertible

Oldsmobile introduced three restyled model ranges for 1957 in November '56. The entire Eighty-eight entry series was named 'Golden Rocket' in honor of the up-coming 50th anniversary. Eighty-eights were the base models and Super Eighty-eights were the mid-range models for the year. Starfire Ninety-eights were the four-model top range. Standard equipment on the senior lineup included all the Eight-eight features, plus electric windows, special emblems, power steering, power brakes and Jetaway Hyda-Matic transmission. Interior trim included cloth, morocceen and leather.

The 'Starfire' name went back to a two-seater concept car that Oldsmobile toured with the GM Motorama Car Show back in '53, named in honor of the Lockheed F-94B Starfire jet fighter. This particular version had a bright turquoise fiberglass body, and turquoise-and-white leather interior. Although this original concept never found its way into production, the Oldsmobile continued to search for a car of this type, to give Ford's 'personal luxury car', the Thunderbird, some direct competition. But the Starfire model introduced in '57 was a car of a completely different stamp, being a full-sized luxury convertible. The name subsequently became synonymous with big, flashy, sporty cars, and it wasn't until 1961 that the Oldsmobile Division considered reconfiguring the car to be a serious rival to the (now four-seat) T-bird. Even in '57, the Starfire was available with a range of J-2 performance options. This technology had been designed to re-assert Oldsmobile's power superiority over Chevy and their small-block V-8. The top J- 2 option (the triple-carb Rocket with 300 horsepower) wasn't recommended for road cars, but Lee and Richard Petty drove the multi-carbureted engines to NASCAR success, until the racing authorities banned this type of induction.

Olds subsequently pulled out of factory-backed racing. With J. F. Wolfram as Divisional CEO, Oldsmobile was the fifth best performing automobile manufacturer in 1957, gaining a 6.2% share of the US market. The company was now solidly positioned as the sales leader in medium-priced cars.

Their concept cars continued to tour with the GM Motorama car show. This year, their big idea was the Mona Lisa, a Starfire Holiday Coupe.

1957 Pontiac Bonneville Convertible

Big things were happening at Pontiac in 1957. Semon E. "Bunkie" Knudsen became Pontiac's new general manager in '56, the son of former GM President William S. Knudsen, he became the youngest general manager at GM at the ripe old age of 43. The change marked the quiet insurgence of Pontiac into the youth car market. Knudsen hired Pete Estes (who became Chief Engineer at Pontiac in 1957) and John DeLorean to head up the design and engineering teams, together with a series of huge performance advances, their work was to positively shape Pontiac cars for years to come. The Strato-Streak V-8 was launched in 1955, it was lighter than the GM Straight 8, and cheaper to manufacture than the Oldsmobile Rocket V-8.

There was a huge emphasis at Pontiac on research and development, and a constant striving for engineering advances. The company funded an influential Product Engineering department.

During this period, they experimented with four-speed floor-shift, rear-mounted transaxles, and supercharging.

A unique, limited edition Custom Bonneville Convertible was released in January 1957. It was Pontiac's first Bonneville. The car was in the Custom Star Chief sub-series (series 28) in which there were four body styles. The production was limited to 630 cars and these were released to the Pontiac dealership of over 1,500 outlets, ostensibly to enable dealers to display the benefits of the modern Pontiac and especially the new fuel injection system. What it actually did was bring buyers to the showrooms in their droves and set the stage for Pontiac being recognized as GM's "Performance Motor Division". The car was only available with a convertible top and the Bonneville V-8, an enlarged Strato-Streak, which developed an estimated 300 plus horsepower at 4800 rpm. The cars were all fitted with the new Rochester mechanical fuel injection system. The Bonneville came with a host of "power" features as standard like brakes and seats.

Engine: Overhead valve V-8, cast iron block

Displacement: 347 cid

Horsepower: (estimated) 315 at 4800 rpm

Transmission: Strato-Flight Hydra-Matic

Compression Ratio: 10.25:1

Fuel System: Rochester mechanical fuel injection

Number of Seats: 6

Weight: 4285 lbs

Wheelbase: 124 inches

Base Price: $5782

1957 Pontiac Custom Star Chief Two-door Catalina Hardtop

Engine: Overhead valve V-8 cast iron block

Displacement: 347 cid

Horsepower: (Hydra-Matic) 270 at 4800 rpm

Transmission: Hydra-Matic

Compression Ratio: 10.0:1

Body Style: Two-door Catalina Hardtop

Number of Seats: 6

Weight: 3750 lbs

Wheelbase: 124 inches

Base Price: $2901

Number Produced: 32,862

The illustrious Pontiac name was introduced by the Oakland Automobile Company in 1926, and introduced their first V-8 right back in '32. It was to become GM's excitement division, particularly under the guidance of the John De Lorean-Pete Estes magical partnership of design and engineering.

In 1957, the company was ranked number six in the car producers' league, with a 5.4 per cent of the US market. The Star Chief two-door Catalina Hardtop was another Custom Star Chief 1957 model in the same series 28 as the Bonneville convertible. Much less exclusive than the Bonneville, 32,862 examples of the two-door hardtop were built in this model year, and sold for virtually half the price. Indeed, by 1956, hardtop sedans were available from virtually all US carmakers, but Pontiac's were particularly snazzy.

When the model was introduced by Pontiac, the advertising slogan for the car was 'Looks Like Pontiac Cornered the Market on Firsts' claiming that the car was built with six-dozen innovations. The car was subjected to a 100,000- mile Marathon run to road test the new features that included the 'cloud-soft ride', 'cat-quick wheel response', and the 'lusty brilliance of its all-new V-8 engine'. The engine in question was the 347 cid Strato-streak V-8 that was fitted to the Super and Star Chief models. It had grown for the third time in three years. The body design was also a fashion first for Pontiac, with 'off the- shoulder', star-flight styling. Pontiac claimed that the car was 'America's Number One Road Car'. Four stars on the rear fender trim identified the series, together with front fenders scripts, chrome semi-cylindrical trim at the back of missile-shaped inserts and full wheel discs.

1957 Studebaker Golden Hawk

The rot had really set in at Studebaker by 1957, despite the best efforts of Harold Churchill. Ranked at thirteenth place for the third year in a row, the company only managed to produce 63,101 units in this model year.

Desperately seeking sales, Studebaker added the budget-price Scotsman range, and a new range of station wagons. Bucking the general trend, the Hawks had their best year ever, selling over 19,000 units. The Hawk quartet of '56 was re-configured for '57 into just two models, the Silver and Golden Hawks. Sky, Power and Flight Hawks disappeared in cost-cutting exercise, the cars were expensive to manufacture and needed buoyant sales to be viable. In effect, this was an admission that Studebaker no longer felt able to compete with the larger manufacturers with their larger model rosters. The '57 Golden Hawk replaced the '56 Sky Hawk and Golden Hawk models.

1957 Silver Hawks were available only as a pillared coupe with a range of engines (both six-cylinder and V-8), whereas the Golden Hawks only came as a hardtop model with a hot 289 cid V-8.

The Packard engine of the 1956 Golden Hawks was no longer available to Studebaker, due to the sale of the Utica, Michigan Packard engine plant to Curtiss-Wright. The car was now fitted with the largest 289 cid Studebaker

V-8, and a Paxton supercharger was fitted as standard. In fact, the supercharger was necessary to maintain the power output. Flight-O-Matic automatic transmission was also fitted as standard. The model body styling was also revised. The tailfins were enlarged from '56 and highlighted with a coved rear panel that was painted in a contrasting color.

A Golden Hawk 400 model was also available, with full leather interior and flared armrests. A magnificent all-white example of this model is on show at the Studebaker National Museum in South Bend, Indiana. The museum is located at the location of Henry and Clement Studebaker's original blacksmith shop, where the whole Studebaker story began.

Engine: Golden Hawk V-8, overhead valve

Displacement: 289 cid

Horsepower: 275 at 4800 rpm

Transmission: Flight-O-Matic

Compression Ratio: 7.8:1

Body Styles: Two-door Hardtop

Weight: 3400 lbs

Wheelbase: 120.5 inches

Base Price: $3182

Number Produced: 4,356

1958 Buick Limited

Engine: V-8 overhead valve, cast iron block	
Displacement: 364 cid	
Horsepower: 300 at 4600 rpm	
Transmission: Dynaflow	
Compression: 10.0:1	
Body Style: Two-door Convertible	
Weight: 4603 lbs	
Wheelbase Length: 127.5 inches	
Price: $5125	
Number Produced: 839	

The year 1958 wasn't a good year for Ford with the introduction of the horrid and unlucky Edsel, but things were even worse over at Buick. Sales in the year plummeted to fewer than 250,000 and the company fell to number five in the sales league. The designers at Buick seemed to have completely missed the plot. As the economy went into a deep (but thankfully mini) recession, and unemployment hit over 5.5 million, car buyers either stayed away in droves or bought smaller economy models, Buick chose this year to launch the big, expensive, luxury top-of-the-line 'Limited' series.

Buick resurrected this grand old nameplate for a trio of cars that were designed to vie with Oldsmobile for glitter, and even challenge Cadillac at the luxury end of the market. A four-door hardtop, two-door hardtop and two-door convertible comprised the full model range in the Limited series.

The most distinctive feature of the styling was twelve diagonal chrome louvers set in the rocket-like trim motive of the rear fenders, together with massive louvered taillights set in squared off tail fins. There was also a heavy sweeping chrome sidespear on either side of the car that swept backwards from the headlamp assembly all the way back to the rear wheel arch. The rear fender was a truly massive chromed construction complete with positively sculptural over riders.

The front fender, also heavily chromed, featured the classic Buick checkerboard effect, whilst the two double headlight assemblies appear to be heavily lidded.

The series was completely disastrous, and far more 'Limited' than Buick had hoped. Across the three models (both hardtop and convertible) only 7136 units were produced. Production of the range was cancelled at the end of the year, as Buick's designers completely revised the entire line. In fact, none of the 1958 model names survived this apocalyptic year.

1958 Chevrolet Impala Convertible

The worst recession year of the post-war era was 1958, and car sales fell by about 30% on the three previous boom years. Americans were seriously concerned that the Russians appeared to be pulling ahead, technologically speaking, in the Cold War – claiming that their nuclear missiles could now reach anywhere in the world. However, with an excellent selection of reasonably priced models, Chevrolet regained poll position in the sales stakes, about 1.14 million compared to just under one million Fords. Imported cars were also becoming more of a factor, cornering an unprecedented market share of almost 10%. These were mostly from Japan, and the first Toyotas (Toyopets) and Datsuns arrived for sale for the first time in America in 1958.

1958 saw the introduction of two luxurious 'Impala' models to the already up-market Bel Air series. For Chevrolet, the cars were launched with relatively stiff retail prices, and were an attempt to reach upmarket. Despite the generally poor trading conditions, the innate class and style of the Impalas meant that they sold very well. The Impala name itself was almost immediately destined for positively iconic status amongst the cognoscenti. A hardtop sport coupe, and a hardtop convertible were simultaneously introduced – both two-door vehicles. The most powerful engine option for the Impalas developed 300 horsepower from a V-8 engine block. The trim features for the new cars included Impala script insignia, broad ribbed sill panels, large dummy chromed air scoops just ahead of the rear wheelwells and triple taillamp arrangements. Inside the car, there were competition style two-spoke deep hub steering wheels, and several Impala script emblems.

Engine: Six-cylinder with overhead valve, or V-8 with overhead valves were both available for Impalas

Displacement: 235.5 and 283 respectively

Horsepower: 145 at 4200 rpm, 185 at 4600 rpm

Weight: 3522 lbs (2841 lbs with a V-8)

Price: $2724 ($2841 with a V-8)

Number Produced: Note: After 1957, Chevrolet production totals are only available by series, no breakdown is available for each series, model or engine type. 142,592 Sports Coupes were sold by the company in 1958, plus 55,989 Convertible models

1958 Dodge Coronet

Engine: Red Ram V-8 overhead valve	
Displacement: 325 cid	
Horsepower: 245 at 4400 rpm	
Compression Ratio: 8.0:1	
Body Style: Two-door hardtop	
Weight: 3540	
Wheelbase: 122 inches	
Base Price: $2644	
Number Produced: Dodge produced 77,388 Coronets in the year, but no figures for the different models were available	

The 1958 Dodges, including the Coronet, used the 1957 swept-wing body shell with minor restyling, except that the grille and headlights were completed revised. The grille now consisted of side-to-side horizontal bars with a single low fender and two shorter bars at the sides that housed the parking lights. The headlights followed the general market trend by adopting the quad-light format. The revised model offered the public plenty of style at a relatively reasonable price, and sold 77,000 units in the year.

The main preoccupation at Dodge in this model year was power. The engine was now the 'wedge' shaped single rocker head design. This became standard across the Chrysler Group, and was cheaper to produce than the Hemi V-8s. In fact, six engine options were available to all Dodge models, ranging from the Inline six-cylinder right through to the super-powerful electronic fuel injection V-8, fitted with the Bendix EFI assembly. However, although this model could develop 333 horsepower at 4800 rpm, only six such cars were built in the year, and were later recalled for the unreliable injector to be replaced with an ordinary carburettor.

Surprisingly, the demand for six-cylinder engines actually increased in the year. Maybe it was the general recession that made buyers more economical.

However, 94.4% of the 1958 cars were fitted with automatic transmission, and 62.5% had power steering. General production fell disastrously in the auto industry as a whole, but Dodge production was particularly badly affected, falling by more than 50%, and retaining only a 3.1% of the market. The 65,000 employees of the company only accounted for approximately two cars each during this dreadful year. Labor problems with the workforce in 1958 had a knock-on effect to 1959 production, and even the introduction of exaggerated styling couldn't gloss over ridiculously long delivery times. The 1958 Coronets were assembled in three Dodge plant locations – six-cylinder powered models in Detroit and Newark, V-8s in Detroit, Newark and Los Angeles.

1958 Ford Edsel Citation

Engine: V-8 overhead valve, cast iron block	
Displacement: 361 cid	
Horsepower: 303 at 4600 rpm	
Compression Ratio: 10.5:1	
Wheelbase: 124 inches	
Base Price: $3766	
Number Produced: 930	

FoMoCo had realized in the late '40s that they needed a truly up-market car to offer their buyers, who would otherwise move up and away to other brands. A Special Products Division was set up in 1955, under the leadership of R.E.Krafve, to develop such as car. The plan was to use components from other parts of the group, but to develop a distinct car in its own right. FoMoCo stylist Roy Brown wanted to create a car with a unique, immediately recognisable appearance. The project was nick-named the 'E-car' (for experiment).

After several delays, the new product was scheduled for a fall 1957 launch, and a name was finally agreed upon – Edsel - after Henry Ford's son, who had died in 1942. The Ford family were initially unsure about using the name, but were talked into it by company chairman Ernest Breech. Ranger, Pacer, Corsair and Citation were chosen as the model names for the series. The Ranger was the base model, and the Citation was the most expensive, 'senior', car in the range.

Eighteen models were launched initially, including five station wagons. The model names for these were Roundup, Villager and Bermuda. Production started in 1957 in four different assembly plants. The Edsel was billed as 'The Newest Thing on Wheels', and there was a massive three-year long publicity build-up to the launch date. Four thousand cars were sold on the September 4 launch date, but within a month, this had slowed dramatically and the cracks were already beginning to show in Ford's big experiment. Having said all that, the Edsels did set an all-time record for deliveries of a newly introduced, mid-price car.

From a manufacturing point of view, the 1958 Citation and Corsair shared a single body shell, but the Citations were trimmed to a higher level. Three models were introduced with this name, a two-door hardtop coupe, a four-door hardtop sedan and a two-door convertible.

1958 Lincoln Continental Convertible

Lincoln introduced the Continental in 1940, and the series quickly became one of the most celebrated lines ever. Indeed, the 1940 convertible model is considered by many to be the most beautiful American production car every built. It was longer and lower than the other model year Lincolns, and introduced the 'continental spare tire' look. The original Continental cars were powered by a V-12 engine.

Lincoln's top-of-the-range series was completely re-styled in 1958, from the Mark IIs of '56 and '57. These had been a very exclusive and expensive range, and sold in modest numbers The Mark III was the largest unibody car ever built, and became the longest American production convertible ever built at a stunning 229 inches in overall length. The chassis was dispensed with, and the suspension, driveline and engine units were fastened to the body structure. It was also fitted with the biggest contemporary American engine – the huge 430-cid V-8. Lincoln offered air suspension, but only 2% of the units sold were ordered with this option. But the cars had lost some of the elite reputation of the series. The Mark II convertible retailed for $10,000, which was nearly $3,000 more than the Cadillac Eldorado Biarritz, but the Mark IIIs were considerably re-positioned in the market, being nearly $1,000 cheaper.

The low-flying lines of this super-long car emphasised its great length, as did the lower body moldings, chrome trim and criss-cross pattern aluminium grille. The elegant streamlining was maintained by the 'disappearing top'.

The fabric hood and glass rear window were fully retracted into the truck, which was hinged at the rear to allow this power-operated system to work smoothly. The luxurious feel of the model was emphasised by full leather upholstery fitted as standard.

The other 1958 Lincoln series were the base level Capris and the mid-range Premiere. The Capri was also completely restyled for this model year, whilst the Premiere used the same body with slightly more trim. All three 1958 models were fitted with the same standard equipment. Lincoln production fell to 17,134 units in this recession year, which translated to position fifteen in the league of model year production.

Engine: Overhead valve V-8, cast iron block

Displacement: 430 cid

Horsepower: 375 at 4800 rpm

Compression Ratio: 10.5:1

Weight: 4927 lbs

Wheelbase: 131 inches

Base Price: $6223

Number Produced: 3,048

1958 Plymouth Fury

Engine: Fury v-8 overhead valve
Displacement: 317.6 cid
Horsepower: 225 at 4400 rpm
Transmission: Three-speed manual with Torqueflite automatic optional
Compression Ratio: 9.25:1
Body Style: Two-door Hardtop Coupe
Weight: 3510 lbs
Wheelbase: 118 inches (206 inches overall)
Base Price: $3032
Number Produced: 5,303

Plymouth had made it back to third position in the producers' league in 1957, and retained this place in '58 with just over a 30% share of the market.

The 1958 Fury was a Belvedere sub-series containing only the high performance sports coupe. Introduced back in '56, the Fury was designed to heat up Plymouth's new performance-orientated image, equipped with a 240 horsepower V-8, and special gold-anodized side trim (this trim persisted into the '58 range). Like the rest of the Plymouth line, the Fury hardtop emerged from the 1957 styling bigger, bolder and better-handling than the first model, complete with a 290-horse V-8. The '58 Fury was a limited edition two-door hardtop with Fury rear fender, nameplates, bumper wing guards, padded interior, and front and rear foam seats, effectively a revised and super-charged Belvedere. Just as '57, the Belvederes remained the top Plymouth range, Plazas were the base models, whilst the Savoys comprised the medium-priced series.

The '58 Furys were now powered by an overhead valve Fury V-8, and an optional Bendix fuel injection (EFI) system was also available for the big-block Golden Commando V-8, but these were later recalled and reconverted. These Golden Commando 'big block' wedge head V-8s were capable of 0-60 miles per hour in 7.7 seconds and could run the quarter-mile in 16.1 seconds.

The '58 model Fury is the car immortalized by Stephen King in his book Christine, but there are several inconsistencies with reality. In the King's book (as in John Carpenter's 1983 film version), Christine is red and white, but the real model was only available in buckskin beige with gold trim. In the book, King refers to the car as having four doors, but the '58 models were only available in the singe two-door version. He also mentions that car as having hydramatic transmission with a transmission lever, but the car actually had TorqueFlite, push-button transmission.

1958 Pontiac Bonneville Custom Convertible

The year 1958 was GM's 50th anniversary, and Pontiac celebrated by announcing a special 'Golden Jubilee' trim for a special line of Star Chief Custom cars. They announced their '58 line-up as 'The Boldest Advance in Fifty Years'. In this year of economic downturn, Pontiac held onto its sixth position in the league of auto producers.

Following its limited production model introduction of 1957, Bonneville was expanded to offer a top-of-the-line luxury line-up for '58. Produced with a unique one-year-only body, the cars showcased the final style statement of long-time GM design guru Harley Earl, and were loaded with chrome, wild twotone exteriors and lavish interiors. The Sport Coupe seats were upholstered in 'Lurex and jewel-tone Morrokide', the convertible in leather. Even the carpets had lurex threads to add more glitter. The dashboard positively gleamed with chrome dials and general sparkle, and had rare inside sliding sun visors. These slid down on a track from the headliner and almost filled the entire windscreen, but they proved to be impractical, so appear only for this model year.

Other fascinating options were a Pontiac dash-mounted compass, Autronic Eye automatic high-beam headlight dimmer, Memo-Matic power memory seat and Wonderbar AM radio. Ironically, the Safeguard speedometer was also available with an excess speed buzzer and warning light! To keep the base price of the cars as low as possible, virtually every extra, however small was listed as a plus-cost option, even the oil filters.

The 1958 Bonnevilles became a line name instead of a singe model designation: the Bonneville Series 25 Super Deluxe range. A second model, the Bonneville Custom two-door Sport Coupe was also introduced. Ragtop and Hardtop were fitted with a new and even bigger Star Chief V-8 for the year, which (with the NASCAR certified Tri-power carburettor option) were capable of delivering a massive 310 horsepower. Fuel injection was also available as an option on both cars. A Tri-power equipped Bonneville convertible was selected as the Official Pace Car for the Indianapolis 500 Mile Race in May 1958, whilst the 300 horsepower hardtop engine was road tested doing 0-60 mph at 7.6 seconds. I.

The Bonneville came with a host of power features like seats and brakes that were optional on the rest of the range.

Engine: Star Chief Base V-8, overhead valve

Displacement: 370 cid

Horsepower: 240 at 4500 rpm (synchromesh)

Base Price: Custom Sport Coupe $3481

Number Produced: 9144 Coupes, 3096 Con.

1959 Cadillac Eldorado

Engine: V-8 overhead valve, cast iron block
Displacement: 390 cid
Horsepower: 345 at 4800 rpm
Compression: 10.5:1
Body Style: Two-door Eldorado Biarritz Convertible (Two-door Eldorado Seville hardtop also available)
Weight: 5060
Wheelbase Length: 130 inches (225 inches overall length)
Price: $7401
Number Produced: 1320

The prevailing style of automotive styling in this year could unkindly be called 'Wurlitzer Jukebox' – and Cadillac was the supreme proponent. If the Chevrolet Impala of the same year was extreme, the Eldorado still managed to out-glitz and out-chrome its rival completely. Of course, it also weighed in at nearly $2,000 more, pushing the price of the Biarritz Convertible to over $7,400. In fact, the 1959 model attracted only 1320 wellheeled customers, but has passed into automotive history as a truly iconic car that has worshippers and detractors in almost equal number.

The car was designed by the outgoing Head of Styling at Cadillac - Harley Earl as his swan song before leaving the company later that year. The car was lower than its predecessor, but it was the body styling that really identified it.

The tailfins were the largest ever to appear on a Cadillac, together with the bullet taillights and other details that so clearly reflected the national fascination with the space race. In fact, contemporaries thought the car looked positively 'Martian'. The Eldos were criticised for poor handling, but at nearly nineteen feet in length, they had been designed for good looks and luxurious cruising rather than hard driving. Despite this, the car could achieve 0-60 mph in eleven seconds, with a top speed of nearly 120mph, courtesy of an enlarged V-8 engine block.

Cadillac itself retained tenth position in the league of cars sales in the US, selling just over 140,000 units. The increased competition from imported cars may have somewhat affected sales, but Cadillac is still the biggest volume manufacturer of exclusively luxury cars in America.

Ironically, just as Cadillac's jukebox look reached its Zenith in the '59 Eldos, the death of Buddy Holly, Ritchie Valens and J.P. 'The Big Bopper' Richardson together in a single air crash signalled the end of the 'Happy Days' rock 'n' roll era, and the flamboyant styling it had inspired.

1959 Chevrolet Impala Convertible

If the nineteen fifties was the decade of the tailfin, then 1959 saw the absolute apogee of this style in several models, including the Cadillac Eldorados (q.v.) and Chevrolet Impala Sport models of that year. The appendages on the Impalas were more 'batfin' than tailfin, and looking liked something off the Batmobile itself. Both cars were simply outrageous, dripping with chrome, with acres of glistening glass and unbelievably lustrous paint jobs in the most in-your-face colors possible. They were the ultimate in rock 'n' roll glamour and excess, cars for fun, dating and cruising. The model name remains in use by Chevrolet to this day, but the cars on which it appears are completely lacking in verve and personality compared to the originals. This dreariness is reflected in the fact that a first class example of the '59 model can command a price tag of three to four times that of the current model. The confidence and joie de vivre of the Impalas reflected the fact that the automotive industry itself was bouncing back with a bumper production year, up nearly half a million units on the previous twelve months. The 'upper crust' that the cars were designed for had given up on the economy models, and was seduced into purchasing 65,800 Convertible Impalas over the life of this model style.

The Impala range was considerably extended in 1959, to include a four-door passenger sedan, a four-door Hardtop Sport Sedan and a re-styled Nomad Station Wagon. It had now moved into a higher market position than the Bel Air range, and the cars were, typically, a couple of hundred dollars more expensive. Like the Bel Airs, Impalas were all available with both six-cylinder and V-8 engines.

A large range of extras was available for the Impalas, everything from six types of V-8 engine (including Turbo-Fire and Special Super Turbo-Thrust) to Magic-Mirror deep lustre acrylic paint and Safety Master brakes.

Engine: Sport Fury V-8, overhead valve

Displacement: 317.6 cid

Horsepower: 260 at 4400 rpm

Transmission: three-speed manual transmission, with PowerFlite automatic as an option (Golden Commando V-8 as standard)

Compression Ratio: 9.0:1

Weight: 3670 lbs

Wheelbase: 118 inches

Base Price: $3125

Number Produced: 5,990

1959 DeSoto Adventurer, Sportsman

First introduced to the DeSoto model line in February 1956, the Adventurer two-door hardtop coupe was a limited production speciality car.

Technically a Fireflite sub-series, the car was heralded as the 'Golden Adventurer' because of its ritzy gold trim accents. The car had a special high performance engine developing 320 horsepower at 5200 rpm. This original Adventurer also had a dual exhaust, and custom finish.

A convertible Adventurer was introduced in the following year, and the Adventurer engines were fitted with dual four-barrel carburettors. This engine provided one horsepower per cubic inch of displacement. In effect, they were vying for power with the Chrysler 300s. In the 1958 model Adventurer, a fuel injector (manufactured by Bendix) was offered as an option. The cars could now perform at 345 horsepower at 5000 rpm.

By 1959, there were still two models in the Adventurer, the two-door Sportsman hardtop, and the two-door convertible coupe. All Adventurers carried the model script on both front wings, and were even more richly trimmed in gold, with fold color sweep inserts affixed, grilles finished in gold, and gold trimmed hubcaps. The hardtop of the Sportsman was finished with simulated Scotch-grain leather. Standard equipment was the same as for the Fireflite series from which this line had evolved, but the car could now develop 350 horsepower at 5000 rpm, as opposed to the Fireflite's 325 at 4600 rpm. Sadly, the company's fortunes were going the other way, with an output of only 45,734 units in the year.

In fact, the Adventurer went down in history as the final model produced by DeSoto in 1961, the company's last production year, providing all of the meagre 3,104 units sold

Engine: V-8 overhead valve, cast iron block

Displacement: 383 cid

Horsepower: 350 at 5000 rpm

Compression Ratio: 10.1:1

Weight: (Sportsman) 3980 lbs

Wheelbase: 126 inches

Base Price: (Sportsman) $4427

Number Produced: (Sportsman) 590 (Convertible) 97

1959 Ford Edsel Corsair Convertible

Engine: V-8 overhead valve, cast iron block	
Displacement: 332 cid	
Horsepower: 225 at 4400 rpm	
Compression Ratio: 8.9:1	
Fuel System: Two-barrel Model PB9E9510-8	
Weight: 3790 lbs	
Wheelbase: 120 inches	
Base Price: $3072	
Number Produced: 1,343	

The 1959 Corsair was the senior model in the Edsel range. But in reality, the car was a Ranger with slightly more trim and a larger engine. It really exemplified the demise of the Edsel, no longer did it have the longer wheelbase and unique design of the 1958 Corsairs, or completely different engine options.

Whereas the availability of four Citations had meant that only two Corsairs were available in '58, four versions of the car were offered in '59, reflecting the discontinuation of the more expensive line: a four-door sedan, a four-door hardtop sedan, a two-door hardtop coupe and a two-door convertible. The new Corsairs were actually over $500 cheaper.

The marque was now a member of the Mercury-Edsel-Lincoln Division, headed up by Ben D. Mills. These '59 models were also manufactured in the single Edsel plant at Louisville, but the production problems that had dogged the earlier versions seem to have been somewhat ironed out.

But any improvement came much too late to help the Edsel. Every problem a car could have seemed to be stacked against it. The name of the line was unappealing. The styling was bizarre and universally derided. The economic downturn had reduced car sales overall, and Edsel in particular. The build quality had been suspect. The final coupe de grace was the fact that dealers no longer wanted to carry the Edsel models.

The Edsel was discontinued in November 1959, just after the introduction of the 1960 models (now just two series, Ranger and Villager). Its illconceived launch was reputed to have cost Ford over $350 million. Although he had assured the public that the Edsel was to be a permanent member of the Ford family, by November Henry Ford II seemed philosophical about the disaster, 'We couldn't sell them, we were losing money, so we made a decision, 'Let's quit'... I'd rather admit the mistake, chop it off, and don't throw good money after bad.'

1959 Ford Edsel Ranger Sedan

The 1959 Edsel line-up was cut down drastically to just ten models – with just three model names remaining, Ranger, Villager and Corsair. The Ranger and Corssair were based on the Ford Fairlaine, and the Villager was a standard Ford station wagon, available in six- and nine-seat versions. The much criticised Edsel styling of '58 was somewhat toned down, but although the cars were essentially more conventional, they were still recognisable. The dodgy horse collar grille was retained, for example. For economic reasons more power train and body parts were taken from the rest of the Ford range.

Edsel production was now limited to just the Louisville plant, and the range was effectively priced 'down-market' of the '58 model range. The economy aspects of the car were featured, and the ad slogan for the model year was 'Makes History by Making Sense'. This was hardly sparkling copy.

The Ranger models were now equipped with V-8s as standard, although they could still be ordered with the original six-cylinder engine. A bigger V-8 (361 cid) could also be specified.

But demand for Edsels continued to fall, from 63,110 in '58 to just 44,861 in '59. This represented 0.8% of the total market. This was in marked contrast to the 200,000 Edsels Ford had hoped to sell each year. In reality, this was also the final year for true Edsel production, as the final models offered in 1960 were effectively Fords with a very light facelift. Ironically, the 1960 car was selected as the year's best buy by Automotive News. Sadly, as this article went to press, the cars had already been discontinued.

Engine: V-8, overhead valve, cast iron block

Displacement: 292 cid

Horsepower: 200 at 4400 rpm

Transmission: Two-speed Mile-O-Matic

automatic

Compression Ratio: 8.8:1

Fuel System: Two-barrel Model B9A9510-A

Body Style (Ranger series): Two-door sedan, Four-door Sedan, Two-door Hardtop Coupe, Four-door Hardtop Sedan

Number of Seats: 6

Weight: 3775 lbs

Wheelbase: 120 inches

Base Price: $2684

Number Produced: 12,814

1959 Ford Skyliner Two-door Hardtop Convertible

Engine: V-8 overhead valve, cast iron block
Displacement: 292 cid
Horsepower: 145 at 4000 rpm
Transmission: three-speed manual, automatic overdrive, Ford-O-Matic and Cruise-O-Matic
Compression Ratio: 292 cid
Fuel System: Holley two-barrel carburetor
Weight: 4064 lbs
Wheelbase: 118 inches
Base Price: $3346
Number Produced: 12,915

When the Skyliner was introduced as a midyear model in the seminal Ford year of 1957, it was the first truly convertible hardtop in American automotive history, and a first for Detroit.

Introduced as part of the completely redesigned Fairlane 500 series, the Skyliner was completely unique with its fully retractable metal roof. All the cars were built with V-8 engines. The Skyliner sold 20,766 units in its first year.

Many car connoisseurs consider the 1959 'Ford Family of Fine Cars' to be the most attractive range ever styled and built by FoMoCo. Indeed, the lineup for the year was awarded the Gold Medal for Exceptional Styling at the Brussels World Fair. With elegance, and understated styling the cars were classy and tasteful, and looked like they could take off into interstellar orbit. The re-style was more of a toning down on that of the previous year, and much chrome trim was dispensed with.

The Galaxie was introduced as the most up-market car in the Ford range for 1959, and the Skyliner was absorbed into this category. The main difference between the new Galaxies and the Fairlane 500s was the styling of the top. The Galaxies had a standard top with a thunderbird style 'C' pillar. The combination resulted in one of the best-looking models ever released by Dearborn.

Generally, automotive manufacturers produced over 30% more cars in 1959, after the mini recession of '58. Even so, Ford trailed Chevy's model year production.

1959 Oldsmobile Dynamic Eightyeight Holiday Hardtop

If American car manufacturers had moved into an automotive space age in 1958, then the '59 models were the epitome of this outrageous style. The completely revised '59 Oldsmobile designs, designated the 'Linear Look', heralded a range of all-new bigger and more streamlined models. It was a substantial move away from the 'chromesmobile' excesses of '58. All '59 Oldsmobile models were now powered by the Rocket V-8 engine. The GM divisions now shared a basic bodyshell, but the styling was different for each different marque.

The Dynamic Eighty-Eights has stunning 'go-faster' lines that emphasised their great length (218.4 inches overall) and, in the final flourish of the age of tailfins, theirs were fairly substantial (though nothing to rival those of their Eldorado GM stable mate). The particularly thin-pillar roofline of the Oldsmobile hardtops prompted the 'SceniCoupe' tag for these models. There was a total of six models in the range, including two- and four-door Holiday Hardtops in the new 'flat top' styling. The four-door model was the company's second most popular offering in this year, whilst the two-door hardtop accounted for 18 per cent of the total Oldsmobile production. Standard equipment for the series included an oil filter, turn signals, air scoop brakes and a safety spectrum speedometer. An innovative new option was the pull out 'Trans-Portable' transistor radio.

Super Eighty-eights were the middle series for '59, and could be distinguished by their dual exhausts and air cleaners (although both of these were also optional on the Dynamics). Olds retained the ninety-eights as its successful top series format, built on a slightly longer wheelbase.

Oldsmobile had been at number four in the league of automobile producers in '58 - their 50th anniversary year – but fell back to fifth position in '59. Even so, they maintained a 6.9 per cent market share, continuing to make 'good cars for good people', whilst introducing technical innovations.

Lee Petty continued to race Oldsmobiles at Daytona (though minus factory support). He took a two-door hardtop to a photo finish in the first Daytona 500-Mile stock car race.

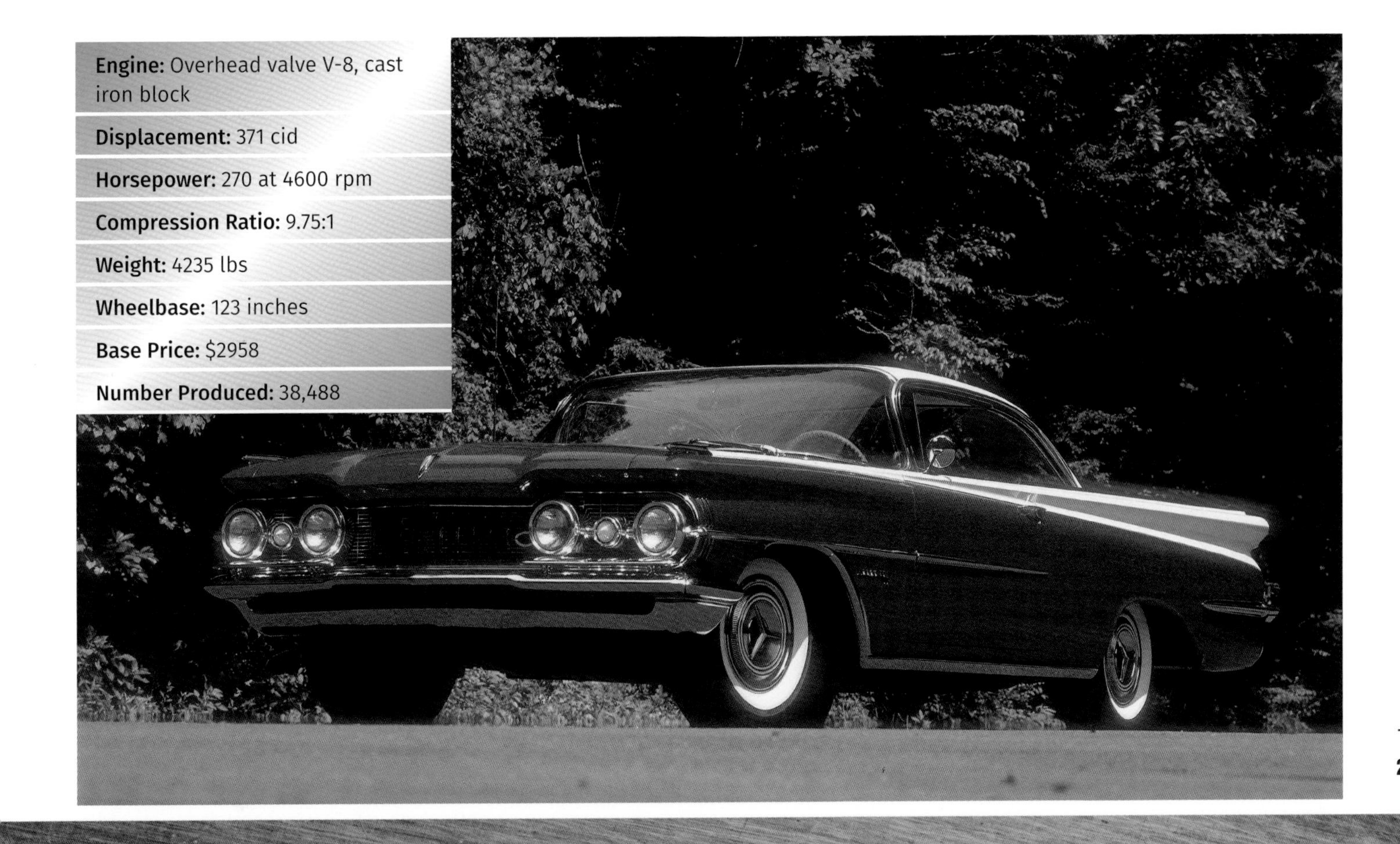

Engine: Overhead valve V-8, cast iron block

Displacement: 371 cid

Horsepower: 270 at 4600 rpm

Compression Ratio: 9.75:1

Weight: 4235 lbs

Wheelbase: 123 inches

Base Price: $2958

Number Produced: 38,488

1959 Plymouth Sport Fury

Engine: Sport Fury V-8, overhead valve
Displacement: 317.6 cid
Horsepower: 260 at 4400 rpm
Transmission: three-speed manual transmission, with PowerFlite automatic as an option (Golden Commando V-8 as standard)
Compression Ratio: 9.0:1
Weight: 3670 lbs
Wheelbase: 118 inches
Base Price: $3125
Number Produced: 5,990

Plymouth sales leapt a full 11.6 per cent in their 30th anniversary model year of 1959, holding third position in the industry sales charts. Although its actual share of the market was reduced – overall production was raised by over 30 per cent. The unsuccessful fuel injection option was deleted from the list of available options. General Manager Harry E. Cheesbrough marked the production of the company's eleven-millionth car.

All Plymouths were heavily re-styled for 1959, gaining huge tailfins and a tire bulge on the trunk lid. Technologically, not much had changed for the 1957-58 models, though.

Furys became a separate series in the 1959 production year, rather strangely grouped together with the Sport Suburban station wagons. They were marketed as higher-level Plymouth offerings and came only with V-8 attachments. The cars were equipped with all the Belvedere options, plus several unique to the new series, such as a deluxe steering wheel with horn ring, lockable glove box, cigar lighter and disc wheel covers.

A Fury four-door hardtop was converted into the latest in a series of Chrysler turbine-engined cars, making a 576-mile cross-country reliability run.

Sport Fury was the highest Plymouth range of all, and was designated as their only 'premium' series. It was a hybrid luxury/performance model. Two body styles were manufactured, a sports coupe and a convertible, available only with 'Sport Fury' or 'Golden Commando' V-8 power. The Sport Fury V-8 developed 260 horsepower at 4400 rpm, as opposed to the 230 generated by the regular V-8. Golden Commando 395 V-8 produced 305 horsepower at 4600 rpm. The huge torque generated by Golden Commando necessitated a beefed-up transmission.

Standard equipment for the model included everything the Fury had, plus swivel front seats, sport deck lid tire cover and custom padded steering wheel.

Right and below: The Sport Fury Convertible was new for '59, with huge tailfins to emphasise its improved performance.

MONTANA
FURY 59

1959 Pontiac Catalina Convertible

Engine: V-8 overhead valve, cast iron block	
Displacement: 389 cid	
Horsepower: 280 at 4400 rpm	
Compression Ratio: 10.0:1	
Weight: 3970 lbs	
Wheelbase: 122 inches	
Base Price: $3080	
Number Produced: 14,515	

Major styling changes followed Harley Earl's retirement from GM, nowhere more so than at Pontiac. Their handsome 'Wide-Track' models and reputation for performance cars really appealed to the '59 buyers and motor journalists alike, and the company leapt to fourth position in the car producers' league (with an output of 383,320 cars, representing a 6.9 per cent market share). Motor Trend Magazine voted Pontiac as the 1959 'Car of the Year'. The new styling included longer, lower bodies with more interior room, crisp twin-fin rear fenders, 'V' contour hoods, increased glass area and a new twin grille theme. This latter element was destined to become a hallmark of the marque.

In fact, all of the GM models shared a basic bodyshell in this model year, but styling was so different from division to division, that the finished products remained completely distinctive.

The old Chieftain model line was renamed 'Catalina' which was now used as a series, rather than a body style designation. Pontiac's new name for the pillarless hardtop coupes was 'Vista'. The cars were cleanly trimmed, with simple 'sweep spear' moldings and undecorated projectile flares. Standard equipment for the series included turn signals, electric wipers, dual sun visors, dome lamps, a cigarette lighter, dual headlamps, front and rear ashtrays, coat hooks, dual horns, tubeless tires and an interesting-sounding instrument panel Snak Bar.

The '59 instrument panel was even glitzier than that of the previous model year, and being virtually all bright metal could literally dazzle drivers on a sunny day. 'Buckety' front seats in a patchwork of many colours were also available, and the overall effect was dramatic.

Seven body shapes were available in the range from the two-door convertible in our photograph to a nine-seat four-door station wagon. The latter had a foldaway third row of rear-facing seats. A couple of El Camino-like pick-ups, 'El Catalinas' as they were nicknamed were also built using the series chassis. Nearly all the '59 Catalinas were fitted with Hydra-Matic (221,622 versus 9,939).

The 1959 Pontiacs were successful in the year's stock car races, and victorious at the Daytona 500 and Darlington 500. The 280 horsepower Catalina Sport Coupe (two-door hardtop) was road tested doing 0-60 miles per hour at 8.8 seconds, or 16.9 seconds over the quarter mile.

1960 Buick LeSabre Convertible

Buick arrived in the '60s, having experienced a difficult end to the '50s, and their fifth decade of production. They had found themselves completely out-of-step with the market, offering massive cars with unpopular styling just when the recession-struck customers were looking for smaller and more economic cars. The wild styling of the '59 line had lacked any frame of reference to the earlier Buick models. It had been coolly received in the showrooms, and the company had slipped down the auto producers' league. From a production of nearly three-quarters of a million cars in 1955, production plunged to fewer than quarter of a million units by the end of the decade, or 4.2 percent of the auto market. They began the '60s in ninth position, which was the company's lowest since 1905.

Buick decided to take radical measures, and dumped the names of their entire product range. Goodbye Roadmaster, Special (until 1961), Limited and Century. Hello LeSabre, Invicta and Electra. A new President, Edward D. Rollert was also committed to resolving the quality problems that the company had experienced. Buick was committed to retaining the unpopular basic body of '59 for the '60 models, but a heavy restyle blunted the paper-fold edges, and smoothed the lines of the new models into a much softer look, whilst clipping the wild fins. For 1960, Buick introduced 'Three magnificent series: Buick LeSabre – Buick Invicta – Buick Electra'.

The LeSabre Convertible headed up Buick's entry-level model range, and gained slightly improved sales. The two-door LeSabre sedan was their priceleader from the seven models in the range. Classic Buick ventiports appeared in trios on the front fenders, and a bright lower body molding accented the car's other bright metal. The cars were equipped with a luxurious Custom interior that included deep-pile carpeting, a padded instrument panel and optional deluxe door handles, window cranks and armrests. All of the cars had Mirromagic adjustable speedometer and instrument dials, electric windscreen wipers, cigar lighter, dual sunshades, Step-On parking brake, dual horns and a single-key locking system. Mechanically, the car was fitted with finned aluminum brake drums for 'complete reliability' and responsive turbine drive transmission. Power steering and brakes were optional. The car exemplified the best of Buick for 1960, with 'The serene feel of security at Buick's All-time Best'.

Engine: V-8 overhead valve, cast iron block

Displacement: 364 cid

Horsepower: 250 at 4400 rpm

Compression Ratio: 10.25:1

Wheelbase: 123 inches

Base Price: $3145

Number Produced: 13,588

1960 Chevrolet Corvette

Engine: V-8 overhead valve, cast iron block

Displacement: 283 cid

Horsepower: 230 at 4800 rpm

Compression Ratio: 9.25:1

Base Price: $3872

Number Produced: 10,261

Opposite: The 1960 Corvette was available in eight exterior finishes, black, white, turquoise, blue, silver, green, red and maroon.

The 1960 Corvette was the best yet, not because of substantial change, but carefully considered modifications. A complete restyle had been scheduled for this model year, but never made it from the drawing board due to financial constraints. Chevrolet produced 10,261 Corvettes in 1960, cracking the breakeven point for the first time in its eight-year life. After Ford enlarged its Thunderbird to a four-seat, square body, Ed Cole had hoped that disaffected two-seater enthusiasts would come over to Corvette, but not so many of these materialized.

Ed Cole was still the car's engineer, and William L. Mitchell was in charge of design. He had become senior Vice President of GM in 1958 at the age of 46, having started as a stylist in the company 22 years earlier. His father had sold Buick cars for the group, and filled his office with his son's fantastic drawings of classic cars. When he got the senior appointment, Mitchell's first action was to change the name of the department from styling (which implied a superficial add-on) to design (essential to the car's identity). His motto was 'Be Brave'.

The existing '58-'59 body was retained for 1960, with its sweet and cheesy grin and classic cove in silver or white. Half were sold with a detachable hardtop, slightly more than half with the optional four-speed manual gearbox.

In his quest for reduced weight, engineer Duntov attempted to replace as many steel parts as possible with aluminium and set to work trying to produce aluminium cylinder heads. But these proved too unreliable to be used on the production model. However, his work on the handling was much more successful. The heavy-duty springs were deleted, the diameter of the front antisway bars was increased, and one was added at the back of the car.

A RPO (production racing option) was offered on the car, with quickened steering and sintered-metallic brake linings inside drums cast with large cooling fins. The top engine option was fitted in this option package, with 290 horsepower. In the production models, tachometers, sun visors, dual exhaust, carpeting, outside rearview mirror and electric clock were fitted as standard.

Racer Briggs-Cunningham entered three 1960 hardtops into the 24 Hour Le Mans race, known as the 'Sting Rays', that had been specially prepared by Alfred Momo (assisted by Chevrolet's Frank Burrell and Zora Duntov). The team reached eighth position. One car reached 151 miles per hour on the four-mile long Mulsanne straight. This racing season established the Sting Ray as the racing Corvette, a beautifully balanced car to the eye and on the road.

Below: The Corvette looked just as good from the back!

LAND OF LINCOLN
000 000
19 ILLINOIS 60

1960 Chevrolet Corvair Wagon

Engine: Horizontally opposed six-cylinder, overhead valve, aluminum block

Displacement: 144.8 cid

Horsepower: 80 at 4400 rpm

Compression Ratio: 8.0:1

Semon E. 'Bunkie' Knudsen was Chevrolet General Manager in this year, and he steered the company back to number one position in the auto producers' league. The introduction of the compact Corvair was the big news at Chevrolet for 1960, and it may well have heralded the origins of the most significant American model range of the post-war era. Initially, the car was only available as a four-door sedan, but the Corvair turned out to be a revolution in American car design, with its rear-mounted, air-cooled, aluminium six-cylinder engine (echoing the German Volkswagon), front luggage trunk, fully independent coiled spring/swing axle suspension and clean styling. The length of all the cars in the range was contained at 180 inches overall. By comparison the compact rivals from Ford and Plymouth (the Falcon and the Valiant) were merely scaled-down versions of larger models.

Initial sales were disappointing, but the introduction of two-door models, and the luxurious Monza Club Coupe in the middle of the year sent sales shooting up. The luxury interiors and bucket seats of the Monza, combined with its agile handling, effectively established the market niche for the entire Corvair line, as economical sporty cars, rather than small family compacts. The range quickly developed a loyal following amongst auto enthusiasts, and the Corvair was nominated as 'Car of the Year' by Motor Trend Magazine. The Corvair models continued to multiply, and by '61, there were a total of eight options in three ranges (500, 700 and Monza 900), including two four-door station wagons, a van-type sports wagon and three half-ton trucks. Promoted as the lowest price Chevys, but badged as Corvairs, the series sold nearly 300,000 units in the 1961 model year.

The Lakewood 700 station wagons were part of the mid-priced six-cylinder Corvair range, and wore '700' nameplates on the cowl side of the fenders. The Lakewood was fitted with all the base equipment for the range, with plusher interior fittings, and exterior chrome decal. The model appeared in both the 500 and 700 Corvair ranges. With the back seat folded, the station wagon had 58 cubic feet of load space. The 500 version of the Lakewood appeared just once for '61.

Despite the sales success of the Corvair range, its handling was considered so unpredictable, that it very unflatteringly inspired Ralph Nadar's expose of the dangers posed by contemporary American cars in his book 'Unsafe at Any Speed'. The book itself influenced a raft of US auto safety legislation that continues to be introduced to this day.

1960 Chrysler 300 F

Chrysler improved their production by 25 percent in 1960, but remained at twelfth position in the auto producers' league. On the corporate level, Lester L. Colbert replaced William C. Newberg as President of Chrysler around midyear in 1960, but was forced to resign in a matter of months.

A new Chrysler era began in 1960, as production was switched to welded monocoque unit bodies. The 'Uni-Body' cars were styled with a new trapezoidal grille, sloping tailfins, boomerang-like taillights, a false spare wheel cover on the trunk lid and white sidewall nylon tires. But the major changes were on the engineering side. A major feat being the introduction of ram-induction manifolding, which acted like a turbocharger on the performance of the car.

The 'sixth in a famous line' edition of the letter series 300 was introduced in this model year, continuing its tradition as a high-performance muscle car. The huge engineering advances applied to the car seemed to return the model to the fierce heritage of the first Letter cars, built for performance, not parading.

Besides its all-new styling and unibody construction, big engineering improvements were also evident, with the unveiling of a Ram-Tuned induction manifold option. Two different horsepower versions were available in 1960, 375 for road cars, or an optional RPO 400. A limited number of nine ten cars (including a very rare convertible) were built with the 400 horsepower engine and a French 'Pont-a-Mousson' four-speed gearbox. Inside the car, the extraordinary AstraDome dashboard made its appearance – a bizarre glass bubble illuminated by avant-garde electrical luminescence. All New Yorker features were included on the car, plus power swivel bucket seats. The 300F Special Gran Tourismos continued the racing career of the Letter cars by winning the first six places in the Flying Mile competition at Daytona, with driver G. Ziegler achieving a top speed of nearly 145 miles per hour. B. Shaw and H. Faubel were also team drivers. Between them, the Chrysler drivers obliterated the times set by a 300B at Daytona in 1956. These were the 'Pont-a-Mousson', short Ram equipped models, the only Letter cars fitted with factory air conditioning. Perhaps they can be fairly described as the ultimate Chrysler 300 Letter car of all time.

Engine: Overhead valve V-8, cast iron block

Displacement: 413 cid

Horsepower: 375 at 5000 rpm

Compression Ratio: 10.1:1

Wheelbase: 122 inches

Base Price: $5811

Number Produced: 248 (plus 964 Hardtops)

1960 Dodge Dart D500

Engine: V-8	
Displacement: 381 cid	
Horsepower: 310 at 4800 rpm	
Compression Ratio: 10.0:1	
Wheelbase: 118 inches	
Base Price: $2607 (plus $418 for D-500)	
Number Produced: (all Darts) 306,603	

Dodge introduced a new crest in 1960, but more importantly, the structure of their entire model range was revised with new model names and target markets. In addition to the full-size Doge models, Chrysler also introduced the new 'downsized' Dart series, with a shorter, 118-inch wheelbase.

The newly introduced low-priced Dart was Dodge's big seller for 1960, and doubled their sales for the year (367,804 as opposed for 156,385). Dodge ran the model head-to-head with Plymouth, and almost achieved their rival's output for the year, hindered as their opponents were by their rather old-fashioned 'jukebox on wheels' styling. The Darts were built on the same wheelbase length as the Plymouth models.

Dodge's basic rationale in entering the Darts onto the market had been to compete directly with the 'big three' manufacturers (Ford, GM, and Chrysler).

The Dart was also the forerunner to their compact models of 1961. The Seneca was the entry-level model, and boasted Chrysler's new overhead valve 'Slant Six' engine, which was destined to be a corporate mainstay for three decades.

The other Dart engine option was the 'A' block, small series V-8. The Darts had a unitised body/chassis that was designed to give the car greater rigidity. Body styling remained recognizably Dodge, but the Darts were slightly more subdued in their trim than the other Dodge models. The side styling was simple with a single horizontal chrome strip, beginning at the front wheel well and ending at the rear bumper. The headlight 'eyebrows' disappeared, to be replaced by chrome bezels. The taillights were also surrounded by chromed bezels. Dodge also offered an unprecedented, long line of luxury extras, including air conditioning.

The Darts were tiered in three levels of trim. The Seneca was the base model, the Pioneer intermediate, and the Pheonix the top-of-the-range. All three tiers were available with the D-500 performance option, a 361 or 383-cid V-8 engine with cross-ram induction and dual four-barrel carburettors giving a supercharged effect. 1960 was the epitome of its power, the engine was detuned in '61.

1960 Ford Edsel

Engine: V-8 overhead valve, cast iron block

Displacement: 292 cid

Horsepower: 185

Compression Ratio: 8.8:1

Number Produced: 777

The Edsel line-up had been drastically cut back for '59, down to ten models, but the advertising slogan of the year 'Makes History by Making Sense' had fallen completely flat and only 44,861 cars were built in the year. A substantial number, as it turned out, compared to the meager 2,846 cars sold in 1960.

The 1960 Edsels were put on sale on October 15 1959, now limited to seven car options, plus two Villager series station wagons. The ad slogan for the year was 'New – Nifty – Thrifty'. In fact, the Edsels of this year were little more than face lifted Ford Fairlane. The bodies were completely new to Edsel, with a lower silhouette that was also longer and wider. They came straight from Bud Kaufman's Ford design studio. Many engineering advances were also introduced, including horizontal coolant flow in the radiator. New styling was fairly conclusive to the distinctive appearance of the marque, as using an elaborate die-cast grille meant dropping the unfortunate horse-collar grille that is now so synonymous with the car. The Edsel side trim for '60 was also borrowed, strongly resembling that of the 1959 Pontiacs. The unusual headlight treatment made them look as though they were floating in front of the chrome grid. A one-piece front bumper was used for the first time, and the rear bumper was also a single piece. In fact, it was so difficult to tell the cars apart from their Ford parents that counterfeit Edsels are now produced. Ironically the lack of success for the '60 Edsel range means that the cars are now rare and valuable. Standard power for all 1960 models was the 292-cid V-8 with the six-cylinder engine available as a delete-option in all models except the convertible. The larger 300-cid V-8 was also an option at extra cost. The cars were quite successful in their position just above the low-priced three, and gave good value for money. Automotive News awarded the car their plaudit of 'Best Buy' for 1960 domestic cars.

Despite Henry Ford II's assurances to the remaining Edsel dealers that the brand would be a permanent member of the Ford family of cars, Ford announced that Edsel production would cease on November 19 1959. The reasons given were poor sales and steel shortages. They had said that they were going to support the model for the long term, but obviously decided that this would mean sending good money after bad. All the 1960 models were actually constructed in 1959, at the Louisville Kentucky, factory.

So the Edsel line came to a disappointing end, after surviving for only three model years, and an output of 110,847 units.

1960 Ford Ranchero

Engine: Falcon Six-cylinder, overhead valve

Displacement: 144 cid

Horsepower: 85 at 4200 rpm

Compression Ratio: 8.7:1

Body Style: Car-based truck with flat bed

Wheelbase: 109.5 inches

Base Price: N/A

Number Produced: 21,027

The Ranchero was offered between 1957 and 1979. It was sold as a completely unique concept, the first car with a truck bed – 'More than a car! More than a truck!' 'A double-duty beauty'. In fact, there had been early hints of the phenomenon early as early as 1930 with the Hudson pickup, and again in 1940 with Studebaker's coupe-pickup. There had also been several European examples of car-trucks over the years, but their American counterparts were certainly much larger and more powerful.

One truly unique feature of the Ranchero models is that they were always patterned after a selected Ford car line. This was changed complete five times over the model lifetime of twenty-three years. It was originally patterned on the Ford Fairlane, and had all the engine of its introductory year, including the Special 352-cid V-8 fitted to the Thunderbird. Trim for the truck were offered in both the Standard and Custom options and included many luxury features.

In fact, the Ranchero was so successful that it spawned the GM competitor, the El Camino, which promptly outsold the original by a factor of more than two in its first year. But the downsized version of the Ranchero released to the market in 1960 immediately turned the tables, and Chevy vacated the arena for the following three model years.

But by 1960, Ford believed that the market was looking for a smaller, lighter and more practical vehicle – that would also be cheaper. So from 1960- 66, the Ranchero followed the Falcon bodystyle design, and became a compact Ranchero pickup. The Falcon model was part of the complete redesign that affected all Fords in 1960, and was the company's contribution to the compact car field. These early Rancheros were great economy haulers, boasting 30 miles per gallon from the 144-cid six-cylinder engine, but the performance was by no means sparkling. Offering much more bang-for-the-buck was the 260-cid small-block V-8 engine option, which gave really great performance in this relatively small vehicle. Even with the straight-six option, the Falcon Ranchero also had the ability to haul a payload of 800 pounds. The trucks made equally good sense around the ranch and around town, and they sold very well in this incarnation.

1960 Ford Starliner

All three main Ford lines were completely redesigned from the ground up in 1960, and introduced to the public in October 1959. The square-bird Thunderbird retained it existing body. At this time, the new thinking in the US auto industry wasn't tending towards high performance, but to the compact market. Interest in performance kicked in later in the decade. But apart from the little Falcon, that promptly ran away with the compact market, all of the Fords other new models were sold on their size and roominess.

Only the engines and drivelines were retained from the previous Ford model year, the rather controversial styling was completely new, and was one of the smoothest model year designs ever to come from the Dearborn drawing boards.

The new models were longer, lower and wider than earlier cars, and trim was restrained compared to many market offerings for this year. All the cars featured a single chrome strip that swept from the top of the front bumper all the way back to the small horizontal fin at the rear of the car. Large semicircular taillights were housed in an aluminium escutcheon panel directly above a large chrome bumper. At the front of the car, a new meshed grille housed dual headlights.

The Galaxie model was essentially FoMoCo's sporty people mover. It was available with powerful engines and sporty styling. The cars were also dragstrip performers, and regularly outpaced their Thunderbird stable mates. They were considered one of the first muscle cars of the decade. The Galaxies were introduced in 1959 as the crème de la crème of the Ford range, although it was hard to tell apart from the Fairlane's in this model year (even by price). But the cars were given completely new bodies in 1960, and were easier to tell apart from other Fords by their unique rear quarter-panel ripples.

The re-designed Galaxie and Galaxie Special series were the top trim level cars for 1960, and included the Starliner (two-door hardtop) and convertible Sunliner.

The five Galaxie and Galaxie Special models were available with both standard six-cylinder and 292-cid 'Y' block V-8 engine options, and were categorized as two series according to engine type. A four-door Town sedan, two-door Club sedan and four-door Town Victoria were also available in the series.

Engine: Ford Six-cylinder

Displacement: 223-cid

Horsepower: 145 at 4000 rpm

Compression Ratio: 8.4:1

Wheelbase: 119 inches

Base Price: $2610

Number Produced: (six-cylinder and V-8 models) 68,641

1960 Plymouth Sonoramic Commando

By the mid-fifties, Plymouth had an excellent reputation as a very reliable family car was long and well established. But this was becoming fairly ordinary, so the car-buying public began to look for new selling points from the auto industry, technical advances and styling changes. Plymouth entered a difficult decade for the company in 1960. Their sales plummeted during the sixties, with only brief recoveries. The rather ugly tailfins of the 1960 Savoy, Belvedere, Fury and Suburban Plymouth models ensured that they were rather coolly received by the market. But things didn't improve with the introduction of the compact models in '62, which the buying public shunned in favor of larger Ford and Chevy models. However, 1960 was a significant year for Plymouth in several regards. It was the final model year for the famous, unattractive 'fins' (or stabilizers as they were called in the sales literature. The cars were also the first manufactured with the 'unibody' construction. This offered weight and material savings over the then conventional body-on-frame construction. It also stiffened the body.This is a practically universal construction method used to this day.

Technically, 1960 also saw the introduction of two legendary engines. The slant six, and the Sonoramic Commando V-8, complete with the cross-ram induction intake manifold, the Plymouth version was available on the 383 and 361-cid 'B' engines. These were first introduced as race tune engines for use in the NASCAR series, then as 'Street Hemi' production options. The package also included heavy-duty suspension and oversized brackets. Richard Petty raced these to three Plymouth victories at the NHRA Nationals in Detroit. His winning time for the quarter mile was 14.51 seconds at 87.82 mph.

Our photographed car is the two-door Belvedere hardtop sedan, which was also available with the slant six-cylinder engine and the standard Plymouth V-8.

The Sonoramic Commando option added a further $389 to the base price of the model, plus a further $211 for the necessary Torqueflite transmissioin. The passenger car Belvederes had the model script on the rather bizarre front fender 'coves' and three shield medallions on the tailfins.

Engine: Sonoramic Commando V-8

Displacement: 383-cid

Horsepower: 330 at 5000 rpm

Transmission: Torqueflite

Fuel System: Dual four-barrel carburetor

Weight: 3505 lbs

Wheelbase: 118 inches

Base Price: $2545

Number Produced: 14,085

1960 Pontiac Ventura

Engine: V-8, overhead valve, cast iron block	
Displacement: 389 cid	
Horsepower: 283 at 4400 rpm	
Transmission: Hydra-Matic	
Compression Ratio: 10.25:1	
Body Style: Two-door Hardtop	
Weight: 3865 lbs	
Wheelbase: 122 inches	
Base Price: $2971	
Number Produced: 27,577	

Starting with the 1959 models, the image of a sporty, youthful car with appeal across the spectrum of car buyers emerged at Pontiac, in sharp contract to its previous 'dull but reliable' image. This resulted in a six-year period in which the company's output was dominated by low-slung, 'Wide-Track' full-size performance cars that performed well in the sales league, and on the track.

Pontiac's major styling changes for the 1960 line included undivided horizontal bar grilles, straight full-length side trim moldings, and a new deck lid that seemed to almost rest on the fenders. The base Catalina models, of which there were seven, had plain beltline moldings. Standard features included turn signals, an oil filter and five tubeless tires. Ironically, the divided grille that had been evident in 1959 was to return and become a trademark of the division. The Catalina range was Pontiac's sales leader for 1960. The new cars were introduced to the public on October 1 1959.

The Ventura model range cars were effectively custom-level Catalinas. They were offered in two options – a four-door Vista Hardtop and a two-door hardtop, both models sold fairly equal numbers, which totalled 56,277. Venturas shared the short wheelbase of the Catalina models. They also had plain belt moldings. Venturas had all the Catalina features, plus custom steering wheel, electric clock, deluxe wheel discs, full carpeting, triple-tone Morrokide seats, right-hand ashtrays and special décor molding.

Nearly all of the cars were fitted with Hydramatic automatic, only 2,381 were built with synchromesh gears.

Pontiac produced 396,716 cars in 1960, to reach fifth position in the automakers' sales league. This gave them a 6.6 percent share of the market. 1960 was also the year when Pontiac began to assert itself in the NASCAR racing series, of which they won four. Jim Wangers also drove a 1960 Pontiac to the NHRA 'Top Eliminator' title. Mickey Thompson, on the other hand, fitted four Pontiac engines in his Challenger I World Land Speed Record car and drove it at 363.67 mph... How things had changed.

1961 Chevrolet SS 409

Engine: W-head V-8	
Displacement: 396 cid	
Horsepower: 409 at 5800 rpm	
Transmission: Muncie four-speed manual	
Compression Ratio: 11.25:1	
Fuel System: Single Carter four-barrel carburetor	
Body Style: Two-door Hardtop Sports Coupe	
Number of Seats: 5	
Weight: 3,737 lbs	
Wheelbase: 119 inches	
Base Price: $3,500 (approximate)	
Number Produced: 142	

The production Chevrolets for '61 followed the General Motors pattern in adopting a brand new downsized body, but they also retained the engineering, contrary to the other GM divisions. The front grilles were generally rather flat and square, whilst the Impalas were instantly recognizable at the back by their triple taillight treatment and crossed racing flags insignia in the center of the rear deck.

The company manufactured 1,604,805 units in the year, to keep them at number one in the auto producers' league. This included a modest 142 Impalas equipped with Super Sport equipment, but though they were few in number, these cars really carried the torch for Chevy's continuing love affair with the muscle car. They showcased the car's top specification for unrivalled looks and performance, and charmingly described themselves as having been designed 'for young men on the move... (who) won't settle for less than REAL driving excitement'.

Chevrolet had been the first US car manufacturer to offer real power to the masses with 'the hot one' back in 1955. The 409 was the inheritor of this mantle of power, identified by its massive horsepower and identified by the capacity of its engine the mighty 396 cid V-8 developed 409 horsepower. The car was an immediate hit, and the car quickly made its way into popular culture, celebrated in Brian Wilson's Beach Boy lyrics in 1962 ('She's real fine, my four-oh-nine' – complete with engine noise on the backing track). The original engine had been conceived in '58, carved from the classic Chevy 'W-head'. But the block was actually recast to increase the bore, and new forged aluminium pistons were fitted in pairs. A forged steel crank was fitted, together with reinforced connecting rods, and aluminum bearings. A massive Carter AFB fourbarrel pumped air into the engine, and a Delco-Remy ignition sparked the heady air and gas mixture. This all boosted output to 360 horsepower. A stiffer sway bar at the front, sintered-metallic brake linings, heavy-duty springs and shocks, power steering and brakes all contributed to the feel of a really hot car, and were included in the LPO (Limited Production Option). Zero to 60 miles per hour took only seven seconds.

The exteriors of the cars were equally impressive, their image super cool, with Special Super Sport Trim, tri-blade spinner wheelcovers and narrow band whitewalls. Inside, the cars were equipped with a padded dashboard, and a grab bar across the glovebox. A 7000-rpm tachometer on the steering column was fitted as standard.

Right: The sleek coupe-bodied SS 409 was a headturner from all angles.

Below: The car's 396-cid V8 produced 409 horsepower hence the its name.

RTV·110

Impala
OHIO

1961 Dodge Polara D-500

Engine: V-8 overhead valve, cast iron block

Displacement: 383 cid

Horsepower: 325 at 4800 rpm

Compression Ratio: 10.0:1

Weight: 3765 lbs

Wheelbase: 122 inches

Base Price: $3252

Number Produced: 14,032 built in 1961

The 'new for '61' Dodges didn't sell at all well. Their styling was controversial, and some dealers even rejected the division's offerings for the new model year. Sales went down to just over a quarter of a million cars, falling away by over 100,000 units from the previous year and Dodge dropped three places in the league of auto producers to ninth position.

The downsized Darts remained as Dodge's bestseller for '61, when a heavy facelift made the car look more like the company's senior models. Their rather bizarre 'reverse slant' tailfins were a dubious Dodge idea that appeared for this year only. Dodge also introduced the compact Lancer in a three-level model range in '61.

The Polara (RD1) was now the only full-sized Dodge series on offer, with the classic 122-inch wheelbase. Six models were offered in this complete range. This senior line assumed a Dart look-alike appearance with an identical grille, though its taillights were deeply recessed. These were contained in the wraparound sweep of the tailfins, and looked like nothing less than jet plane exhausts. The model trim featured chrome moldings around the front windshield and rear window and a split chrome side strip with a contrasting aluminum insert. An odd 'Flight Sweep' trunk lid wheel cover was also optional.

The standard engine fitted to the line was the 265 horsepower 361-cid V-8, with an optional 383-cid version. The D-500 V-8 and Ram-Induction D-500 V-8s were also available as optional extras for the Polara. Both featured an unusual intake system with two Carter AFB four-barrel carburettors mounted on a 30- inch long intake manifold. The carburettor mounted on the right (over the valve cover) actually fed the left bank of the engine and vice-versa. The extremely long manifolds produced incredible low-end torque. This was known as the 'Sonoramic' long ram engine.

Polaras have a great following today, they look like the essence of '60s space travel on wheels, with their bizarre styling and sputnik-like ornaments.

The equipment nomenclature of the line thoroughly reinforced this impression, with its 'MirrOMatic' rear-view mirror, 'Satellite' revolving clock, 'Tower Bank' front seats and 'Astrophonic' radio.

1961 Ford Galaxie Sunliner

Engine: FE series V-8
Displacement: 390 cid
Horsepower: 401 at 6,00 rpm
Compression Ratio: 10.6:1
Wheelbase: 119 inches
Base Price: $2,847
Number Produced: 44,614

The year 1961 saw the third major restyle of the full-sized Fords in as many years. From the beltline down the cars were completely new, but the upper body structure was retained from the previous year. The new front-end styling was highlighted with a full-width concave grille with a horizontal dividing bar. At the back, the Galaxies were fitted with tiny tailfins. A similar chrome strip was used to that in the previous year, and was complemented by a ribbed aluminium stone guard on the Galaxies.

By '61, and Ford and Chevrolet were becoming increasingly competitive. Chevy was in front, and Ford strove to catch up by introducing all kinds of technical innovations, such as fuel injection in '57. But the Automobile Manufacturers Association ban on factory racing involvement (designed to promote greater car safety) badly affected Ford in the late fifties. This was because the company followed the AMA edicts to the letter, whereas other producers, such as Chevrolet and Pontiac, continued to support their performance programs as though nothing had happened.

But by 1960, the 'Blue-Oval' boys were back in the race and began designing bigger and better engines for the high-performance cars they wanted to re-introduce to the market. Going through several stages of V-8, Ford finally arrive at their first 400 horsepower engine via a dealer-installed 6V induction system. This consisted of three Holley two-barrel carburetors, which could be dropped straight into the space vacated by the 375/390 horsepower engine with single-carburetor intake. This option was available for an extra $260.

Among the first drivers to take advantage of this big beast was drag racer Les Ritchey, who went racing at the 1961 NHRA Winternationals, with a top speed of 105.5 miles per hour at the wheel of his 401 horsepower Ford, despite its great weight of two tons.

This big new engine option was offered on all 1961 Fords, except for the extensive range of stations wagons. The engine elevated the car from being a big cruiser available with all the usual kit (fender skirts, continental kit and special hubcaps) into a performance beast with 'No Lag. No Drag. Just flashing power'. The Sunliner was truly one of Ford's 'Lively Ones' complete with modern Borg-Warner transmission, beefed-up suspension and strengthened brakes.

1961 Pontiac Catalina Super Duty 389

Bunkie Knudsen moved to Chevrolet in 1961, and Estes took over at Pontiac. Under his able management, the division continued to grow both in sales volume, and performance reputation. The Tempest, a radical new compact model introduced in 1961, projected the company to number three in the auto producers' league with its great success. Traditionally, this was a hot spot for competing manufacturers, but Pontiac managed to dominate this rung on the ladder for most of the decade.

The Super Duty performance cars ended Pontiac's formerly staid image forever, preparing the ground for the GTO models that came later in the sixties. Jim Wangers, winner of the 1960 'Top Eliminator' title, was a Pontiac promotions guy who drag- raced the company cars in his spare time. He and group of keen young Pontiac engineers were instrumental in bringing the Super Duty race-car options together. By their very nature the Super Duty Pontiacs were limited-production special options. The Catalinas were most often fitted out with the racing package, being the smallest and lightest cars full-size cars in the Pontiac stable.

The 1961 version used the 389 cid (6.1 liter) Super Duty engine, and took Pontiac to success in 21 of 52 NASCAR Grand National stock car races. But the omnipresent threat of the big Chevy 409 engine meant that engine power was increased to 421 cid (6.3 liters) in 1962, becoming the ultimate Super Duty engine option. It was strengthened with forged pistons and four-bolt main-bearing caps. The strength was no more than the engine required, as it developed a reputed output greater than 500 horsepower. In 1963, the engineers added a McKellar solid-lifter camshaft with dual-valve springs and transistorized ignition. This brought the rating right up to 550 horsepower. To maximise the effect of this huge power output, the cars were radically lightened, with reduced-weight front ends, aluminium bumper brackets, trunk lids and radiator supports. Even the glass windows could be changed to plexiglass as a dealer-fitted option. Most radical of all were the so-called 'Swiss Cheese Catalinas' with holes drilled into their chassis. This option reduced the weight of the car to an extraordinarily low 3325 lbs (considering the weight of the extra go-goodies). A Swiss Cheese Super Duty car with full providence would now be worth in excess of $100,000. So the Super Dutys reached the epitome of their developmental success, just in time for GM to impose a factory-sponsored racing ban.

1962 Buick Skylark

Engine: V-8 overhead valve, aluminum block

Displacement: 215 cid

Horsepower: 190 at 4800 rpm

Compression Ratio: 11.0:1

Base Price: $2787

Number Produced: 34,060

Production climbed in 1961 to a 5.28 percent market share, and eighth in the industry production rankings. But things went much better in '62, Buick sold more cars than in any year since 1956 – 415,892 units for a 6 percent share of the market, and rose to sixth in the league of producers. Buick introduced the first production V-6 (developing 135 horsepower) on their entry level Special series, which gave them a price advantage in the burgeoning small domestic car market. It would be the first in a long line of Buick V-6s. The car won 'Car of the Year' for 1962.

The Skylark name had first appeared at Buick in 1953, on a limited edition model. From being part of the Special series in '61, Buick launched the name as a complete two model series for '62, a hardtop and a convertible, both with improved power. The Skylark range would ultimately serve as the basis for the mighty Buick GS line. It was their top compact range for the year, in a year that compact models were taking a bigger slice of the American market. Buick launched six series in all Special, Special Deluxe, Skylark, LeSabre, Invicta and Electra. The car's styling reflected that of the senior Buick models, and followed market trends with the installation of bucket seats. In fact, this trend was reflected in 14.3 percent of the cars bought in this year as sporty 'personal cars' became more popular. Their styling was slightly refined for the year, and a power top convertible model was added to the range midyear.

The two-door hardtop Skylark was a true hardtop with no side window, and hugely outsold the convertible, more than four to one. Buick had abandoned torque tube drive in 1961, and now offered a four-speed manual option on the V-8 Skylarks (although three-speed was fitted as standard). At the same time, Buick changed the position of the starter. The new custom exterior styling included front and rear bumper guards, turbine wheelcovers, classic Buick ventiports (though these were now lozenge-shaped, rather than round), wrapped taillamp housings and a lower body bright rocker. A Skylark badge was positioned on the front fender. Skylark interiors were all-vinyl, and equipped with a heater and defroster as standard. Other luxuries included padded cushions, a deluxe Skylark steering wheel, front and rear carpeting, dualhorns/sunvisors/ armrests and a cigarette lighter. 1962 was also the year in which Buick launched its bid to get back into the power race with the Wildcat 401 cid engine.

1962 Dodge Polara 413 Max Wedge

Engine: V8 overhead valve, cast iron	
Displacement: 413 cid	
Horsepower: 410 or 420	
Compression Ratio: 11:1 or 13.5:1	
Weight: 3315 lb	
Wheelbase: 116 inches	

Things remained depressed at Dodge in '62, but general production remained at a similarly low level, and the company stayed at ninth in the auto producers' league. C.E. Briggs was the general manager of the division.

But there were chinks of light in the company year, perhaps the brightest was the development of the 'Max Wedge' performance engine. This was actually street slang for Dodge's Ramcharger 413 V-8. 'Max' for maximum performance and 'Wedge' for its shape. The Max Wedge was introduced in '62 as Dodge's answer to the Chevy 409 and Pontiac's 421 Super Duty engines. All of these were now in cut-throat competition on the super-stock race circuit. Chrysler was the last of the 'big three' to join up, but with Tom Hoover now in charge of the competition performance program, they didn't lag behind for long. The first thing Hoover did was to turn his attention to the newly revamped Dart/Polara models. The Polara 500 was the top trim level production Dodge for 1962. It shared body and chassis components with the Dart series, and was built on the same 116-inch wheelbase.

Hoover also used the division's existing 413-cid V-8 as the basis for a completely revamped super-stock competition powerhouse. Longer exhaust valves were fitted, and a forged steel crankshaft. Magnafluxed forged steel rods, lightweight forged aluminium pistons and redesigned cylinder heads were also added to reduce weight and increase output. Unfortunately, it proved impossible to install oil seals, so the engine proved to be a terrible oil burner. Induction was provided by a unique cross-ram aluminum intake with twin Carter carburettors positioned diagonally. Cast iron manifolds were adjusted to provide open operation in race situations. Reviewer Roger Huntington heralded the manifolds as a work of art.

The Ramcharger 413 was available in two forms, one rated at 410 horsepower, the other at 420. Transmission was provided by either a heavy-duty Borg-Warner manual gearbox, or a Torqueflite three-speed automatic. Beefed-up braking and suspension were also fitted to the beast. Parts for the latter were borrowed from Dodge's super-tough police car models.

The Max Wedge was also designed for street use, and Motor Trend Magazine's road test of August '62 produced a top speed of almost 110 mph. In fact, the Polar 413 actually became the first production stock car with a factory option engine to break the quarter-mile-in-12-seconds barrier.

Right: External styling not to everyone's taste but interior color coding scores a point.

Below: The heart of the matter was the potent 413 Max Wedge V8 putting the car's performance on par with the Chevrolet SS 409.

lights
wipers

1962 Oldsmobile Starfire

Engine: V-8
Displacement: 394-cid
Horsepower: 345 at 4600 rpm
Transmission: Hydra-Matic automatic
Compression Ratio: 10.50:1
Fuel System: Rochester 4GC four-barrel
Body Style: Two-door Convertible
Weight: 4488lbs
Wheelbase: 123 inches
Base Price: $4744
Number Produced: 7149

Oldsmobile were at number five in the automakers' league for 1962, on a production of 428,853 units. Oldsmobile was somewhat upstaged in the performance arena of the late '50s by the introduction of the Chrysler Hemi and high-winging Chevrolet V-8, but continued to manufacture solid, popular and comfortable cars. The company also re-asserted itself in the high performance stakes with its introduction of the tri-carb-based J-2 option that was offered on most models.

Like so many other US automakers, the early sixties Oldsmobile models looked as though they had just landed from outer space. Mid-1961 saw the introduction of the Starfire convertible that started the swing to big sporty cars. The car was fitted with standard 'buckety' front seats, and hard-to-see tachometer. The car was equipped with abundant bright trim, this had been limited in the other GM ranges for '61, but was more in evidence at Oldsmobile. Starfire became a separate range for '62, after the positive reception to the initial model in the previous year. The Holiday hardtop coupe was added to the big bucket-seat ragtop, and immediately outsold it, virtually five to one. The two cars were Oldsmobile's attempt to corner a greater share of the personal/ luxury car market. To compete with the other cars in this market, they were deliberately well-equipped with all the hardware of the less expensive Olds models, plus sports console, Hydra-Matic automatic with console shifter, power brakes, power steering, brushed aluminum side trim, a dual exhaust with fiberglass packed mufflers and leather upholstery.

The big technological news at Olds for '62 was the introduction of America's first production turbocharged V-8 engine. It was launched in mid-season as the 'Turbo Rocket', and fitted to the new Jetfire coupe. Featuring 'fluid injection', the 215-cid engine-block pumped out 215 horsepower, and had lively acceleration. But initially, the engine had some reliability problems that somewhat curtailed sales.

1962 Pontiac 421 Super Duty

Pontiac had made it to number three in the auto producers' league by 1962, and had every intention of staying there. Their other mission for the year was to clean up at the nation's drag strips. The Catalina chassis grew again, by about one and a half inches, this time. As well as the ordinary production models, Pontiac intended to build Super Duty Catalinas to further their performance ambitions. The '62 Super Duty models were fitted with a 421-cid engine, and yes, that is a bigger umber than the Chevy 409s. Although they were on public sale, the cars cost a small fortune, the engines alone coming in at $1300. But the cars were unbeatable. However, they were never really built for street use, 'These cars are not intended for general passenger car use, and are not supplied by Pontiac Motor Division for such purposes' read a corporate disclaimer. Like the Max Wedge Mopars, the Super Dutys were factory super-stock racing-cars. Rather bizarre, as to be race legal the car had to be on sale to the general public - without any modification.

The previous Super Dutys of '60 and '61 had been equipped with a 389-cid engine, which had proved a big success. But things moved quickly in this competitive world, and the 1960 win at Daytona (when Fireball Roberts lapped the track at an astounding 155mph) seemed a long time ago when Pontiac was considering how the revise the car. First fitted to the car in late 1961, the 421-cid/405 horsepower engine was Jim Wangers' racing dream. The fusion of this big engine and a specially built chassis produced a purpose-built screamer. The block sported four bolts, a forged-steel crank and forged-aluminum pistons. This developed fearsome compression. A special cam with solid lifters, two big Carter AFB carbs and conduction system were also included. As for the body, Pontiac was the first manufacturer to click that offering weight saving components would enhance performance. This resulted in replacement aluminum panels, hood, front bumpers, inner fenders and radiator brackets being offered.

Engine: Super Duty V-8

Displacement: 421-cid

Horsepower: 405 at 5600 rpm

Transmission: Borg-Warner T-10 four-speed manual

Compression Ratio: 11:1

Fuel System: Two Carter AFB four-barrel carburetors

Body Style: Two-door Hardtop

Number of Seats: 6

Weight: 3800 lbs

Wheelbase: 120 inches

Base Price: $4400

Number Produced: 200

1963 Chevrolet Impala SS

Engine: (Base V-8) Overhead valve
Displacement: 283 cid
Horsepower: 195 at 4800 rpm
Compression Ratio: 9.25:1
Wheelbase: 119 inches
Base Price: $2774
Number Produced: 399,224

Chevrolet launched their widest ever range in 1963, and sold three out of every ten cars in America, over 2,300,000 units (including Corvette and Corvair). This output was 700,000 cars more than that of closest rival, Ford. Bunkie Knudsen was still at the helm of the division. This was a vintage year for factory drag racing options, and the Chevelle launch year. At the other end of the model range, the full-size Impala was the 'upper crust' Chevy model. It was the only '63 model to receive new sheetmetal, and looked elegant and uncluttered. Impala had started life in the late fifties as a model in the Bel Air series, and soon became instantly recognisable by their triple taillight treatment. A large range of options were available for the car, including the RPO Z11 race ready package.

Impala was Chevy's largest seller, and their plushest car had most of the equipment standard on lower lines. They also had bright aluminium front seat end panels, patterned cloth and leather grained vinyl upholstery in color-coordinated materials over thick foam seat cushions. A sport-style steering wheel was fitted, and other extras included an electric clock, parking brake warning lamp, and dashboard face panels in textured bright metal. The Super Sport (RPO Z03) trim was uprated for '63, with the addition of special chrome spinner hubs, extra emblems, matching cove inserts, bucket seats and a locking storage compartment (when the transmission was upgraded to four-speed or Powerglide automatic). Super Sports also had the option to fit the Turbo-Fire 409-cid V-8, and most of these fitted in '63 were installed on this model. This was just one of six V-8 options that were available, producing between 250 and 430 horsepower, all the way from mild to wild. The biggest engine was available only with the Z11 package, and was the prototype for the spectacularly successful 396-cid V-8 of 1965.

Motor Trend Magazine road tested two Impala SS sport coupes with V-8 engines at both ends of the mild and wild scale – the 250 horsepower 327 and the newly introduced 340 horsepower 409. Both cars were installed with Powerglide automatic, power steering and power brakes. The tester found the automatic gears tricky on both cars, and slow to shift, but liked their performance in the acceleration tests. Of course, the 409-cid engine was much quicker, covering a quarter mile in 15.9 seconds, reaching 88 mph.

1963 Chevrolet Corvette Sting Ray Z06 Fastback Coupe

The 1963 Corvette turned out to one of the greatest ever, a brilliant synthesis of styling and performance. The cars were called 'Sting Rays' to emphasise that they were different to all the Corvettes that had gone before. The Sting Ray was the joint concept of design guru Bill Mitchell, and engineering genius Zora Duntov. Duntov had been a racing driver, and was very aware of the handling and performance of the car, while Mitchell wanted to inject the glamour of the European sports cars into Corvette. The contemporary Jaguars in particular are credited with influencing the sinuous, cat-like lines of the car. He himself drove an E-Type.

The Corvette received a major restyle in 1963, for the first time since its introduction ten years earlier. Two models were available for the first time. The traditional convertible, and the first Corvette coupe, introduced for the first time in this model year. This 'boattail' fastback would be the only Corvette ever to have the unique 'split' rear window, now highly coveted by classic car collectors. It gave the car the appearance of a kind of fish spine running the entire length of the car. Apart from this unusual feature, the rear deck view resembled that of the '62 body, but the rest of the car was unrecognisable from its predecessors. The headlights were hidden in an electronically operated panel. This was not a gimmick, but a serious attempt to improve the aerodynamics of the car. Corvette and Jaguar were two of the few marques to have grasped this concept in the early '60s. Corvette was the first manufacturer to introduce hidden headlights since the DeSoto of 1942. The recessed fake hood louvers were of the more superficial styling type, however. Mechanically, this was the year when Duntov introduced the famous ladder-frame chassis and suspension layout. The center of gravity for the car was lowered from nineteen to sixteen and a half inches (so that ground clearance was reduced to only five inches) and weight was evenly distributed between the front and rear axles. This gave the car both excellent road holding and ride quality. The interior had circular gauges with black faces. Windshield washers, carpeting, an outside rearview mirror, dual exhaust, tachometer, electric clock, heater and defroster, cigarette lighter, and safety belts were all fitted as standard.

The Sting Rays also hugely improved Corvette's performance in the market place, and the car sold over 22,000 units in the first year of this model.

Engine: V-8 overhead valve, cast iron block

Displacement: 327 cid

Horsepower: 250 at 4400 rpm

Compression Ratio: 10.5:1

Base Price: $4257

Number Produced: 10,594

1963 Chevrolet Nova SS

Engine: Overhead valve, Six-cylinder

Displacement: 194.4 cid

Horsepower: 140 at 4400 rpm

Compression Ratio: 8.5:1

Wheelbase: 110 inches

Base Price: $2472

Number Produced: 24,823

In Chevrolet's biggest range year of 1963, the Novas were the base models, and were becoming one of the most important models in the Chevy line-up. The Chevy II/Nova line had been introduced in 1962 as the Chevrolet repost to Ford's rather conventional Falcon compact range. The substantial difference between the two ranges was that the Falcons were essentially down-scaled versions of the larger Fords, whereas the Novas came straight off the drawing board. Chrysler, AMC and Studebaker were also in the compact race, with their Valiant/Dart, Rambler American and Lark ranges. Although Chevy already had its compact Corvairs, they wanted a car that could compete head-to-head with the other compacts on the market. The original economy Chevy models featured simple, rather square styling with unitized body and front sub-frame construction. The cars were known as the GM 'X' bodyline, and were originally manufactured by Fisher Body. The fenders were designed to 'bolt on' to facilitate easy replacement. Eleven basic models were arranged in four series, according to the engines option used. The 100s were pretty Spartan, but the top-of the-range Nova 400s were more richly trimmed both inside and out. All of the cars were fitted with completely new four- or sixcylinder engine options, except the 400s, which only came with the six-cylinder engine. There was also a factory-offered kit for dealer installation of a V-8 in two sizes (283 or 327 cid).

Detailed refinements and new freshness for its basically simple lines were the major exterior changes for the 1963 Chevy II/Nova models. On the inside, the cars were awarded a completely new upholstery and trim. There were still three series in the range, the Chevy II 100, Chevy II 300, and Chevy II Nova 400; ten models were now on offer. A new Super Sport (SS) option was also introduced, exclusive to the Nova 400 Sports models for an additional $161. '63 was the only year in which Chevy offered a 'drop-top' option for the Nova SS, as this was discontinued in 1964. This means that this is now one of the most valuable Novas, even though the car was only fitted with the larger 194 cid six-cylinder engine. Junior cars in the series were also fitted with the 'Iron Duke' four-cylinder engine.

The Chevy II/Nova outsold all the domestic compacts in 1963, and remained in manufacture until 1979.

1963 Chrysler New Yorker

Despite the fantastic reputation of several of its models, including the Letter cars and the New Yorker, Chrysler were struggling in '63, with falling sales in its fourth decade. They were at eleventh in the auto producers' league of output. Chrysler had finally abandoned tailfins in 1962, and there was a complete styling revamp for 1963. A tasteful slab-sided look evolved, with a minimal amount of trim. Chrysler touted this as its 'crisp, clean custom look'. But sales actually dropped and the company remained at eleventh position in the auto producers' league. A five-year or 50,000-mile warranty on the drive-train components was introduced as a marketing tool.

The 1963 New Yorkers displayed the crisp, new custom look expounded in the years advertising. They shared the 'C' body with the Newports and 300s. The new styling look gave the hood, fenders and rear deck a flat, wide look that contributed to the overall integrated styling theme. A thin beltline molding ran from front to back of the car, and the usual New Yorker trim bars were now found on the front fender, behind the wheel opening and below the New Yorker nameplate. The grille was divided into two halves, by a chromed center bar, and egg crate inserts on either side of this, harking back to the models of the mid-fifties. The number of trim bars was reduced to six. In the interior, the car was embellished with deeply quilted luxurious Jacquard fabrics and soft vinyl trim. The New Yorker line included five model options.

Our featured car is the Salon option, a limited edition four-door vinyl roofed hardtop with a luxurious interior, special side trim and Salon nameplates. The model was introduced as the top option in the New Yorker range, selling only a modest 593 examples. It is said that only 11 of these are still registered. The car had pillarless styling, unusual for a four-door car, with no 'B' pillar between the front and rear doors, and no framing around the window glass. In fact, there was absolutely nothing between the vent windows and the 'C' pillar at the back of the car. The Salons were fitted with leather interiors as standard, together with a load of power options including power steering, brakes, windows and door locks. Inside the car, Auto Pilot, air conditioning, seat belts, six-way power seats, an FM radio and cup holders were all fitted for a luxurious feel. Aficionados of the car say that it was the finest American car available for sale in '63.

Engine: New Yorker V-8, overhead valve

Displacement: 413.8 cid

Horsepower: 340 at 4600 rpm

Compression Ratio: 10.0:1

Wheelbase: 122 inches

Base Price: $5344

Number Produced: 593

1963 Ford Galaxie 500 427

Engine: Thunderbird High-performance V-8

Displacement: 427 cid

Horsepower: 425 at 6000 rpm

Compression Ratio: 11.5:1

Base Price: $3268

Number Produced: 33,870

Ford completely revised their full-size car range for the fifth year in a row.

As in the previous year, the base model cars were devoid of any sculptural lines except for the beltline feature line. Each model name was carried in script on the fender immediately behind the front wheel opening. Once again, the taillights were large round units mounted at the top of the rear fenders with a stamped aluminium panel being used on the Galaxie 500 series.

There are several Galaxies model years that describe the top range as '500s'. The number does not stand for 500 horsepower or 500 cubic inch displacement. The leading rumor for the origination of the notation is the 500 miles as in the 500-mile stock car races in which the Galaxie models competed.

The '63 Galaxie was a classic from the first day it rolled off the showroom floor, the car's distinctive lines and styling are timeless and are now among the most collected Fords of any era.

The Galaxie 500 series was the top trim level for the model in this year. The series contained sedans, hardtops and convertibles. The model was slightly more elaborately trimmed that the simply-styled base-model Galaxies, having an attractive full length upper and lower body side molding, and trimmed with chromed 'A' pillar and window moldings. There were also two horizontal chrome strips on the side of the car. Between the two chrome pieces, just in front of the taillights were six cast 'hash marks'. The 500 XL models were available as two- and four-door hardtops, convertible and midyear fastback model. The fastback featured a sporty new roofline with no post and Starliner type looks. The standard equipment on these cars was a deluxe offering with standard bucket seats and console, spinner wheelcovers, wall to wall carpeting, courtesy lights and contoured deluxe seat upholstery.

The 427-cid big-block V-8 was the most powerful engine option for the year. The high-performance race with other US manufacturers was in full swing by now, and this was the most powerful engine in Ford's entire history. It fully reflected FoMoCo's determination to sell more cars via racing successes. Two versions of the engine were introduced, a single four-barrel version at 410 horsepower and a dual four-barrel carburettor at an astounding 425 horsepower. Transmission types were the Synchro Smooth manual column shift, a four-on-the-floor manual, Ford-O-Matic Drive automatic and Cruise-O-Matic three speed automatic.

1963 Ford Thunderbird 'M-Code' Sports Roadster

The Thunderbird was Ford's answer to the Chevrolet Corvette and debuted onto the market in 1955. It started out as a two-seat personal car, but was re-built as a four-seat model in 1958. This was the last time it would appear as a true two-seat car until 2002, but the Sports Roadster of 1962/63 was a partial return to two-seat styling. It seemed the perfect answer to the nostalgia felt for the original T-birds. The model was an attempt to revive the sporty appearance of the original 'personal' cars. It featured a fibreglass tonneau package (designed by Bud Kaufman), which covered the back seat of the car to give a two-seat appearance. The tonneau covered the twin front headrests and was 'flowed' from these to the back of the car to make the styling more aerodynamic. The convertible top could operate with the tonneau in position. The introduction of the Sports Roadster meant that the car buyer could purchase a sporty two-passenger convertible and four-passenger car in one and the same car.

For its second and final model year, the car received new doors and front fenders, and a bodyside crease. The Sports Roadster was fitted with a dash-mounted grab bar for the passenger, and four dazzling Kelsey-Hayes wire wheels. The rear fender skirts were deleted to allow clearance for the knock-off wheel centers. For performance fans, Ford offered a special 'M-code' 390-cid FE V-8 rated engine, rated at 340 brake horsepower. It was equipped with three Holley two-barrel carburettors and an aluminum manifold to keep the carbs level and at the same height. The 'M-code' option was rare, and only 145 Thunderbirds were ever built with this engine, including thirty-seven '63 Sports Roadsters.

Although the idea for the Sports Roadster was intriguing (it is usually attributed to Lee Iacocca), a combination of the model's high price and difficult tonneau-installation limited its sales. The price was so high, that the buyer could have a Cadillac Convertible for only $162 more. The Kelsey-Hayes wire wheels have also been blamed for the demise of the car, not only because they added expense (a whopping $372.30), but because they were blamed for causing wheel failure, extremely dangerous at speed. The best-known customer of the Sports Roadster, Elvis Presley, had a blow out accident at high speed that immediately hit the headlines. Ford discontinued the model in '63.

Engine: 'M-code' Thunderbird Special Six-Barrel V-8

Displacement: 390-cid

Horsepower: 340 at 5000 rpm

Compression Ratio: 10.5:1

Base Price: $5563

Number Produced: 37

1963 Mercury S-55

Mercury cars had first been introduced in 1939 as intermediate models, priced between Ford and Lincoln. The joint Lincoln-Mercury division was founded by Ford in 1945, and so it wasn't until the 1949 models appeared that Mercurys were able to shed their image as glorified Fords (at least for the moment). The youth-appeal of the cars was hugely improved when James Dean drove a Mercury in the cult movie *Rebel Without a Cause.* However, Fords and Mercurys shared body shells throughout the fifties, and early 60s so it was difficult for the marque to define itself. Later advertising also stressed the classy heritage of their stable-mate Lincolns.

Basic Monterey styling for the year was reminiscent of the 1961 models, with six taillights located in the rear deck panel, and a concave vertical bar grille housed four chrome-trimmed headlights. The Monterey Custom was the intermediate model for the year, and had additional chroming and standard features.

The 1963 Mercury Monterey S-55 was the top of the Mercury model line for this year, and was offered in four body styles – the two-door convertible, two-door Marauder Fastback Coupe, four-door Breezeway Hardtop sedan and the two-door Breezeway Hardtop. The unique and practical Breezeway roof, first offered in 1957 on the Turnpike Cruiser and reintroduced in this model year, featured a roll-down back window. The standard engine for the line was the Marauder Super 390-cid V-8, with four-barrel carburetor, but the Marauder 390 V-8 with two-barrel carburetor was available as a no-cost option. The Marauder Fastbacks were true muscle cars that were raced to victory in the 1963 NASCAR series, with 427-cid V-8s. They had been designed with aerodynamics and speed in mind. Transmission options for the production models were three or four-speed manual or Multi-Drive Merc-O-Matic automatic. S-55s were identified by marque insignia in front of the rear fender, chrome bars and special wheel covers.

The Mercury advertising slogan for '63 was 'Shift to the Real Performer! Go Mercury!' The S-55 was a real sizzler that delivered the luxury of the Monterey Custom series plus sports car features – individually adjustable bucket seats, bright-metal floor console with shift lever and storage compartment. Red and white courtesy lights were positioned inside the doors. When the doors were opened the interiors were revealed as deeply luxurious, complete with vinyl upholstery, front and rear armrests and a padded dashboard.

Engine: V-8 overhead valve, cast iron block

Displacement: 390-cid

Horsepower: 250 at 4400 rpm

Compression Ratio: 8.9:1

Number Produced: 1203

1963 Pontiac Grand Prix

Engine: V-8 overhead valve, cast iron block
Displacement: 389-cid
Horsepower: 303 at 4600 rpm
Compression Ratio: 10.25:1
Weight: 3915 lbs
Wheelbase: 120 inches
Base Price: $3490
Number Produced: 72,959

The Grand Prix model was introduced to the Pontiac range in 1962, replacing the Ventura model (although Ventura-Catalinas were still available as a trim option). This first Grand Prix model was the top of the line sporty Pontiac for the year, and was available with a range of engine options. Sixteen Super Duty Grand Prix cars were built in the first model year, equipped with 421 V-8s inducted through twin Carter four-barrel carburettors. The Super Duty engine option could develop in excess of 400 horsepower, and were usually equipped with a four-speed manual transmission. The Grand Prix cars were identified by clean side styling, ornamented with a checkered flag badge in the rear concave section of the side spears. The cars also had a rocker panel molding, anodised grille insert and nose-piece and special rear end styling. The factory-fitted equipment included all the lower-model range Bonneville equipment (except the courtesy lights) and solid color Morrokide upholstery, bucket seats and a center console complete with a tachometer. A single model was offered in '62, a two-door hardtop. 30,195 were sold.

The '63 Grand Prixs were handsomely restyled from bumper to bumper. The new model brought glamour and performance to the division. The new car had a clean look with no side trim, and a grille that emphasized negative space with bright accents and enclosed parking lights. In fact, this grille was unique to the bucket-seated hardtop. The car also had grilled-over taillamps, that were mounted on the deck lid and a concave rear window treatment. Standard equipment continued to include the full list of Bonneville equipment, together with extra Grand Prix fittings. The standard engine for the model was the 389- cid V-8 engine fitted with the optional Tri Power carburetion (i.e. three twobarrel carburetors). The car boasted a smart interior, with bucket seats, console and Morrokide upholstery as well as a wood-grained dash and steering wheel. The elegant black-faced instrumentation and chrome bezels made the dash elegant and tasteful.

The Grand Prix badges were now mounted on the sides of the rear fenders and rocker panel moldings. A single two-door hardtop was still the only model on offer, but sales more than doubled at 72,959 for the year.

1963 Studebaker Avanti

Studebaker had introduced the Lark in 1958, and the car proved to be a big success. However, the board could not agree on how to spend the profits. President Harold Churchill wanted to invest in the future, using the money to keep Studebaker at the forefront of small car development. But the remainder of the directors wanted to diversify. The disagreement led to Churchill's early replacement by Sherwood Egbert in '61. It was under his direction that the company launched the Gran Turismo Hawk and Avanti models.

The Avanti was Studebaker's almost unique fiberglass-bodied sport coupe, and was the first completely new-bodied model that the company had introduced since 1953. It was built on a modified Lark Daytona convertible chassis. In 1961, Egbert commissioned Raymond Loewy to design the car. Egbert flew himself to Palm Springs to meet the celebrity industrial designer. The car was a triumph, with smooth lines, an under-the-bumper radiator air intake and wedge-shaped design the model hallmarks. All '63 Avantis had round headlight enclosures. Avanti interiors were equipped with four aircraft inspired seats. Two power units were available for the car, the Avanti R1 V-8 that developed 240 horsepower, or the Avanti Supercharge R2 V-8, capable of 289 horsepower. The latter was a $210 option.

Daniel Jedlicka wrote an Esquire article about the car (after the demise of Studebaker) in 1969, called 'Instant Classics'. In it, he wrote of the Avanti 'Raymond Loewy styled it and liked it even better than his slick '53 Studebaker coupe. Ian Fleming bought one. The roof was trimmed with a steel boxlike frame attached to a hefty roll bar and windshield support. There were aircraft type rocker switches mounted in the roof, a Paxton supercharge V-8 and a fibreglass body with tremendous impact resistance. It hit 170 mph at Bonneville'. It was all true.

Despite the success and acclaim generated by the Avanti, Studebaker's position continued to be precarious, and the company lost money. Egbert stepped down from his position in '63, due to failing health. Production was now centralized at the Hamilton, Ontario plant in Canada. The writing was on the wall for the Avanti, and production was discontinued in December '63.

Engine: Avanti Supercharged R2 V-8
Displacement: 289-cid
Horsepower: 289
Compression Ratio: 9.0:1
Weight: 3148 lbs
Wheelbase: 109 inches
Base Price: $4445
Number Produced: 3834

1964 Chevrolet El Camino

Engine: Six-cylinder with hydraulic valve lifters	
Displacement: 230 cid	
Horsepower: 155 at 4400 rpm	
Compression Ratio: 8.5:1	
Base Price: $2,267	
Number Produced: 36,615	

Just as the Camaro followed in the footsteps of Ford's Thunderbird, and trounced it, so the El Camino followed the Ford Ranchero by a year, and established itself as the car-derived truck, outselling its rival virtually two to one in its first year. The El Camino first appeared in 1958, and followed a long Chevrolet tradition of morphing its cars into commercial vehicles. The lovely boxy roadster pickups of the '20s were particularly evident of this tendency. Through a long line of trucks, business coupes, coupe pick-ups, the El Camino finally arrived. Essentially, it was a two-door station wagon with the load space roof removed. The vehicles were cheap for the manufacturers to originate, as virtually all their components were already tooled up and ready to go. If Ford and Chevrolet were unsure that the car-truck hybrid would find a natural market, they were soon fully justified. Both vehicles offered the driver all the creature comforts with brilliant practicality, looked hot and moved well. What more could anyone want? Over its (almost) three decades of production, the El Camino outsold the Ranchero twice over, and outlasted its rival by nine production years. The model sold 1,056,424 opposed to Ford's 23-year total of 508,000.

By the time the El Camino came onto the market, there was considerable competition in this market area from all kinds of makers – Studebaker offered a coupe-pickup in 1940, Hudson had one back in '34.

Ford had downsized their Ranchero in 1960, to a Falcon-based model. These vehicles surged into a market that had been vacated by the El Camino in '61, '62 and '63. But 1964 marked a sharp reversal in fortunes. GM introduced their Abody intermediate sized models, and immediately developed a new Chevellebased El Camino. This model put the Ranchero in second place for the balance of its career. Weight and wheelbase were down, but the 1200-lb cargo box was more capacious. The new El Caminos and Custom models (an extra $80) were offered with both six-cylinder and V-8 engine options. They were offered with every Chevelle option, apart from the Super Sport package – that would come later in '68.

1964 Ford Mustang 260

Engine: V-8, overhead valve, cast iron block	
Displacement: 260 cid	
Horsepower: 164 at 4400 rpm	
Compression Ratio: 8.8:1	
Wheelbase: 108 inches	
Base Price: $2368	
Number Produced: 97,705	

The big news at Ford for 1964 was the introduction of the Mustang in April. As this was late in the model year, the cars are often known as '64-1⁄2s'. Lee Iacocca is attributed with its introduction, despite the reluctance of the Ford top brass. The Ford Studio team presented the original design in response to a brief set out by Iacocca. The cars were Ford's response to the improving economy of the early '60s, when consumers were looking for smaller cars complete with luxury and good performance. The car was designed with such clean, attractive lines that it was awarded the Tiffany Award for Excellence in American Design, the only car ever to be so honored. It was the right car at the right time, and went on to have spectacular success, selling over half a million examples in its first year of production. A million were sold in the first two years. An unbroken record to this day. This unprecedented achievement can be attributed to its clever combination of desirable features and striking design aimed at a large segment of the buying public. Not only did the Mustang spawn an entire class of family sports car, but an entire cohort of youthful Americans became known as the Mustang generation. In some regards, this is misleading, as it was the fact that the car appealed to such a large cross-section of age groups, which was the crux of its desirability.

The car was received with such a wave of universal enthusiasm that demand outstripped production for almost a year. Its combination of comfort, convenience, luxury and economy engendered a pride of ownership unmatched in this price bracket. Although the Mustang was the very first of the 'pony' muscle cars, it was first designed and marketed to be a low cost 'personal luxury' sports car. The initial production did not have serious performance engines. The base engine option was the six-cylinder 170-cid, developing a rather weak 101 horsepower. The base V-8 displaced 260 cid and developed 164 horsepower. The top V-8 option, the Challenger High-Performance (Hi-Po) engine was introduced in June '64. This powerplant developed a punchy 271 horsepower, and came with a special handling package, including fourteen-inch red line tires. Only 7273 Hi-Po Mustangs were built.

The original Mustang would diversify into an enormous range of variants, factory special editions, performance options and custom-built models. The streets of America were soon full of them.

Right: This Skylight Blue Mustang is equipped with the base option 260-cid V8.

Below: The optional Ford 289-cid V8 developed 210 horsepower with a dual-barrel carbureter and 220 with a four-barrel.

HAF 64
MUSTANG

1964 Mercury Comet Caliente

Mercury celebrated its Silver Anniversary in 1964, and was at number nine in the auto producers' league with an output of 298,609 cars. The Comet models were produced by Mercury between 1960 and 1967. The original Comet series was based on the Ford Falcon platform of 1960. The sister cars shared many technical features, but the Comet had its own distinct bodylines, exterior ornamentation and interior trim. The first convertible Comet was introduced in the '63 line-up.

The Mercury Comets were extensively re-styled for 1964, with a Lincoln- Continental-style grille. The same theme was repeated on the rear deck panel. A wraparound trim piece was seen on the tips of the front fenders. Three thin, vertical trim slashes were on the sides of the front fenders. The signal lights remained embedded in the front bumper. Overall, the new squared-off bodyline, slightly longer than the earlier models, made the cars far more futuristic looking.

The 1964 range introduced new names to the Comet line-up, the Comet 202, Comet 404 and Caliente: three distinctive packages with the same bodylines and drivetrain components. The Caliente was the sports model Comet, with bucket seats. Six-cylinder and V-8 engines were available for the model, as were manual and automatic transmission options. The Cyclone was the ultimate Comet package with the V-8 engine as standard equipment and 6000rpm factory dashboard tachometer.

'Every bit as hot as it looks!' was how the sales literature described the Caliente series. The model trim consisted of a wide, full-length molding on its sides and a nameplate on the lower front fenders. The interior of the car was embellished with a padded instrument panel with walnut grain trim and deeploop carpeting as luxury standard features. Caliente hardtops and convertibles were available only in solid colors. A third model, a four-door sedan was also on offer.

A team of customized Calientes, equipped with 289 cid/271 horsepower V-8s joined the Mercury muscle car club by travelling for over 100,000 miles at average speeds of over a hundred miles per hour.

Engine: Six-cylinder, overhead valve

Displacement: 170 cid

Horsepower: 101 at 4400 rpm

Compression Ratio: 8.7:1

Body Style: Two-door Convertible

Wheelbase: 114 inches

Base Price: $2636

Number Produced: 9039

1964 Oldsmobile F-85 Cutlass 4-4-2

The F-85, launched in 1961, was the smallest Oldsmobile model in decades. It was a compact companion to the full-size Super 88 Holiday model, and strongly resembled this car. The four-door model, the coupe and the four-door wagons were sold in the $2300-$2900 range. F-85 output totalled 76,394 for this model year. The F-85s rode on a 112 inch wheelbase.

Sportiest of the new F-85 range was the mid-year, Deluxe-trim, bucket-seat pillared coupe called the Cutlass. This progenitor of a well-regarded Oldsmobile model name was priced at $2621 attracted 9935 sales. The car came with only one engine option – the 155 horsepower aluminum V-8.

Convertibles were added to the compact F-85 range in '62, in both base and bucket-seat Cutlass trim. The Cutlass model sold 9893. In 1963, Oldsmobile retained the trim options of the F-85s, but gave the car more of a big-car appearance, as the buyers had requested. The new Cutlass Sports Coupe hardtop was the bestselling model in the range, attracting 29,269 buyers with an attractive base price of $2592. All F-85 models were equipped with a 215-cid aluminum V-8 engine, rated with either 155 or 215 horsepower.

The F-85 went from a compact model to mid-size 'senior compacts' for '64, by adopting a handsome new 'A-body' platform, shared with several GM stablemates – the Buick Special and Skylark, the Pontiac Tempest and Le Mans and the new Chevy from Chevelle. It was completely redesigned, and gained a full eleven inches in length (just three inches on the wheelbase). The Sporty Cutlass now became a separate three-model series, which included a two-door coupe, two-door Holiday hardtop and two-door convertible. The $2784 Holiday hardtop coupe was the bestselling model in the range, with sales of 36,153. Cutlasses also now came with a new cast-iron 'Rocket' 330-cid V-8 making 290 horsepower. The aluminium block was a thing of the past.

1964 was also the introductory year for the 4-4-2 equipment package, at an extra price of $136, RPO code B09. This translated as four-barrel, four-speed and dual exhaust. Other equipment included a heavy-duty suspension, dualsnorkel air cleaner, oversized redline tires and special badges. The cars were also fitted with a heater/defroster, self-adjusting brakes, oil filter, front stabilizer and dual sun visors. Deluxe models also boasted a Deluxe steering wheel, padded dashboard and carpets. Interiors were vinyl or cloth. The F-85 models were built in six Oldsmobile plants at Lansing, Atlanta, Kansas City, Linden, Southgate and Arlington.

Engine: V-8, overhead valve, cast-iron block

Displacement: 330-cid

Horsepower: 210 at 4400 rpm

Compression Ratio: 9.0:1

Number Produced: 36,153

1964 Pontiac GTO

Engine: Tri-Power V-8

Displacement: 389-cid

Horsepower: 348 at 4900 rpm

Compression Ratio: 10.75:1

Body Style: Two-door Convertible

Number of Seats: 5

Weight: 3360 lbs

Wheelbase: 115 inches

Base Price: $3400

Number Produced :(With Tri-Power 389) 8,245

The original GTO package was initially offered for the LeMans sports coupe and convertible, but a hardtop was added to the line-up shortly after the launch. The LeMans were a subseries of the Tempests, which were enlarged for this model year into a senior compact. The base powerplant for the line was a six-cylinder engine assembled by Pontiac Motor Division from Chevrolet-produced components. But this was a world away from the legendary 'Goat'.

The re-designed '64 Tempest body provided the ideal basis for the GTO, which effectively became America's first modern muscle car. Actually, it wasn't the hottest thing on the road, but it represented a powerfully attractive combination of power and affordability that thrust its way to the forefront of this market sector.

Pontiac General Manager Pete Estes had given the development of the GTO his enthusiastic blessing, and somehow managed to keep it away from the anti-performance top brass at GM long enough for the concept to emerge undiluted. Effectively, he bought John DeLorean and fellow engineer Bill Collins enough time to wait for the new '64 Tempest body, with its full-perimeter frame and solid rear axle. Their fundamental idea was simple: take big-block engine

and drop it into the lightest possible chassis, whilst keeping the price within reach of the average baby boomer. But as well as producing a car aimed at the youth market, Pontiac had also managed to produce a car for the general market by using a mid-sized body.

The multi-talented Jim Wangers, drag car racer and marketing guru, also worked with DeLorean on designing an appealing performance package. GTO was a tag borrowed from sports-giant Ferrari, which brilliantly encapsulated the aspirations of the package. The American public loved it with a passion. Pontiac sales manager Frank Bridge predicted that he would be able to sell 5000 examples of the GTO package, but 32,450 were actually ordered, and the production line failed to keep up with the demand. According to Car And Driver Magazine's David E. Davis, the GTO combines 'brute, blasting performance with balance and stability of a superior nature'.

Above: The convertible version of the GTO together with decal and wheel details.

Left: The more powerful Pontiac V8 "Tri-Power" unit was equipped with three two-barrel Rochester 2G carburetors giving at 348 horsepower.

1964 Studebaker Gran Tourismo Hawk

Studebaker introduced the GT Hawks in 1962, and the introductory models had a squared roofline, no fins and rocker panel moldings. They were offered with six-cylinder engines, but these cars were only available outside North America. A 259-cid V-8 engine option was also offered in this market. American/Canadian GT Hawks came with a standard 289-cid V-8, which developed 210 horsepower. The Hawk name appeared on the deck lid and 'Gran Turismo' script appeared on the doors. Only one model was offered for the three years of GT production, a two-door hardtop.

The 1964 Gran Turismo Hawk had new, smooth deck lids and slightly restyled imitation side grilles. Their identifications were the same as in '62- '63, a vertical Hawk emblem in the center of the grille. A stand-up 'S' hood ornament was also fitted, but a decklid appliqués had been removed. The rear deck was now smoothed off, and a new red, white and blue round medallion added. A tremendous list of options were offered for the Gran Turismo, including a half-vinyl roof, rear deck lid radio antenna, aero strut wheel covers, door handle guards, visor vanity mirror, power steering, hill holder, air conditioning, white sidewall tires, tinted glass, transistor AM/FM push-button radio, Strato-Vue rearview mirror, license plate frames and rubber floor mats, locking gas cap, seat belts, windshield washer and adjustable shock absorbers.

As well as the standard engine option, GT Hawks were also offered with a 289 'Power Pack' V-8 that could develop 225 horsepower. Super Hawks had four larger engine options. Two 289-cid options were available, one developing 240 horsepower and the other 290 horsepower. Two 304.5-cid engines, developing 280 and 335 horsepower were also offered as options on this model. This was the final year in which Studebaker manufactured its own power plants.

A total of 1548 Gran Turismo Hawks were sold in this model year. The model was discontinued in '64, and did not appear again.

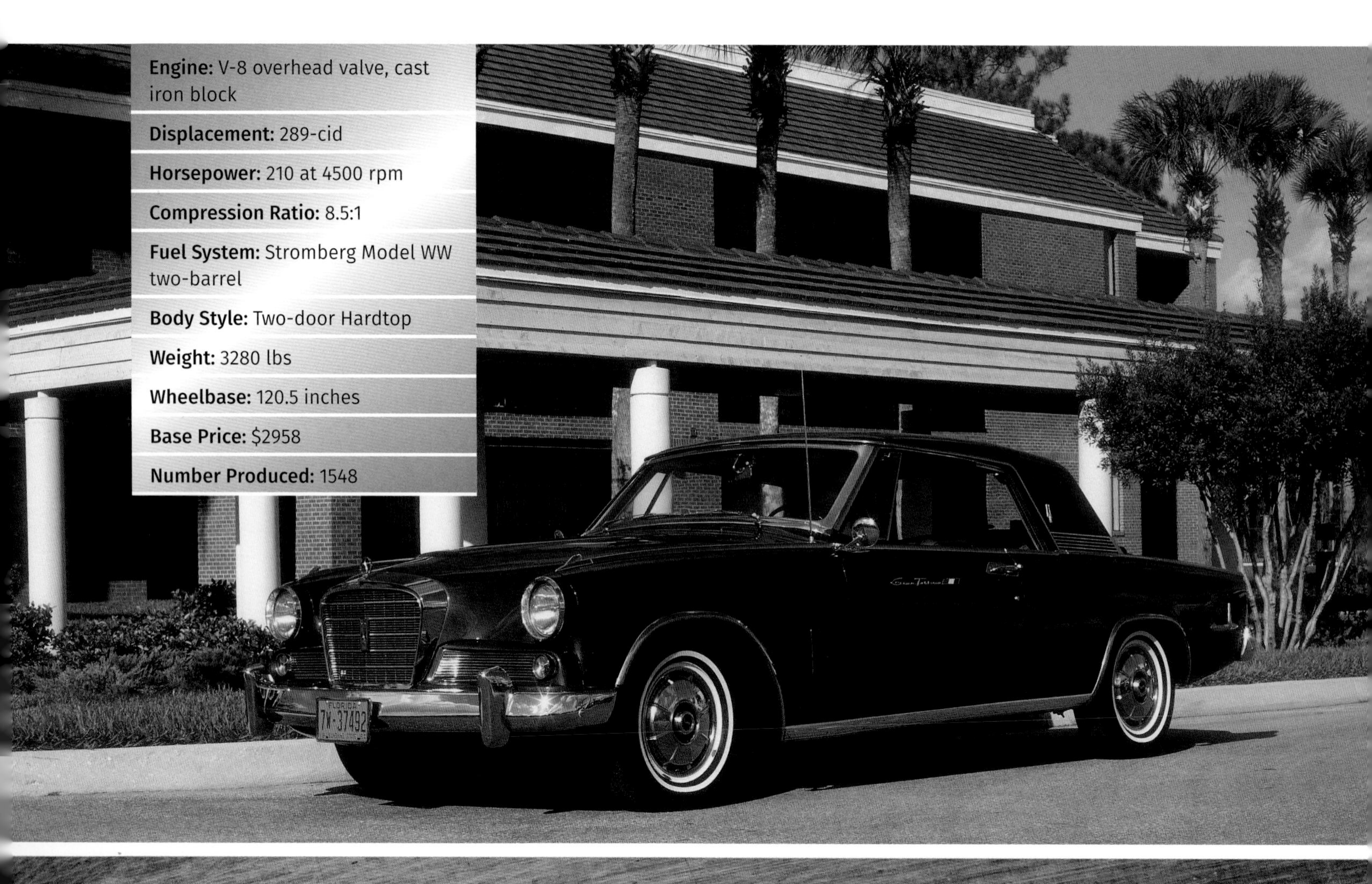

Engine: V-8 overhead valve, cast iron block

Displacement: 289-cid

Horsepower: 210 at 4500 rpm

Compression Ratio: 8.5:1

Fuel System: Stromberg Model WW two-barrel

Body Style: Two-door Hardtop

Weight: 3280 lbs

Wheelbase: 120.5 inches

Base Price: $2958

Number Produced: 1548

1964 Studebaker (Lark) Daytona

As of January 1 1964, all Studebaker production took place at the Canadian plant, and the company was in negotiation with General Motors to buy in engines for their 1965-66 cars. Model year production fell to only 19,748 units, and the company fell to twelfth position in the auto producers' league. Avanti production had been discontinued in December '63, though leftover cars were on sale in '64.

The Lark name was de-emphasised in 1964, and was actually only being used in connection with the early Challenger and Commander models. Several new model names and a revised squared-up design were the major changes for this model year. All the outer panels were new, their design was courtesy of Brooks Stevens. The overall length of the car grew by six inches. The grille became more horizontal, with an egg-crate center and integral headlamps, and a pointy new rear end carried high-set tail and backup lamps. Despite this, sales tumbled.

A new Daytona model was introduced, the four-door sedan. This joined the other three cars in the line-up, a two-door hardtop, four-door station wagon and two-door convertible. The new Daytona models were fitted with wide metal side moldings across the back of the deck lid, and had a standard circled 'S' hood ornament. Four power plants were available for the car, the 64S and SC- 10 six-cylinder engines, and the 64V and VC-50 V-8s.

Studebaker's automotive division was in serious trouble by 1966. Calendar year auto production was down to a meagre 2.045 units. The corporation's other diversified holdings now represented the majority of the company's activities, and kept the corporation going. In mid-1967, the Studebaker Corporation purchased the Wagner Electric Corporation, and combined in the November of that year to form the Studebaker-Worthington Corporation. The car company had now effectively ceased to exist, with production terminating in March '66 ending 114 years of Studebaker tradition, and 64 years of car production. The only automotive survivor of the corporation was the Avanti. Two South Bend businessmen, Nathan Altman and Leo Newman revived the marque after the company left their plant in the town. They created the Avanti Motor Corporation, which was to subsequently pass through several changes of ownership.

Engine: V-8 overhead valve, cast iron block

Displacement: 259.2-cid

Horsepower: 180 at 4500 rpm

Compression Ratio: 8.5:1

Fuel System: Stromberg Model WW two-barrel

Wheelbase: 109 inches

Base Price: $2797

1965 Buick Skylark Gran Sport

American car sales accelerated dramatically in the mid-sixties, partly encouraged by the launch of a host of tasty models by the big automakers and by the growth of the economy. Multi-car households were greatly on the rise, and the industry sold more than nine million cars for the first time in 1965. On the flip side of the market, some old names were on the way out, including Studebaker and Packard.

Buick passed Dodge and Oldsmobile in 1965 to reach fourth position in the automakers' league of production, with an output of 653,838 units. The 'Father of the Skylark', Edward T. Rollert moved to GM head office, and a new Buick General Manager, Robert L. Kessler, was appointed.

The son of the Skylark, on the other hand, took poll position as the epitome of the '65 Buick range. The Gran Sport was the top optional package for both three Skylark models (the two-door coupe, sport coupe and convertible) and the senior Riviera model ranges. The Riviera version was advertised as an 'Iron Fist in a Velvet Glove', the Skylark GS (a midyear introduction) as a 'Howitzer with windshield wipers'. Both cars were fitted with the Wildcat 401-cid V-8 engine 'Buick engineering wraps their potent 401 powerplant and super quick Turbine trans with a reinforced Skylark shell' proclaimed a contemporary article in Hot Rod Magazine. 'The Buick Gran Sport evolves as one of the hottest of factory-produced street/strip hybrids.' Indeed, many people view this model as Buick's first serious attempt to enter the muscle car market, although it also had a reputation for luxury and refinement. The 1965 Skylarks were plush with cloth and vinyl (or leather-grained vinyl) with bucket seats optional. Outside, the cars were adorned with unique coved styling on the exterior. This intermediate Buick range was very much aimed at the new market, which demanded both 'young' styling and high performance.

From a performance point of view, the engines produced 325 horsepower, with the addition of heavy-duty cross-flow radiators and dual exhaust manifolds with oversized pipes.

Engine: Wildcat V-8

Displacement: 401 cid

Horsepower: 325 at 4400 rpm

Transmission: Turbine drive automatic

Compression Ratio: 10.25:1

Fuel System: Carter AFB four-barrel

Body Style: Two-door Sport Coupe

Wheelbase: 115 inches

Base Price: $2622

Number Produced: (for Skylark model) 4501

1965 Chevrolet Chevelle Malibu SS Z16

Chevrolet had launched the Chevelle/Malibu range in 1964, anticipating a general improvement in the market for cars priced and sized below fullsized models – senior compacts. The car was placed between the compact Chevy II and the regular full-sized models, and was designed to take the same market share that the classic mid-size Chevys of the mid fifties had done. Projected demand was so high that Chevy even opened a new factory in Fremont, California. The model was quite square in it styling, but had curved side window glass and a 'wide' look that gave the range a distinctive new look. Eleven models were available in two basic lines called Chevelle 300 and Chevelle Malibu, with a convertible as an exclusive upper level offering. A Super Sport option was also released. The car had SS insignia, and bucket front seats were popular options.

For their second year, the Chevelles were mildly restyled. The nose was 'veed' slightly outwards and a new grille came in. At the rear, Chevy gave the line new taillights. The Chevelle Malibu SS 396 (RPO Z16) was the top-of-the-range car. It was aimed at a young, muscle car-obsessed market. Z16s were blacktop-bruising powerhouses, accurately described as 'one of the wildest pieces of equipment on wheels'. They were introduced as a special midyear package, and featured the 396- cid V-8 with dual exhaust and chrome accents. Mechanically, the cars were constructed on the stronger convertible frame, and fitted with special shocks and a brawny sports suspension with front and rear sway bars. Heavy duty cooling was also installed. The cars were also fitted with hydraulic power-assisted steering for more responsive handling, and power brakes. Fifteen-inch wide simulated mag style wheel covers were included in the package, and enhanced with Firestone gold line tires. Considering the power and cost of the upgrade to the Z16s ($1500), they were quite discreetly styled, and it was tricky to instantly differentiate them from the other Chevelles. The interiors also had an SS-396 emblem on the dashboard, and were equipped with a 160 mph speedometer and an AM/FM stereo multiplex radio.

Only 201 cars equipped in this way were built in '65: 200 hardtops and one super-rare convertible. Surviving Z16s are revered today as one of the most rare and powerful muscle cars ever produced.

Engine: V-8

Displacement: 396 cid

Horsepower: 375 at 5600 rpm

Wheelbase: 115 inches

Base Price: $4091

Number Produced: 200 hardtops, one convertible

1965 Chevrolet Corvette Roadster

The Year 1964 has seen the birth of the 'sporty compact' Ford Mustang, but as the advertising slogan of 1965 proclaimed, 'Corvette is America's one true sports car'. Even so, these first 'pony cars' had more 'youth appeal' than any other car on the road at this time with the exception of the Corvette. The Sting Rays were now offered in two body styles, five engines and three transmissions, so the car came in a good many configurations. These ranged from the plush for the 'boulevardier' to the seriously sporty for the 'aficionado' and several in-between options. The car itself was getting increasingly sophisticated in both appearance and driving enjoyment.

Every Corvette now came with four-wheel disc brakes (although drums were available as a minus cost option), fully independent suspension, retractable highlights and a bucket-seated interior. The big mechanical change for '65 was the introduction of a big block V-8. This was the height of the muscle car era in America and performance (together with 'think young' styling) was the driving force behind the phenomenal success of Detroit in this period. Introduced midyear, the engine that was to become known as the Turbo Jet came with up to 427 horsepower. In fact, it was so big that an impressive power bulge appeared on the hood lid. This 'W' engine had been used for five years already in Chevy saloons, and on the NASCAR circuits in '63. For some strange reason, Chevy had decreed that medium-sized cars like the Corvette should not have engines with a higher displacement than 400 cid, so the engine was sleeved down to 396. To handle the power of the Turbo Jet, the Borg-Warner gearbox was replaced by a home grown GM model. Manufactured at Muncie, Indiana, it became known as the 'Muncie Box'. A beefed up clutch and cooling system were also added. Duntov had to rebalance the car due to the increased engine weight, but managed to do so with characteristic precision.

The car itself had cleaned up looks, with very limited ornament. The look was more uncompromising and rather less sinuous than the early models, more piscine than panther, perhaps. The front fender vents certainly had a look of shark gills, and would become a mainstay of Corvette design. This all led to Corvette seats reaching another new high of 23,562 cars.

Engine: (standard) V-8, overhead valve

Displacement: 327 cid

Horsepower: 250 at 4400 rpm

Compression Ratio: 10.5:1

Wheelbase: 98 inches (175.2 overall)

Base Price: $3212

Number Produced: 15,377 convertibles

1965 Mercury Comet Cyclone

Mercury announced the first Cyclone Comet as a new mid-year model in 1964. It was available in only one body style, the two-door hardtop. The cars were equipped with a special exterior trim, which was very restrained. The car only had decoration on the wheel well moldings, a rocker panel molding and the letters C-O-M-E-T on the rear fin. A special insignia with the word 'Cyclone' and a checkered flag adorned the lower section of the front fenders. The wheel covers were made of stainless steel, which gave the appearance of chrome. The Cyclone was considered the first 'macho' Comet model. Sales literature referred to the V-8 engine 'under the hood, a whiplash of surging power' and of the 'masculine feel of black vinyl in the instrument panel'. Macho or not, the blurb also pointed out that, as a safety feature, the 'bucket seats are contoured to hold you more securely in turns'. Inside the cars, a three-spoke steering wheel, tachometer and console were fitted as standard. Even the engine was decorated with special chromed parts, including air cleaner, dipstick, oil filter, radiator cap and rocker arm covers.

7,454 Cyclones were sold in their first year of 1964. Mercury continued the Cyclone model in 1965, but made several changes to the car. Cyclones were still only available as a two-door hardtop, but styling was overhauled. The checkered flag emblems were moved to the rear fenders. The stainless steel wheel covers remained, but the interior upholstery options were updated. The cars were offered with vinyl roofs, curb moldings, a distinctive grille design and revised tachometer.

The standard Cyclone engine for 1965 was a 200 horsepower version of the small block (289-cid) 'Cyclone V-8', with two-barrel carburetion. An up-rated 289 V-8 engine, the 'Super Cyclone', with 225 horsepower, was also offered as an option. Both engines were offered with the 'dress-up' chrome of the '64 cars. Cyclones were also offered with a 'macho', but simulated, twin air scoop hood in fiberglass.

The cars had a great reputation for reliability, with the exception of the clutch, which sometimes began to slip at around 10,000 miles. Many owners changed these to the heavy-duty Cobra clutch that offered a permanent solution to the problem.

Engine: Cyclone V-8, overhead valve

Displacement: 289 cid

Horsepower: 200 at 4400 rpm

Compression Ratio: 9.3:1

Wheelbase: 114 inches

Base Price: $2625

Number Produced: 12,347

1965 Mercury Park Lane

Engine: Park Lane V-8

Displacement: 390-cid

Horsepower: 300 at 4600 rpm

Compression Ratio: 10.1:1

Fuel System: Ford C5AF-9510E four-barrel

Body Style: Two-door Convertible

Number of seats: 6

Weight: 4013 lbs

Wheelbase: 123 inches

Base Price: $3526

Number Produced: 3008

Mercury was at ninth position in the auto producers' league for 1965, on a production of 346,751 units manufactured. The Park Lane had returned to the Mercury line-up in 1964 as the top-of-the-line series. A wide band of tire-level chrome trim, running across the bodysides, set it apart from the other Mercury models. Its interior featured nylon face, biscuit-design upholstery and large walnut-tone panel inserts. The Park Lane convertible, like the Monterey version, came with a retractable glass rear window in a Breezeway roof.

The 1965 advertising for the Mercury models were described as being 'In the Lincoln Continental tradition', meaning that they were to be raised to new levels of luxury. The write-up also extolled the car's 'beautiful proportions, confident stance, elegant appointments. And its thoughtful touches'. Ford group vice-president Lee A. Iacocca was put in charge of a program to improve Lincoln-Mercury sales. He was keen to identify the line with the prestigious Lincoln reputation, but this didn't really become apparent until the 1969 model line-up.

The 'king-size' 1965 Park Lane models had three new styling features - rectangular rear fender nameplates, chrome gravel shields and a band of molding above the rocker panels. They were equipped with more luxurious interiors than other Mercurys, including a padded dashboard, padded visors, courtesy lights, visor-mounted vanity mirrors plus a trip odometer.

A total of 32,407 Park Lanes were sold in 1965, and four body styles were offered in the range. The Breezeway was a sedan with slanting and retractable rear window. Then there were two Marauder models, a four-door hardtop fastback and a two-door hardtop fastback. A two-door convertible concluded the line-up. As well as the tilt steering wheel, locking differential and crinkle all-vinyl bucket seats, and the abundance of wood grain trim, the most important aspect of the convertible model was the Breezeway glass rear window.

1965 Plymouth Belvedere Satellite

Plymouth was back in the big league by '65, having recovered from the disastrous compact model introduction of '62. They were at number four in the table of auto production, on an output of 728,228 cars. It was at the end of this model year that the fourteen-millionth Plymouth rolled off the production line. The generic Plymouth advertising slogan for the year was 'Plymouth presents the roaring '65s'.

For this model year, the standard 1962-64 Plymouth was restyled into the mid-size Belvederes. In fact, the Belvedere nameplate was no longer used to designate the level of trim and equipment, but was now used to classify the intermediate size Plymouth ranges. They utilized the same platform that Plymouths had used the previous year. The Belvederes were now offered in three model tiers. Three Belvedere I models were the base-level offerings for the range in this model year and had fairly limited standard equipment and trim. The higher level Belvedere II models (there were five of these, ranging from a convertible to a nine-seat station wagon) were far more luxuriously appointed, with carpeting, special trim and upholstery, foam cushions and backup lights. The exterior trim was also more generous.

The sporty Satellites were now the premium level cars in the Belvedere range. They were introduced rather tenuously, as the 'New Way to Swing Without Going Out on a Limb'. The cars were endowed with rich and sporty characters, fitted with all the equipment of the Belvedere I and II models, plus front bucket seats, center console, full wheel covers and all-vinyl trim on two-door hardtop model. The side trim moldings apparent on other models were deleted, but louvers were added to the rear fenders and rocker panel moldings were used, as were trimmed wheel openings and a rear panel beauty strip. Two Satellites were available - hardtops or ragtops.

Belvederes were now fitted with column-mounted gear shifters, instead of the derided push-button models. As well as the slant six, a whole range of V-8 engine options were available for the '65 Belvederes – the 318-cid, 361-cid, 383- cid and 426-S (wedge head). Plymouth also built a relative handful (actually 27) Belvederes powered by the 426-cid/425 horsepower Hemi V-8. At an extra $1105, these cars were aimed solidly at the professional auto racer. Styling for these cars was a revamp of the '64 body, with single headlamps and a crossbar grille superimposed over a rectangular mesh background.

Engine: Plymouth V-8, overhead valve

Displacement: 318-cid

Horsepower: 230 at 4400 rpm

Transmission: TorqueFlite automatic

Compression Ratio: 9.0:1

Number Produced: 1860

1965 Pontiac Catalina Convertible

Stacked headlights continued on Pontiac's all new big '65 models, as did the performance oriented '2+2' package option. 2+2 was only available for the big Catalina convertible and the semi-fastback Sport Coupe hardtop. The package had first been introduced to the Catalina range in 1964, and was the first such performance option to be offered by Pontiac on the full-size models. The 2+2-adapted cars were marketed as a sort of giant five-seat luxury sports car with a long list of technical and luxury choices '...as fine as you want – or as fierce'. The top '64 engine option was a 370 horsepower version of the 410-cid V-8, which came fitted with Tri Power (three two-barrel carburetors). However, the weight of the car topped 4,000lbs, which tended to slow things down, somewhat.

Power for the 2+2 option was up-rated in the following model year, but when comparing even the revised performance of the package with the GTO option, it has to be said that even this sporty package wasn't quite as spectacular in these bigger cars as it had been on the smaller GTO models. Even so, the '65 2+2 option cars came fully loaded with a 338 horsepower, a four-barrel 421-cid V-8 fitted as standard that developed 338 horsepower at 4600rmp. Hardtops fitted with this engine could achieve 0-60 mph in 7.4 seconds, and the quarter mile in 15.8. This engine option was available for an extra price of \$108 - \$174 on other Pontiac models. The 2+2 421 was offered in three different strengths, including a top-rated 421, capable of 370 horsepower equipped with Tri Power as in '64. With this version fitted, the cars could achieve 0-60 mph in just over seven seconds (7.2) and the quarter mile in 15.5. Hydra-Matic automatic transmission came fitted as standard with the package.

The basic problem remained the weight/power ratio. The entire Pontiac range became bigger and wider for 1965, wheelbases lengthened to 121 and 124 inches, and the Wide Track chassis was wider than ever. The 2+2 Catalinas were muscle cars, but only just. Other styling changes for the '65 Catalina models included twin air-slot grilles, V-shaped hoods with a prominent center bulge, curved side-glass and symmetrical 'Venturi' contours with fin-shaped creases along the lower body. Coupes and convertibles with the 2+2 package were trimmed with '421' engine badges on the front fenders, 2+2 numbering on the rear fenders and deck lid, and simulated louvers behind the front wheel cut-outs.

Engine: V-8

Displacement: 421-cid

Horsepower: 338 at 4600rpm

Compression Ratio: 10.5:1

Weight: 3795lbs

Wheelbase: 121 inches

Base Price: (Catalina Convertible) \$3103

1966 Cadillac Eldorado

Engine: V-8 overhead valve, cast iron block	
Displacement: 429 cid	
Horsepower: 340 at 4600 rpm	
Compression Ratio: 10.5:1	
Wheelbase: 224.2	
Base Price: $6631	
Number Produced: 2,250	

Safety was becoming an issue in UC car making by 1966. The Department of Transportation was set up, and The National Traffic and Motor Vehicle Safety Act and The Highway Safety Act were both passed by Congress.

Despite the fact that industry volume eased to 8.6 million cars in '66, Cadillac, the most distinguished name in US car production, enjoyed its first year of manufacturing more than 200,000 units. This enabled the company to reach eleventh position in the league of auto producers. It was the company's fifth record sales year in a row, and it turned out that 1966 was to be the best sales and production year yet. Indeed, productivity was also at a record level. An unprecedented number of cars, (5,570) were produced in one week in December '66, and in one record day in October, no less than 1,017 cars came off the Cadillac production lines.

A minor facelift was given to all the '66 year models at Cadillac, including a new bumper, grille and more integrated taillight housing. White sidewall tires were also fitted to the cars. Variable-ratio power steering was a new option, as were carbon-cloth seat heating pads. The Eldorados were the top-of-the-line model for the year, and Fleetwood type trim was seen again on this year's model. Finless rear fenders had appeared in 1959, on a long, low and luxurious body, and this look was retained for '66. Soft Ray tinted glass was available as an option. The interiors of the cars blended fabrics, leathers and walnut in a tasteful welter of luxury. The Eldorado was available with bucket seats at no extra cost, and leather upholstery was available in a delicious range of colors – green, black, white, midnight blue, gold, vermilion and antique saddle. A sixway power front bench was fitted as standard. Mechanically, the cars were full of power-operated options, including power-operated cent panes, automatic level control, cruise control and power-operated door locks.

Little change had been in evidence with the model year Fleetwood Eldorado and it turned out that the car was to be the final rear-wheel car in this line. Cadillac had decided that it was time for something completely new, and a brand new front-wheel drive Eldo arrived in the following year with completely new 'sports' styling. This car built on the unrivalled heritage and reputation of the marque, and it strengthened Cadillac's dominance of the upper end of the US car market.

1966 Chevrolet Corvair Corsa

Chevy trailed Ford production for the second time in the decade in 1966 – though only by a modest 6000-odd units. Both companies topped an output of 2.2 million cars, despite industry volume contracting to 8.6 million. Over at Corvair, the GM bosses mandated that there should be no changes from the restyle of 1965, so the cars remained little changed, but remained desirable semi-sports cars and still looked terrific. In fact, its smooth-lined restyle package was to influence all the GM styling for years to come. The range had also been given a mechanical overhaul in '65 with the addition of a revised rear suspension (to correct the tendency of the cars to over-steer), and a turbocharged option.

The Corsa had been introduced as the top-of-the-line model in 1965, more richly equipped than the Corvair 500 and Monza lines, but hugely outsold by both of these. The Monzas had effectively saved the Corvair range from extinction on its introduction in 1960. It was outselling every other Corvair range by the following year. Like the Monza, Corsas were only available with 164-cid six-cylinder engines. But the Corsa engines were fitted with four Rochester carburetors that enabled them to develop nearly fifty percent more power.

1966 marked the final appearance of the Corsa option. In its last season, the sporty high-performance Corsa clung to the few visual distinctions that set it apart from the less expensive Monzas. The car's standard interior equipment included bucket seats, carpeting, glovebox light, fold-down rear seat, ashtrays and rear seat foam cushions. The Corsa was equipped with full instrumentation, including tachometer, oil pressure gauge and temperature gauge.

The Corvair range as a whole sold 103,743 units in 1966, but production was slowed down to 73,360. In fact, the writing had been on the wall for the entire range since Ford had introduced its Mustang in 1964, whose raunchy performance was preferred by the public. Chevrolet also contributed to the drying up of demand by introducing the Camaro in 1967. 'Unsafe At Any Speed' – Ralph Nadar's expose of the Corvair's poor handling, hadn't helped sales, and these plummeted to unacceptable levels for a volume manufacturer. The Corvairs limped along until 1969, when the final gold Monza coupe rolled off the Willow Run, Michigan production line.

1966 Chrysler Imperial

Engine: V-8, overhead valve, cast iron block
Displacement: 440-cid
Horsepower: 350 at 4400 rpm
Compression Ratio: 10.1:1
Wheelbase: 129 inches
Base Price: $6540
Number Produced: 1878

Chrysler launched Imperial with a single four-door limousine in 1946, at the then extraordinary price of $3875. The marque was to be used as the home for Chryslers 'extra fancy' models, to distinguish them from less expensive and exclusive models. It became a separate brand of Chrysler Corporation in 1954. The management tone for the line was arrogant to say the least. Dealers were encouraged to 'select' their buyers – 'the type of person who appreciates a fine motor car... one whose prestige and pride of ownership will reflect credit on the car and on your organization'.

The styling also reflected this 'manufacturer knows best' attitude. When the customers were looking for something with flash and glamour in the early fifties, the Imperials were demure and sedate. The Imperial ethos was passenger comfort and quality engineering, rather than up-to-date styling. It wasn't until 1955 that, now separate from the parent company, the cars began to take on a look of their own. The much better looking cars of '55 and '56, with their freestanding taillights were a tremendous boost to Imperial's image. Designer Raymond Loewy encapsulated the company's ambition, to show that a car could have style and class without looking like 'a jukebox on wheels'.

By 1957, the styling was quite extreme, however, and the very long and low cars of this model year (which were built on a 129-inch wheelbases) looked as though they had just landed. For 1960 and '61, the Imperials developed extreme tailfins. Happily, these were a thing of the past by 1964, when Imperial styling reflected the influence of the classy Lincoln Continentals of the early sixties.

The Imperial Crown models were the first introduced back in the forties, and underwent man re-styles over the years. A revised Crown had been launched in '64, so the '65 and '66 updates were mild. The '66 model featured a new ice cube tray-style grille with four headlights recessed into chrome panels. The rear deck style remained unchanged. All the cars were fitted with power brakes/steering and windows, an electric clock and automatic transmission. The LeBaron four-door hardtops were identical to the Crowns, except they had different badges and more luxurious interiors.

There were four Imperial models for '66, and model year production of 17,653 cars was recorded.

1966 Ford Fairlane 427

Engine: Single Overhead Camshaft 8V V-8	
Displacement: 427 cid	
Horsepower: 657 at 7500 rpm	
Transmission: Four-speed manual	
Compression Ratio: 12.1:1	
Fuel System: Two Holley four-barrel	
Body Style: Two-door Hardtop Coupe	
Wheelbase: 116 inches	
Base Price: $2649	

Right: The 427-cid V-8 was the performance option. This car has the optional 'Ram Air' hood scoop.

Below: The 427 engine was in reality only fitted to a small number of Fairlanes, most owners opting for the 289-cid V8.

The Ford Fairlane model range was manufactured by FoMoCo from 1955 through to 1969. The original body design was the full sized Ford body. This body started out as a family vehicle, and evolved into many different models and body styles. The exterior paint and trim options were virtually endless, with a rainbow of paint colors and massive variety of cloth trim for inside.

From 1960-62 the Fairlanes began to decrease in size and the car went from a large to mid-sized model. The first sports coupe model was introduced in 1962, which continued until 1965. These models were bigger than the Falcons, but smaller than the Galaxies and offered the best of both worlds with solid performance and great economy.

In 1966, Ford restyled all the full-sized cars for this model year, including the Fairlane. The model was getting larger again, but still not a full-size Ford. The mid-sized Fairlanes were all new for this model year, wearing swoopier GM-type styling on slightly larger dimensions, though the body weight remained pretty much the same. The new dual stacked headlamp design was popular from the start. The model now boasted many available options and performance goodies to include 427 dual-carburetor engine setups, four-speed manual transmission, bucket seats and a console. A '66 Fairlane built specifically for the dragstrip was produced with a fiberglass hood. Fifty-seven of these were built with stripped down race specifications, and equipped with 427-cid race engines. However, the great majority of 1966 Fairlaines were actually fitted with a standard 200-cid six-cylinder engine, and a 200-horsepower 289-cid as the base option V-8.

This model year also introduced the Fairlane convertible, which was available in GT, GTA, 500XL and 500 models. The GT and GTA models were offered with more of the performance accessories and were most commonly built with 390-cid big block engines and luxury interiors complete with bucket seats and consoles. The GTA was a 'GT' with an 'A' automatic transmission, GTs were fitted with four-on-the-floor manual transmission. Disc brakes weren't introduced until the following year.

FORD

1966 Ford Galaxie 500 7-Litre Series Convertible

Ford finally made it to number one in the production league table for 1966, under the leadership of Donald N. Frey. FoMoCo continued its policy of major annual re-styling for several model lines. While 1965 and 1966 full-sized Fords bear a resemblance to each other, they are quite different cars. The hood is the only interchangeable exterior body component. The '66 models featured more rounded lines than the previous year, though the feature lines were in exactly the same positions.

Galaxie was the intermediate Ford trim for 1966, and offered a full range of seven different series containing nineteen different models. The series were the LTD, the Galaxie 500XL, the Galaxie 500 –Litre, the Galaxie 500 and Custom 500, and the lower trim levels also existed as a parallel V-8 format. Hardtops, sedan and convertible Galaxies were on offer. The model became known for both its clean lines and performance options.

The LTD had a rather different look with a distinctive trim, ornamentation and interior trim. The 500 XL was the ultimate Galaxie in '66 with deep foam contoured bucket seats, full-length console, wall-to-wall carpeting, padded dashboard and visors and special 500 XL exterior ornamentation. Ford also offered an impressive eight different engine choices, all the way from a standard economy six-cylinder to the 'side oiler' 427-cid engine with dual carburetion and 425 horsepower. FoMoCo also introduced a new 428 cubic inch Thunderbird V-8 that was standard equipment on the Galaxie 7-Litre model.

The 7-Litre was the high performance version of the Galaxie 500 XL, equipped with Cruise-O-Matic transmission. The four-speed manual was also available as a no-cost option for those who chose the even sportier driving characteristics it imparted. The cars also had low restriction dual exhausts, a non-silenced air cleaner system and power disc brakes. The model also had a Sport steering wheel, of simulated English walnut. The Galaxie exterior was available in fifteen Super Diamond Lustre enamels, in twenty-three two-tone combinations. Vinyl-covered roofs were also on offer. Inside the cars, forty-two different upholstery choices were available. The '66s had several power options, for steering, windows and seats together with optional disc brakes and air conditioning.

Engine: Thunderbird Special V-8

Displacement: 428 cid

Horsepower: 345 at 4600 rpm

Transmission: Cruise-O-Matic automatic

Compression Ratio: 10.5:1

Number Produced: 2368

1966 Ford GT40

Engine: V-8

Displacement: 427 cid

Horsepower: 485 at 6200 rpm

Weight: 2450 lbs

Wheelbase: 95 inches

Ford was at number one in the 1966 league of autoproducers. This was in part to their successful introduction of production performance cars like the Torino, Mustang and Thunderbird. These were both inspired, and initiated by technological advances learned at the classic American races and at Le Mans. As far back as 1962, Henry Ford II had decided that Ford should participate in the International Competition field.

The prototype Mark II GT40s made their first appearance at the 1965 LeMans, featuring a NASCAR approved 427 cid V-8. Unfortunately, the cars were obliged to retire with transmission problems. Ford and racing partner Shelby American spent the rest of '65 developing the Mark II for the 1966 endurance-racing season. They refined the big block 427, reducing the engine weight by fifty pounds by using aluminium heads with smaller valves and other modifications. Shelby American did most of the work on the chassis and suspension, reinforcing and revising these where necessary. The GT40 body consisted of a semi-monocoque construction of 0.61mm steel, hinged front and rear panel sections and doors of reinforced fibreglass.

Their work was fully justified. The cars not only took 1-2-3-5 at the first 24- hour Daytona Continental race, but 1-2-3 at the '66 LeMans. In fact, the GT40 was the first American car to win this prestigious race. The winning car was driven by Bruce McLaren and Chris Amon, and set a new average race speed of 125.4 mph. It had a top speed of 205 mph. This is where the GT40 earned its reputation as a 'Ferrari eater', proving more than a match for the P3.

Despite failing to win at Monza and the Nurnburgring 1000km race in Germany, Ford took the constructors' championship for prototype and series production cars. This resulted in the GT40 Mark II becoming the vehicle that showed the world that heavier, larger displacement cars could deliver spectacular performance. It was a victory for American talent, and gained a reputation for the GT40 as the being the best supercar of the era. As well as the fifty-plus cars that FIA rules insisted upon, Ford also had to produce at least 1,000 units of the car for public sale to qualify for competition in the production car series. These were known as the GT40 Mark III cars. Several changes had to be made to ensure that the car complied with safety regulations. The engine was changed to the 289-cid used on Shelby 350 Mustangs, but the clutch and transmission were identical to the race set-up.

1966 Oldsmobile 4-4-2 W-30

Oldsmobile entered the supercar sweepstakes in April '64 with a self explanatory name four-four-two – four-barrel carburettor, four-on-the-floor transmission and dual exhausts. The name remained the same for fifteen years, even though the transmission was also made available as both three-speed synchomesh and automatic over this period.

In 1966, Oldsmobile introduced the W-30 option, which later became known as 'Outside Air Induction'. This fantastic package became Olds hottest muscle car equipment. The package consisted of a RPO list of extras, designed to boost the 4-4-2s to be suitable for super-stock drag racing. The package consisted of a 400-cid L69 big-block V-8. This 360 horsepower engine had been introduced in the November of the previous year. It was equipped with three Rochester two-barrel carburetors, designed to keep the air/fuel mixture cooler and denser, to increase the available power. The engine option was not limited to the W-30 models, however, they were also fitted to 2,129 production models, including the F-85 Cutlasses.

The W-30 equipment also included ram-air ductwork that rammed cool air into the Rochester via two authentic large plastic scoops, which were situated in two openings of the front bumper, reserved for the turn signals in ordinary production cars. The hoses took up so much room under the hood that the battery was relocated to the trunk at the back of the car, which had the additional benefit of loading some weight onto the rear wheels. Other '66 W-30 equipment included a four-bladed fan with no clutch, heavy-duty three-core radiator, close-ratio manual transmission and a hydraulic, high-lift, longduartion camshaft. Despite all this heavy gear, no output rating change was listed for the W-30-equipped Oldsmobile.

Only 54 of these W-30 cars were built in 1966. The cars were not actively promoted at all. As soon as they were, production jumped to a far healthier 502 cars in '67. Some equipment changes were made, including replacing the carburetor system with a single four-barrel unit. This was due to a GM edict banning multiple-carb setups. An optional automatic transmission was offered.

The W-30 was most popular in 1970 when 3,100 W-30 equipped cars hit the streets. Oldsmobile continued to offer the package until 1980, but it was substantially diluted by this time.

Engine: L69 V-8

Displacement: 400-cid

Horsepower: 360 at 5000 rpm

Body Style: Two-door Coupe

Wheelbase: 115 inches

Number Produced: 54

1966 Plymouth Belvedere

Engine: Street Hemi V-8, overhead valve with hemispherical combustion chambers

Displacement: 426-cid

Horsepower: 425 at 5600 rpm

Compression Ratio: 10.25:1

Body Style: Two-door Convertible

Weight: 3320 lbs

Wheelbase: 116 inches

Base Price: $2910

Number Produced: 2,759

Plymouth was now number four in the auto producers' league, but on a reduced production output of 678,514 cars. There were quite of lot of interesting developments at Plymouth for '66 - mostly positive. The revived fullsize Fury was re-introduced to the range, with mild styling tweaks. The 'glassback' Barracuda continued to make a valiant attempt to take pony car sales away from Ford's Mustang (without a great deal of success, it has to be said). The Fury VIP model was added to the Plymouth range to give the Ford LTD and Chevy Caprice a run for their money in the luxury hardtop coupe market. The car came with a 230 horsepower 318-cid V-8 as standard, but engine options went all the way up to a 365 horsepower 440-cid V-8. Like Dodge, Plymouth also offered a muscular Street Hemi on selected models, including the top-of-the-line Satellite Belvederes.

The 1966 Belvederes received a major restyling. The new car had a square body, and slab-like fenders. A full-length sculptural depression panel relieved this. The large front wheel opening now curved up to this panel. In profile, the edge of the front fender thrust forward into a wide V-shaped form. The sedan models had a square-angular roof with thick rear pillars. Hardtops retained the cantilevered roof treatment with a thicker base, and this treatment was echoed at the rear of the station wagon models. As in the previous model year, Belvederes were offered in a three-tier model range, the Is, IIs and Satellites.

Belvedere Is (three models) had thin, straight moldings along the lower feature line, a heater and defroster, front seat belts, an oil filter and five blackwall tires. Belvedere II models (of which there were five), were adorned with a wide full-length chrome spear placed above the bodyside centreline. Inside the cars, there was upgraded upholstery and carpeting. Satellites, as usual, had less side trim than the other models, but came with a fancy trunk treatment, rocker panel moldings, bucket seats and console, wheel covers and vinyl trim. As in '65, there were two models – the two-door hardtop coupe and two-door convertible.

1967 Chevrolet Camaro Z/28

Engine: Small-block V-8, overhead valve, cast iron block	
Displacement: 302 cid	
Horsepower: 290 at 5800 rpm	
Transmission: Muncie four-speed manual	
Compression Ratio: 11:1	
Fuel System: Single 800-cfm Holley four-barrel with high-rise intake	
Body Style: Two-door hardtop coupe	
Number of Seats: 5	
Weight: 3070 lbs	
Wheelbase: 108 inches	
Base Price: $3,380	
Number Produced: 602	

Chevy was still at number one in the production stakes in 1967, and intended to stay there. Inspired by the huge success of the Mustang, Chevrolet entered the pony car race in 1967 with its sporty Camaro. Chevrolet was determined that the new model should appeal to as wide a range of drivers as possible. A massive 'building block' system of option packages allowed for the creation of many distinctive vehicles, tailored to many different types of drivers. Two basic body shapes were available, a two-door hardtop coupe and a two-door convertible. A huge range of engines from an inline six-cylinder to a 295 horsepower Turbo-fire V-8 was available for the retail models. The basic price was fairly reasonable at around $2,500, but the extensive list of options could easily push the price for an individual car past $5,000. The Camaro quickly established itself as Chevy's real contender in the sporty compact field, ousting their fast-fading Corvair, which lost its turbocharged option this year. They were soon taking market share off Ford, as sales of the new model achieved sales of over 220,000 in its first year. Unsurprisingly, the car shared the same basic profile with the Mustang, having a long hood on a visually short deck.

The Z/28 competition package Camaro was created for the recently formed Trans-Am racing series, and had a fantastic option package that included numerous performance goodies, including a powerful small-block V-8 and disc brakes. The distinctive Z/28 badging didn't actually appear until '68. In fact, the

car effectively saved the Trans-Am, as the complete dominance of the Mustang in the '66 season was a complete turn-off to the race goers. Chevrolet's Vince Piggins promised the organizers that he would turn up with a racecar for the next season, and then persuaded the board to back his plan to take the Camaro to the races. To be race legal, the car had to be available for general retail sale, and attract at least 1000 customers, and displace no more than 305 cid. This resulted in a huge engineering project, redesigning a hot small-block V-8 especially for the car, with a forged steel crank, L79 big-port cylinder heads, and a Holley four-barrel carburetor. Unofficially, the car was rated at over 400 horsepower. The car was also fitted with F41 stiffened suspension and a Muncie four-speed transmission. The only external signs of the powerhouse within were Corvette rally wheels with bright rims, and wide twin racing stripes on the hood and rear decklid.

Although sales were a modest 602 in their first year of introduction, this was one Camaro model that really caught on, selling 7,199 in '68, production soared to 19,000. The reviews for the car were fantastic, it was described as a 'happy and extremely potent screamer'. Apart from a two-year hiccup in the seventies, the brilliant Z/28 stayed in production until the new millennium.

Above A Z-28 in full flight on raceday.

Left: Trans Am racers were production-based cars with engines of 305 cid or less. Chevy worked a forged steel version of the 283-cid V-8's crankshaft into its 327 V-8 to get 302 cid. Big-port Corvette heads, solid lifters, a hot cam, a baffled oil pan, and a Holley four-barrel on a tuned aluminum manifold were specified. Horsepower was rated at 290 but thought to be closer to 360 in reality.

1967 Chevrolet Corvette L-88

Engine: L-88 V-8, aluninum cylinder heads	
Displacement: 427 cid	
Horsepower: 430 at 5200 rpm	
Compression Ratio: 12.5:1	
Wheelbase: 98 inches	
Base Price: $5675	

Right: This Goodwood Green coupe is one of the earliest L-88s built, and is owned by collector Bill Tower, a former GM engineer. He thinks it was the first car in the series to be built.

Below: Understated engine appearance belies its 430 horsepower.

An all-new Corvette had been planned for 1967, but development problems intervened, and delayed the new model for a year. The Sting Ray put in one more glamorous appearance, looking cleaner and meaner than ever with less decal, and retained its now familiar hood hump. The body was fundamentally that of the original '63 Sting Ray.

The special L-88 model Corvette was primarily designed as a race-ready car for the track, and anyone who tried to use it as a streetcar was destined for a shock. The car was stripped of any of the normal refinements that street Corvettes were loaded with, radio, heater, automatic choke or underhood cooling, even reliable brakes. In fact, Corvette was quite coy about identifying the cars at all in case the unsuspecting ordinary buyer might insist on purchasing one. They had originally planned to identify the cars with special L/88 badges, but dropped this idea. Chevrolet even grossly underrated the car as having 430 horsepower, hoping to steer their racier customers to the L-71 engine (rated at 435 horsepower). They only wanted to sell the cars to genuine racers. In fact, the car could not even be filled with regular gas, it required special high octane fuel of at least 103 octane to prevent engine damage.

What the car did have was a selection of specially designed performance

enhancing equipment, including the (now reliable) aluminum cylinder heads, a huge four-barrel Holley carburetor and the first Corvette ram air hood forcing cool air down into it (the first Corvette to have a fully functional hood scoop). This was the result of extensive testing to identify the flow characteristics of the air coming off the windshield. The L-88s were also fitted with stiff suspension, G81 Positraction differential, J56 power-assisted metallic brakes and an M22 'Rock Crusher' four-speed manual transmission. Racing-style sidepipes manufactured by Stahl were also fitted to the car, as was a transistorized transmission. All of these modifications were supervised by Corvette chief engineer Zora Duntov.

At the Le Mans 24 Hour race of June '67, an L-88 was clocked at 170 mph before a connecting rod disconncected.

The L-88 was manufactured for a total of three years ('67, '68 and '69) and a total of 216 were produced. Rather appropriately, at least one L-88 was amongst a batch of complimentary Sting Rays given to NASA astronauts.

Below: Presumably an aftermarket decal as Chevrolet themselves avoided over badging the L-88.

1967 Chevrolet Impala SS 427

GM celebrated the manufacture of its 100 millionth American-built car in April 1967. The company was now divided between the car and truck divisions. Ralph Nadar's influence continued to pervade the company thinking, with safety becoming an increasingly important issue. An energy-absorbing steering column and seat belts were both now standard. Chevy was now back on top of Ford in '67 (out-producing their rival by over 220,000 units), thanks in some part to the launch of the Camaro.

Chevrolet's Impala SS was still a separate series in '67, but would revert to option package status in 1968. Chevy introduced the big new 427 cid V-8 in this year. Although there was a growing market for compacts in the US, the full-sized Chevys, including the Impalas, held onto sales of over half a million units. The Biscayne, Bel Air, Impala and Caprice models were all built on the full-sized chassis. The basic Impala range for this year included six options, ranging from a two-door convertible to a nine-seat station wagon. By contrast, the Impala Super Sport series had two options, a two-door hardtop and two-door convertible. The cars also came with a choice of engines, an inline six-cylinder or the overhead valve V-8 in various power levels. Turbo Hydra-Matic transmission was introduced in '67, and was available as an option.

The sporting Impala once again featured an all-vinyl interior, with front Strato bucket seats and a division console that housed the shift lever were standard, although Strato bench seating was available at no extra charge. Exterior identification for the series consisted of a black-accented grille, with bright metal horizontal bars, front and rear wheelhouse moldings, a black highlighted body sill and deck lid latch panel. 'SS' grille badges and specific 'SS' full wheel covers were also part of the equipment for the model. The bright metal hood louvers were simulated.

The car impressed contemporary customers with both its 'mobile creature comforts' and performance, 'which was, until recently, reserved for dragstrip specialists'. He continued that the car 'may well be the Ultimate average car – the status symbol of Everyman'. Of course, at around $5000 for the full SS 427 package, not every man could afford the luxurious, fully loaded, feel of the SS. In fact, this price tag rather blew Chevy's image as being one of the 'low-priced three'. Car Life magazine really liked the car, rating the ride good, the acceleration exciting, the handling fair to good, but the braking dreadful (with rear wheel lockup and significant fade).

Engine: V-8 overhead valve, cast iron block

Displacement: 427 cid

Horsepower: 385 at 5200rpm

Compression Ratio: 10.25:1

Wheelbase: 118 inches

1967 Chevrolet Nova SS

Engine: V-8 overhead valve, cast iron block,

Displacement: 326.7 cid

Horsepower: 275 at 4800 rpm

Compression Ratio: 10.0:1

Number Produced: 10,100

The Nova had been dropped, briefly, in anticipation of the larger Chevelle SS, but public demand had resulted in a midyear return for the model, with two new engine options. These were the first V-8 for the Nova, and the 230 inline. The convertible and 3-seat station wagon were dropped, but the 2-seat wagon continued. The Chevy II 300 series trim was deleted. '65 had heralded minor trim changes, which were reinforced in '66. The installation of more serious engines to the car in the '64-'65 period began to establish a serious reputation for Nova SS as a muscle car.

Both the Chevelles and Chevy II Novas were completely restyled in 1966. The hot ticket version of the car was the L79 version of the 327 V-8, which developed 350 horsepower. Officially, this engine option was dropped in 1967, to prevent sales being drawn away from the newly introduced Camaro range. But although the L79 option was officially deleted, production records show that six such cars were built in '67. Only minor styling changes were introduced in 1967, to consolidate the car as one of the most clean-looking and collectible Chevys of all time. A new grille and side trim were the most visible changes to the car. Mechanically, the dual-pot barking system was fitted as standard, but disc brakes were available as an option. Revised for '67, the standard engine option for the Nova and Nova SS was a 187.6 six-cylinder inline. Optional engine variations were the RPO L22 250-cid 155 horsepower six-cylinder block, and the RPO L30 327-cid 275 horsepower V-8. The small, taut Nova Super Sport two-door hardtop coupe (series 117) continued to make an excellent high-performance car when equipped with these special options.

The exterior of the Nova SS was distinguished with a special blackaccented chrome grille, complete with SS insignia low on the driver's side. Above a black painted sill area, there was an elegant chrome accent stripe, lower body moldings, and bright wheelhouse accents extending along the lower fender edges. The model had specific Super Sport full wheelcovers, fender scripts and full-width color accent deck lid trim panels. The car interiors were all vinyl, with front Strato bucket seats, optional Astro-bucket headrests, with bright seat end panels as standard. In cars where the transmission was upgraded to four-speed or automatic transmission, a floor shift trim plate was added. The cars also had a specific Nova Super Sport three-spoke steering wheel.

1967 Dodge Charger 426 Hemi

Engine: Hemi V-8	
Displacement: 426 cid	
Horsepower: 425 at 5000 rpm	
Transmission: Four-speed manual with Hurst shifter, TorqueFlite automatic optional	
Compression Ratio: 10.25:1	
Fuel System: Two 650-cfm Carter four-barrel carburetors	
Body Style: Two-door Fastback Coupe	
Number of Seats: 4	
Weight: 4160lbs	
Wheelbase: 117 inches	
Base Price: $4500	
Number Produced: 118	

Dodge had slipped back to seventh in the auto producers' league in 1967, though production remained pretty steady at just under half a million. Byron Nichols was now the General Manager of the division. With its unique looks and massive powerhouse, the Charge Hemi 426 was a scary beast, especially if you came upon one on the road. Rather more sedately, Nichols described the car as 'a fresh concept in styling and engineering excellence from bumper to bumper'.

The basic body fastback Charger body was pretty much the same one that had been introduced in 1966, with the cars built on Coronet chassis. The road-going version was fitted with front fender mounted turn signal indicators and some extra chrome trim, together with front and rear bucket seats and a cigar lighter. All were fitted with V-8 engines. But these were pretty tame compared to the Hemi 426. These had been designed for one thing and one thing only – racing at the new super speedway in Daytona Beach, Florida. Owned by Bill France, this was the real testing ground for the biggest muscles in Detroit. Chrysler, of course, had a huge advantage in having the milestone hemihead engine in its parts cupboard. The original engine was launched in 1951, but was abandoned in the late fifties in favor of the equally powerful, but smaller wedge-heads. However, the concept was revived six years later, bigger and meaner than ever.

Back in 1962, Chrysler engineer Tom Hoover was asked to recreate the Hemi for the modern NASCAR circuit. They completed the job in February 1964. In fact, the engine proved to be so dominant that Bill France actually tried to ban the Hemi from the race circuit on the basis that it didn't comply with the race rules, not being fitted to a regular production car. Chrysler got round this by offering de-tuned version of the engine to the public – the 'Street Hemi'. This engine was offered on a wide array of Dodge and Plymouth models from '66 – '71, but in the Charger version, the hot performance of the hemi was matched by the stunning looks of the car.

1967 Dodge Dart GTS (383 V-8)

Engines and cars got bigger in the mid-sixties, heavier and more complex. Detroit had learnt the way to power through stock-car racing and gradually introduced what it had learnt to production models. The company's fortunes had also grown in these years, as it climbed higher in the production league table. In 1967, they were at seventh place (on production of nearly half a million cars), but had made a brief stop at fifth in '66. R.B. Curry became the General Manager of Dodge in this year.

The Darts got a completely restyled unibody, 43 percent of these were fitted with the V-8 engine option. The car was now America's largest compact and was styled with curved side glass and delta-shaped taillamps. Polara and Monaco were also extensively redesigned and built larger, riding a 122-inch wheelbase.

The restyle for the Dart was total, from the ground up. The cars retained the chassis of the previous year, but were restyled to look larger than the earlier model (although they were actually half an inch shorter). The full-width grille housed single headlights and featured a vertical bar arrangement with a largescale dividing bar in the center of the concave grille. The side profile was rather more rounded than in the previous incarnation of the car, but carried similar lines. At the back, the taillights were almost square and fitted into a flat section on the trunk lid. The cars were trimmed with chromed windshield and rear window moldings.

The Dart was the base model of the range, the Dart 270 was the intermediate trim level, and the Dart GT was the top-level model. The GT was fitted with all the options of the entry-level models, plus a padded instrument panel, special wheel covers and bucket seats. There were two GT models, a two-door hardtop coupe and two-door convertible. The cars were fitted with two basic engine options, the slant six-cylinder and the V-8 – available with various-sized engine blocks. The convenience options for GT models included an interior console, buffed paint, a tachometer and simulated 'mag' wheel covers.

Engine: Overhead valve V-8, cast iron block

Displacement: 383 cid

Horsepower: 270 at 4400 rpm

Compression Ratio: 9.2:1

Fuel System: Carter two-barrel BDD-4125S

Weight: 3030 lbs

Wheelbase: 111 inches

Base Price: $2860

Number Produced: (All Dart GTs) 38,200 including 21,600 V-8s)

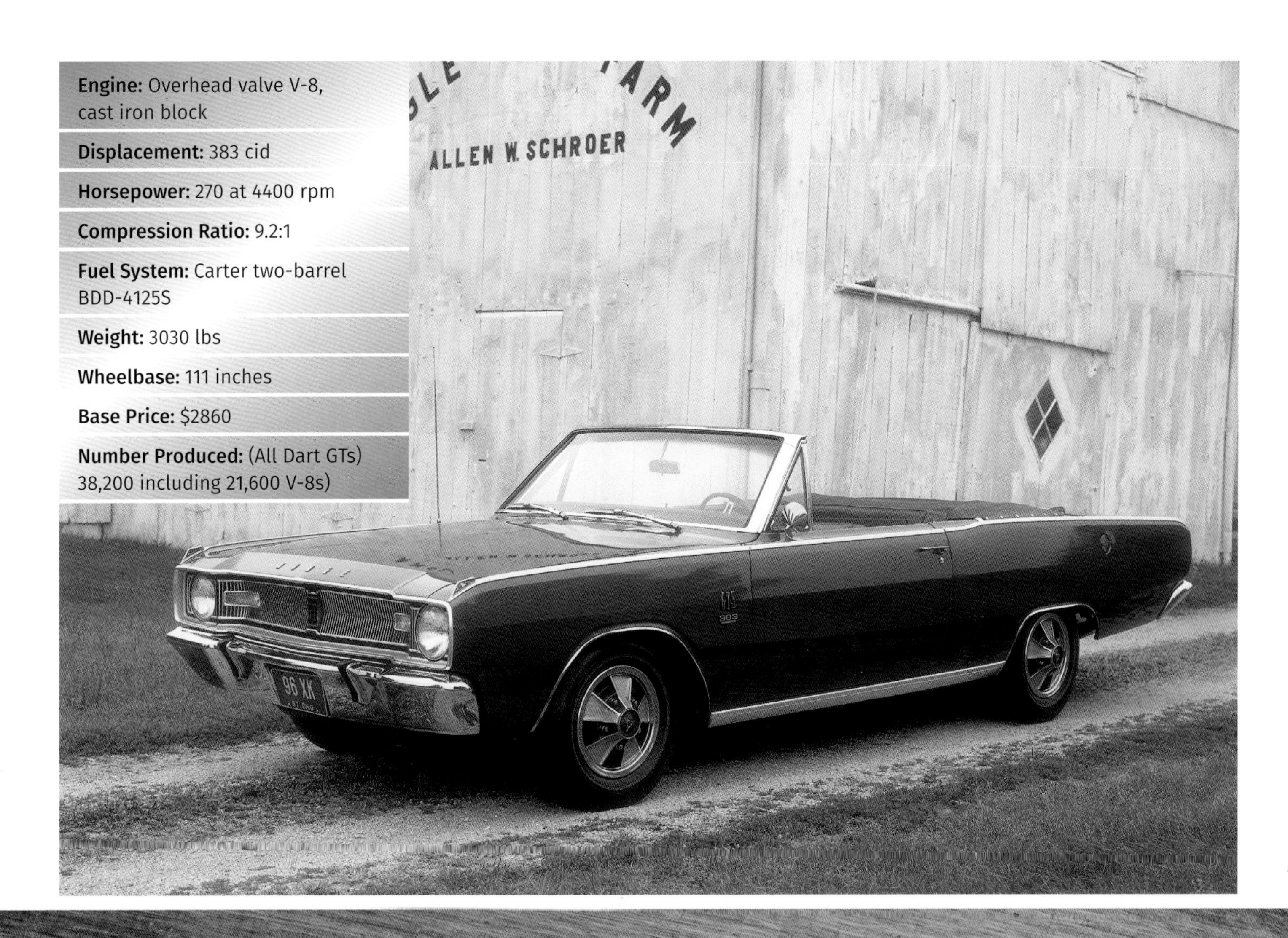

1967 Ford Thunderbird Landau

The Thunderbirds received a complete re-style for 1967. The front end of the car now featured a full-width grille with hidden headlights. As in 1966, the rear end featured a large, single taillight lens with a horizontal trim strip in the center. In addition, there were backup lights, in the center of the strip, which gave the impression of a large round taillight.

Introduced in 1955 as a luxury two-seater Thunderbird, the model range gradually evolved over the years with the open version selling only 7.5% of the total range by 1966. This was the final year that offered a convertible Thunderbird, and some say that the later series had all the luxury, but none of the sportiness of the earlier car.

For the fifth Thunderbird series, and for the first time in Thunderbird history, a four-door sedan, the 'Landau Sedan', was introduced to the model range. It was built on a longer wheelbase than the two-door hardtop and featured unusual rear doors that opened to the front, of 'suicide doors' as they were known. This model seemed to some to mark the descent of the Tbirds into middle age. The production engines at least were now tuned for refinement and not performance, and the bodies looked bulky compared to the sprightly earlier two-seat models.

The base interior now included full five-dial instrumentation plus a glovebox-mounted emergency flasher. Other niceties, such as an AM/FM radio, Stereophonic tape system, six-way power driver's seat, cruise control and seat belt warning lamps were available as options.

1967 turned out to be the fourth best sales year to date for the Thunderbirds. The model sold 77,956 cars out of a total Ford production of 1,742,311 units. This was despite industrial action by the United Auto Workers that knocked out fifty-seven Ford production days. The two-door Tudor models and the new 'Fordor' Landau appeared in most of the advertising for the year, but the most popular model in the range was the Tudor Landau – despite the fact that virtually all of the year's advertising ignored it. While appearances may well suggest otherwise, almost no parts from the '67 Thunderbirds are interchangeable with the '68-'69 models. However, Thunderbird sales then declined until the introduction of the 1972 models, which reversed the decline until the disastrous effect of the oil crisis of 1974.

1967 Mercury Cougar XR-7

Mercury entered the pony car race in 1967 with the introduction of the Cougar. The model was launched with some powerful performance packages.

The Cougar was one of the most handsome automobiles of 1967, and Cougar was a niche marketing success for Mercury. Overall, its probably leant more to comfort than performance. Its styling was more mature, featuring disappearing headlights, wraparound front and rear fenders and triple taillights (with sequential turn signals). The front and rear end styling were similar. Cougars came, equipped with all-vinyl bucket seats, three-spoke 'sport-style' steering wheel, deep-loop carpeting and deluxe seat belts. The standard transmission for the model was floor-mounted three-speed manual transmission.

There was a single base Cougar model, the two-door hardtop coupe. A Cougar XR-7 two-door hardtop coupe was introduced in the mid-model-year. No convertible or fastback models were offered until 1969. Except for a medallion on the quarter panel of the roof, it looked just like the standard Cougar model. It was also embellished with a black-faced instruments set into a simulated walnut dashboard, and a more luxurious, simulated leather, interior.

The Cougar was immediately successful, selling a total of 150,893 cars in its first model outing, accounting for nearly half of Mercury sales for the entire year. Only 7,412 of these were fitted with the optional front bench seat, and only 5.3% with the four-speed manual transmission. The most popular options were automatic transmission (96.2%), power steering (97%), tinted glass (69.7%) and power brakes (65.6%). A GT performance package was also available for a further $323. This included a firmer suspension with bigger shocks, fatter anti-roll bars, power front disc brakes and a 390-cid V-8 (rated at 335 horsepower) that required premium fuel.

Engine: V-8, overhead valve, cast iron block

Displacement: 289 cid

Horsepower: 200 at 4400rpm

Compression Ratio: 9.3:1

Fuel System: Autolite C7DF-9510-Z two-barrel

Body Style: Two-door Hardtop Coupe

Number of Seats: 5

Weight: 3015 lbs

Wheelbase: 111 inches

Base Price: $3081

Number Produced: 27,221

1967 Mercury Cyclone 427

Engine: FE-series V-8	
Displacement: 427-cid	
Horsepower: 410 at 5600rpm	
Transmission: Four-speed 'top loader' manual	
Compression Ratio: 11.1:1	
Fuel System: Single-Holley Four-barrel carburetor	
Body Style: Two-door Hardtop Coupe	
Number of Seats: 5	
Weight: 3600 lbs	
Wheelbase: 116 inches	
Base Price: $4100	
Number Produced: 8	

Mercury offered thirteen different styles for the '66 and '67 Comet ranges. These cars are amongst the most desirable Mercury models ever built, gone was the economy car image of the 1960-63 Comets, and the new V-8 engine selections massively up-rated the available power.

The 427 Cyclone, introduced by Mercury as their answer to Pontiac's '64 GTO phenomenon. This car was responsible for spawning several other muscle car competitors, including the Buick Gran Sport, Chevrolet's SS-396 Chevelle and Oldsmobile's 4-4-2.

Chrysler and Ford had been slow to join the club, but hoped that the Mercury Comet Cyclone would be able to take on their competitors. But the small block V-8 of the 1965 production model was no match for the GM big block big boys. The 225 horsepower Super Cyclone 289 could only manage the quarter mile in a rather sedate 17.1 seconds at 82 mph. The simulated air scoops didn't help much, either. The Cyclone model had originally been introduced to the Comet range in 1964 in an attempt to attract a more masculine, 'macho' buyer, with its up-rated power train and black vinyl dash, but it soon became evident that this

Right: The 427 FE V8 was the top of the range option. Only eight cars were built with this in 1967.

model wasn't muscle enough to see off the serious competition.

1967 Comet customers were offered two new big block engines, the 410 horsepower 427-cid (complete with four-barrel carburetion). Mercury also offered the even bigger and badder Ford 427 FE, complete with 425 horsepower. This engine was the king of the FE-series big-blocks, and dated from 1958. It had been growing progressively over the years, from 352 cid to 390 in '61, 406 in '62 and finally 427 in '63. It had been used to power the '66 GT-40s to a 1-2-3 finish at Le Mans. Company records show that around eight Mercurys were built in '67, complete with this single four-barrel carburetor engine option. The car was designed exclusively with drag racers in mind, and came equipped with a heavyduty clutch, 'top loader' four-speed manual transmission and heavy-duty suspension, battery and extra-capacity radiator.

Above: Styling was designed to appeal to a more macho buyer.

1967 Plymouth Barracuda Formula S 383-cid

Engine: Barracuda V-8

Displacement: 383-cid

Horsepower: 280

Transmission: three-speed manual

Number Produced: 28,196

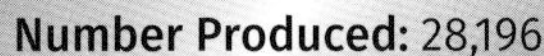

The Barracuda had been launched by Plymouth in 1964, specifically designed to go head-to-head with Ford in the Mustang pony car race. It was a sporty new 'glassback' coupe with a large and distinctive wrap-over rear window treatment that gave the car a fastback shape. It also had a uniquely styled roof and deck, in which the deck lid bulged up to meet the glass. The car also boasted distinct trim features, that included a split, negative space grille with center insert and horizontal outer division bars, wide rocker sill panels and a chrome band across the base of the rear window.

A single two-door sports hardtop Barracuda model was offered in the model's first, second and third years. The news for '65 was the introduction of the Barracuda Formula S package. This included a Commando 273-cid V-8 engine, rally suspension, heavy-duty shocks, Goodyear Blue Streak tires, a tachometer and 'open wheel' covers.

For 1966, the Barracuda was offered as part of the Valiant series. The car was redesigned towards the font, but remained largely unchanged at the rear end and above the belt. It was now fitted with the split Valiant grille opening. The 'S' option package continued to be offered, and the standard engines were either the 225-cid slant six, or 273-cid two-barrel V-8.

The Barracuda models shed their links with the Valiant family in '67, and three distinct body shapes were now on offer, a two-door notchback coupe, a two-door fastback coupe (the 'Sports Barracuda') and a two-door convertible. The 'glassback' look was consigned to the past, and the cars were offered to the public with completely new styling. Barracudas now had curvy, flowing features with extremely restrained, modern trim. Contemporary motoring magazines compared the styling of the car to that of the Buick Riviera, but they were really worlds apart. The new Barracuda had single headlights, concave roof pillars (on the notchback coupe), curved side glass, a concave rear deck panel, wide wheel openings and a sleek fastback.

Standard equipment for the '67 Barracuda included three-speed transmission, and the 225-cid slant six or 273-cid V-8. The interiors were equipped with carpeting, front bucket seats (in the convertible), a fold-down rear seat (in the fastback), a padded dash and all standard safety equipment. The 'S' equipment package was still available in an up-rated form. Whatever the rationale, the '67 Barracuda doubled its sales from the previous model year.

1967 Plymouth Belvedere GTX

The year 1967 Plymouth production models were introduced to the market in September 1966. 638,075 cars were sold, putting Plymouth at number four in the league of production. Robert Anderson was the chief executive officer of the company in this year.

The 1967 Belvedere range grew to include another tier of performance, the GTX models. All '67 model year medium-sized Belvederes and the derivative models of the series were slightly face lifted. The horizontal grille blades were thinner and house side-by-side headlights with small grille extensions between them. The parking lamps were moved into the bumper, and the taillamps were redone. A new economy station wagon model was introduced to the range, simply known as the Belvedere.

Belvedere I models had a cigar lighter, padded dash, two-speed wipers with washers, back-up lights, front and rear armrests and rocker panel moldings. The Belvedere II line had all of the above plus foam front seats, parking brake warning lamp, wraparound taillights, carpeting, wheel opening moldings and full-length side moldings. The Satellite models had additional extras including front bucket seats with console (or center armrest seat), deluxe wheel covers, glove-box light, fender-top turn signals, upper body accent stripe, courtesy lights and aluma-plate full-length lower body trim panels.

The high performance GTX had all this and more, including a 'Pit-Stop' gas cap, Red Streak tires, dual hood scoops, dual sports stripes, heavy-duty threespeed TorqueFlite transmission, brakes, suspension and battery. The standard engine for the GTX cars was the 440-cid four-barrel V-8 engine. They inherited the benefits of the racing heritage of the Mopar big-block engines, and were the muscle cars to beat on the streets.

Two GTX models were available, the two-door hardtop coupe and two-door convertible. 720 GTX cars were fitted with the Street Hemi option package in this year, 312 of which were fitted with four-speed transmission.

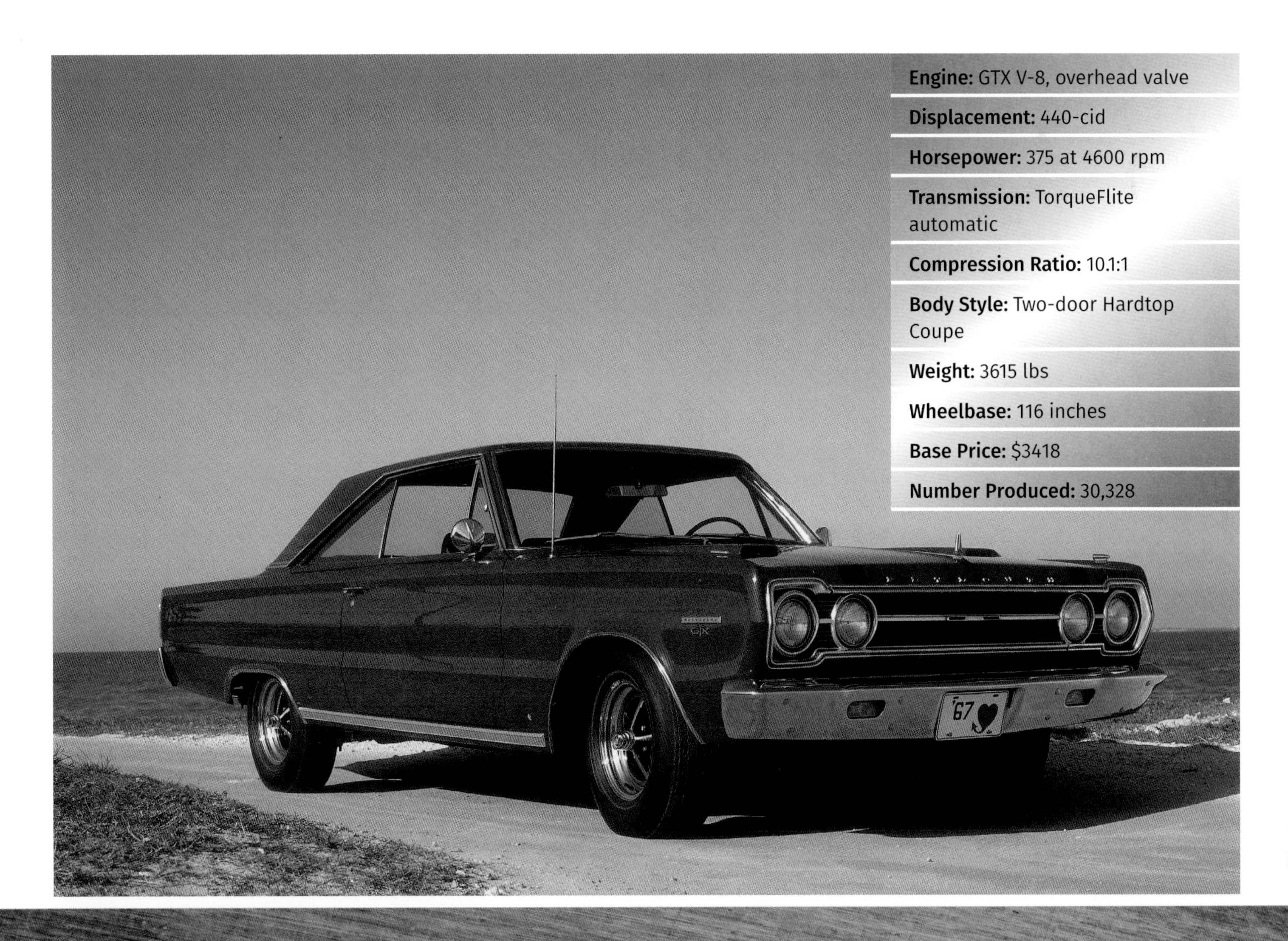

Engine: GTX V-8, overhead valve

Displacement: 440-cid

Horsepower: 375 at 4600 rpm

Transmission: TorqueFlite automatic

Compression Ratio: 10.1:1

Body Style: Two-door Hardtop Coupe

Weight: 3615 lbs

Wheelbase: 116 inches

Base Price: $3418

Number Produced: 30,328

1967 Pontiac Grand Prix

Engine: Grand Prix V-8
Displacement: 400-cid
Horsepower: 350 at 5000 rpm
Compression Ratio: 10.5:1
Weight: 4040 lbs
Wheelbase: 121 inches
Base Price: $3813
Number Produced: 5856

Pontiac was at its habitual third place in the auto producers' league for 1967, on a production of 857,171 units. '67 was a big year at the company with the introduction of the Camaro-clone Firebird model, the continued presence of the GTOs, and the expansion of the Grand Prix range to include a convertible. Interesting '67 Pontiac conversions included the 'Monkeemobile' based on the GTO phaeton, made for The Monkees TV show. Racecar driver/builder John Fitch also marketed his hot performance Firebird – the Fitchbird.

The Grand Prix models underwent a transformation for 1967. Following the trend for hideaway headlights, these were now completely concealed behind the grille, and could not be seen at all, except when they were lit. The turn signals were equally discreet, positioned in recesses in the front wings. The cars were also equipped with elegant fender skirts and lower body accent moldings, on their face-lifted 'Poncho' bodies. For identification, the cars had GP letters on the left hand side of the grille and on the rear fender. Grand Prix models were built on a massive 121-inch wheelbase and the overall length of the car was almost 215 inches (nearly eighteen feet...). The Grand Prix now weighed over 4,000 pounds, and typified the mid-sixties trend towards the bigger, heavier, accessory-laden cars being built in America.

Naturally, cars of this great bulk needed equally hearty power to compensate for the extra weight. The standard engine fitted to the model was the 350 horsepower 400-cid V-8 (coming just under the wire for engine size, according to GM directives). Despite this, the sheer mass of the car meant that it was never going to be truly quick. The interior of the car was sporty, however, equipped with front Strato Bucket seats and a console (plus all the government required safely features). From the long range of options, Grand Prix could be as well equipped as a Cadillac.

The '67 Grand Prix range consisted of two models, a convertible (as shown in our photograph) being added to the two-door hardtop coupe (which lost its vent windows this year) in an effort to boost sales for this line-up. Unfortunately, the convertible attracted only 5,856 buyers, and so ended up being a one-year option only.

1968 Ford Mustang 428 Cobra Jet

This was another of the specialist high-performance cars initially brought to the market by enthusiastic dealers. Robert F. Tasca ran a Ford dealership in East Providence, Rhode Island, and is credited with having invented the tag, 'Win on Sunday, sell on Monday'. Tasca put his money where his mouth was, and sponsored several individual cars, plus a fully-fledged drag race team. The latter won championships at the NHRA Winternationals in 1964 and '65. Tasca Ford became a byword for everything 'hot' produced by the blue oval, under the watchful eye of performance manager, Dean Gregson.

This reputation led Tasca to feel somewhat disappointed in Ford's revised Mustang GT, introduced in 1967. This was a larger version of the original car, fitted with the big-block 390 cid FE-series V-8. Tasca made his own revises to the car. He used a 428-cid Police Interceptor short-block engine with reworked cylinder heads and carburetion. This 428 version Mustang could cover the quarter mile in thirteen seconds, and earned the soubriquet 'KR' – 'King of the Road'. Hot Rod magazine writer Rod Dahlquist polled his readers with the question – should Ford make a production version of the KR? Several thousand affirmative responses came in, and Henry Ford II duly set the ball rolling. The Dearborn technical people followed Tasca's recipe for success, but also included a 390 GT cam, PI rods and low restriction dual exhausts. Production started on the strip-ready models on December 13 1967, slightly later for the street-going models. These were launched on April 1 1968. The car was offered as a $420 Mustang option.

Ford decided not to use the 'KR' name for the car, but chose 'Cobra Jet' instead. It was offered as a fastback, coupe and convertible, and had a long list of extras available including power front disc brakes, strengthened suspensions and a black-striped ram-air hood, four-speed or C6 automatic transmission, F70 tires, chromed quad exhaust tips. GT-styled bright steel wheels and competition-style flip-open gas caps were fitted as standard. The 335 horsepower car was Ford's new shot at muscle car prominence, and they raced it to top honors at the NHRA Winternationals in Pomona, California. At the time, it was probably the fastest regular-production sedan ever built. The strip-ready model covered the quarter mile in a scorching 13.56 seconds. A total of 2827 were built, 2253 fastbacks, 564 hardtops and 10 convertibles.

1968 Plymouth Road Runner

Engine: Hemi V-8	
Displacement: 426 cid	
Horsepower: 425 at 5000 rpm	
Compression Ratio: 10.25:1	
Wheelbase: 116 inches	
Base Price: $4,400	
Number Produced (Hemi): 840	

Plymouth launched the Road Runner in 1968 as a mass-market muscle car. It remained in production until 1974. There were plenty of rivals to the car, but they tended to be expensive and exclusive, and out of reach of the very young pockets they were aimed at. Most of the cars that could do the quarter mile in 100mph were at least $3,300 – nothing below this price mark could cut the mustard.

Plymouth were looking to give more bang for the buck, and aimed to build a car that could reach the quarter mile at 100 mph, for less than $3,000. Luxuries and non-essentials were pared down to an absolute minimum. Resources were concentrated on the performance package instead, the 'gogoodies'. The suspension was stiffened and a A833 four-speed manual column change gearbox was fitted. Despite the sporty image of the car, bucket seats were not included. But outside, the car was pretty glamorous with a blacked out Sports Satellite grille and GTX hood. All '68 Road Runners were either hardtops or coupes, a convertible option was offered for a single year in '69.

But Plymouth also made the car fun. It was named after a cartoon character! The Road Runner had decals of its namesake, and a droll horn that played the recognizable 'beep beep beep beeeep beep'. Even so, it was the muscle

Below: The 426 Hemi V8 was available for an extra $715 making the car the fastest production sedan available at the time.

car image and power of the car that made it a success – it was only $2,900 and full of go. Plymouth engineers had concocted a 335 horsepower V-8 from off-the-shelf equipment that enabled them to offer lots of car for the power. In fact, it was heralded as the first car since the GTO to be offered specifically at American youth. Demanded for the car skyrocketed, 44,599 were sold in '68, and the car became the second most successful production muscle car in America, behind the Chevy SS 396 Chevelle.

Above: The Plymouth Coupe's good looks appealed to its young customers.

Below: The Warner Bros Road Runner decal from the cartoon show of the same name.

Despite the modest price of the car, options could be added to build it up to be either more luxurious or more powerful. The 426 Hemi engine was offered, but at the hefty price of $715. Fitting the Hemi also necessitated fitting Torqueflite automatic transmission, at a further $139, and a whole lot of beefed up equipment, including a K-member subframe and further-stiffened suspension. This raised the price of the car to around $4,400. Even so, the car was rated as probably the fastest production sedan on offer. About 840 coupes and 169 hardtops were built with this option.

1968 Plymouth GTX

Engine: GTX V-8, overhead valve	
Displacement: 440-cid	
Horsepower: 375 at 4600 rpm	
Transmission: TorqueFlite automatic	
Compression Ratio: 10.1:1	
Body Style: Two-door Hardtop Coupe	
Weight: 3520 lbs	
Wheelbase: 116 inches	
Base Price: $3329	
Number Produced: 17,914	

Plymouth remained at number four in the producers' league of 1968, on a production of 790,239, and had an 8.1 percent market share. G.E. White was the chief executive officer of the company in this year. Retail sales for the Chrysler-Plymouth group hit an all-time high.

The GTX models were now part, and top of the Plymouth Intermediate model range, which also included the Belvedere, Satellite, Road Runner, Sport Satellite and Satellite Sport Wagon. Including the two GTXs (a two-door hardtop coupe and two-door convertible), the Intermediate line-up comprised sixteen different models that ranged from two-door convertibles to nine-seat station wagon options. The Road Runner was a newly introduced lightweight pillared coupe that was essentially a no-frills Belvedere equipped as a fullblown muscle car. A modified version of the car was driven to success by the Sox and Martin drag-racing team. The GTX models remained the powerful and expensive Intermediate cars, but considering their serious equipment, they were still quite reasonably priced. They were considered 'special' level cars.

All Intermediate cars were all given a sleek new body. The square look was gone, and the cars now had smooth-flowing 'coke bottle' lines that emphasised the low profile and width of the cars. The grille was different for each intermediate line. The taillamps were housed in sculptured fender extensions with a sideways 'U' shaped appearance. The rear deck lid was somewhat high and slightly 'veed'. GTX models were equipped like the Sport Satellites, and had the same grille style, but also had many standard extra. They were fitted with 440-cid 'Super Commando' V-8 engines, heavy-duty brakes, suspension, battery and shocks, together with a fully-functioning dual scoop hood and GTX identification.

Despite their already sparkling performance, 446 Street Hemi packages were installed into Belvedere GTX hardtop and convertible models. 234 of these were also equipped with four-speed transmissions.

1968 Pontiac Firebird 400

Pontiac introduced the Firebird in 1967, as the Chevy Camaro-clone of the division. Both cars were designed to go head-to-head with Ford's Mustang. The first car was made at Lordstown, Ohio in early January 1967, and was officially released in the February of that year. Two basic models were launched (the two-door hardtop coupe and two-door convertible), but the cars were marketed in five 'model-options' created by adding regular production options (RPOs) in specific combinations. All Tempest and GTO powertrains were available for the car, a total of four major engine types (two six-cylinders, and two V-8s). All others were described as optional equipment. Sales were excellent, with 82,560 model units being sold.

Styling for the second year Firebird models was almost identical to the original cars, and continued to be offered as two-door hardtop coupes and twodoor convertibles. Body styling was sculptural with twin grilles of a bumperintegral design, front vent windows, and three vertical air slots on the leading edge of rear body panels. Federal safety side-marker lights made use of the Pontiac logo. The only major change was that the vent windows were now replaced with one-piece side door glass. The cars were also fitted with outside mirrors, side marker lights, E70-14 black sidewall wide-oval tires with a Space Saver spare. Cordova vinyl tops were available, as were wire wheel covers. Inside the car, apart from the standard GM safety features, front bucket seats continued to be fitted as standard, as was vinyl upholstery. The dash was simulated burl woodgrain. Astro-ventilation was now offered on the cars.

The Firebirds were now offered with seven different engine options, ranging from a base 175 horsepower six-cylinder, to a fire-breathing Ram Air II 400-cid V-8, rated at 340 brake horse power. This top option (available for an extra $616) could blow even the Mustang GT off the strip, and was fitted with twin functional hood scoops and a de-clutching fan. The V-8 options were revised to be more emissions-friendly. Technical changes to the '68 model included biasmounted rear shock absorbers and multi-leaf rear springs.

'68 Firebird sales were even better than in the launch year, with the model attracting a total of 107,112 buyers. This was despite the number of cars now competing for 'pony car' sales. Every major US car manufacturer was now in on this act. The Firebird sales were a significant contribution to Pontiac's total model year total of 910,977.

Engine: Firebird 400 V-8

Displacement: 400 cid

Horsepower: 335 at 5000 rpm

Compression Ratio: 10.75:1

Base Price: $3397

Number Produced: 90,152

1969 AMC Hurst SC/Rambler

AMC was born in 1954 after the merger of Western independents Nash and Hudson. It specialised in affordable transport for ordinary American families, and did well during the leaner times at the end of the '50s. The company was in ninth position in the 1969 car producers' league, with an output of 282,809 units.

AMC was in fact about the last car manufacturer in the US to join the power-hungry muscle car club. They launched their first two hot models in 1968, but with the help of the Hurst Performance division, and some of their components (including a Hurst four-speed gearbox) they launched the Hurst SC/Rambler as a '69 midyear model. It had been launched at the Chicago Auto Show in March '69. The car was based on the lightweight AMC Rogue body. The car was promptly nicknamed the Scrambler, and became one of the flashiest fun machines of the early muscle car era. Following the AMC tradition, the car had a wild patriotic paint job of red blue and white. The wheel centers were also blue, as was a big arrow on the hood, directing air into the rather bizarre boxy scoop (helpfully labelled 'AIR') sucking cold air down into the carburettor to maximise performance. It turned out that the car was the final Rambler American (now known simply as the Rambler), a famous nameplate that dated as far back as 1902. The Scrambler retailed at a competitive price of $2998, which was stunning value for such a high performance, high-spec auto. Average wages for a full-time worker were now $7095. AMC only intended to manufacture 500 examples of the car, but demand far outstripped this meagre supply, and they actually sold over 1500 in the model year.

One road test clocked the car doing 0-6- at just over six seconds, the quarter mile in about 14 seconds, while the car was travelling at 100.9 miles an hour. The car retained its compact outside dimensions and overall styling in line with AMC's new policy of maintaining design continuity from year to year for its lowpriced models. But the car had all kind of performance upgrades with new accelerator cable linkage, power front disc brakes, a dual exhaust system, a heavy duty cooling package and beefed-up suspension with stiffened springs and shocks. Inside, it was fitted with sporty bucket seats, a sport steering wheel and a Sun tachometer. Outside, the performance heritage of the car was emphasised by dual racing mirrors and racing style hood tie-down pins and spoke sport wheels. Sadly, although the car was a great success of its type, and many critics admired its performance, it turned out to be a one hit-wonder.

1969 Chevrolet Chevelle SS 396

Having been introduced in 1964, Chevelles soon became one of the most successful medium-priced 'senior compact' in the US. Chevrolet had introduced the car in advance of an anticipated increase in demand for cars of this type, and the model was soon selling well. The Malibu option became one of America's most popular cars.

The SS396 started out as a little-known, limited-edition model created to test the waters for a super car version of the Chevelle, and to give the Pontiac GTO some competition for this market. 201 were built in 1965, fully loaded with performance options and expensive enough to put off all but the most hardened muscle car fans. The 1966 edition was slightly less powerful than its predecessor, but priced more competitively to appeal to a wider public. It certainly worked, and over 72,000 Super Sport Chevelles hit the streets that year. The familiar SS badge finally appeared in 1970.

By 1969, the Chevelle was Detroit's best-selling muscle car, having beaten Pontiac's GTO. The car outsold its rival by around 14,000 units. The SS 396 model lines that could be equipped with the RPO Z25 now included a Malibu sport coupe, convertible and even an El Camino. The base engine for the 1969 SS393 was the 325 horsepower L35 396. Other options included the 350 horsepower '396' engine (known as the RPO L34), and the top-choice 373 horsepower L78. The L78 396 was fitted with the L89 aluminum-head option for this year only. The thinking behind this modification was to reduce the weight of the big block engine, rather than increase the power. However, the COPO (central office production order) system was used to circumvent the GM edict against the largest engines, and a 425 horsepower powerhouse, the L72 427 V-8, was fitted to around 300 '69 Chevelles. Front disc brakes were also added for this year. Externally, the cars were fitted with five-spoke sport wheels as standard (replacing the steel rims and hubcaps of earlier years. In the interior, a sport steering wheel, bucket seats and a console were offered as optional extras. SS 396 badges were proudly positioned on both front fenders, but a lot of stripes and chrome trim and black painted shadows were discontinued for this year's styling treatment.

The lower body could be painted in a contrasting color, and the D96 upper-body stripe could be added as an option. Two rather toxic special colors were also optional, Hugger Orange and Daytona Yellow for an additional $42.15.

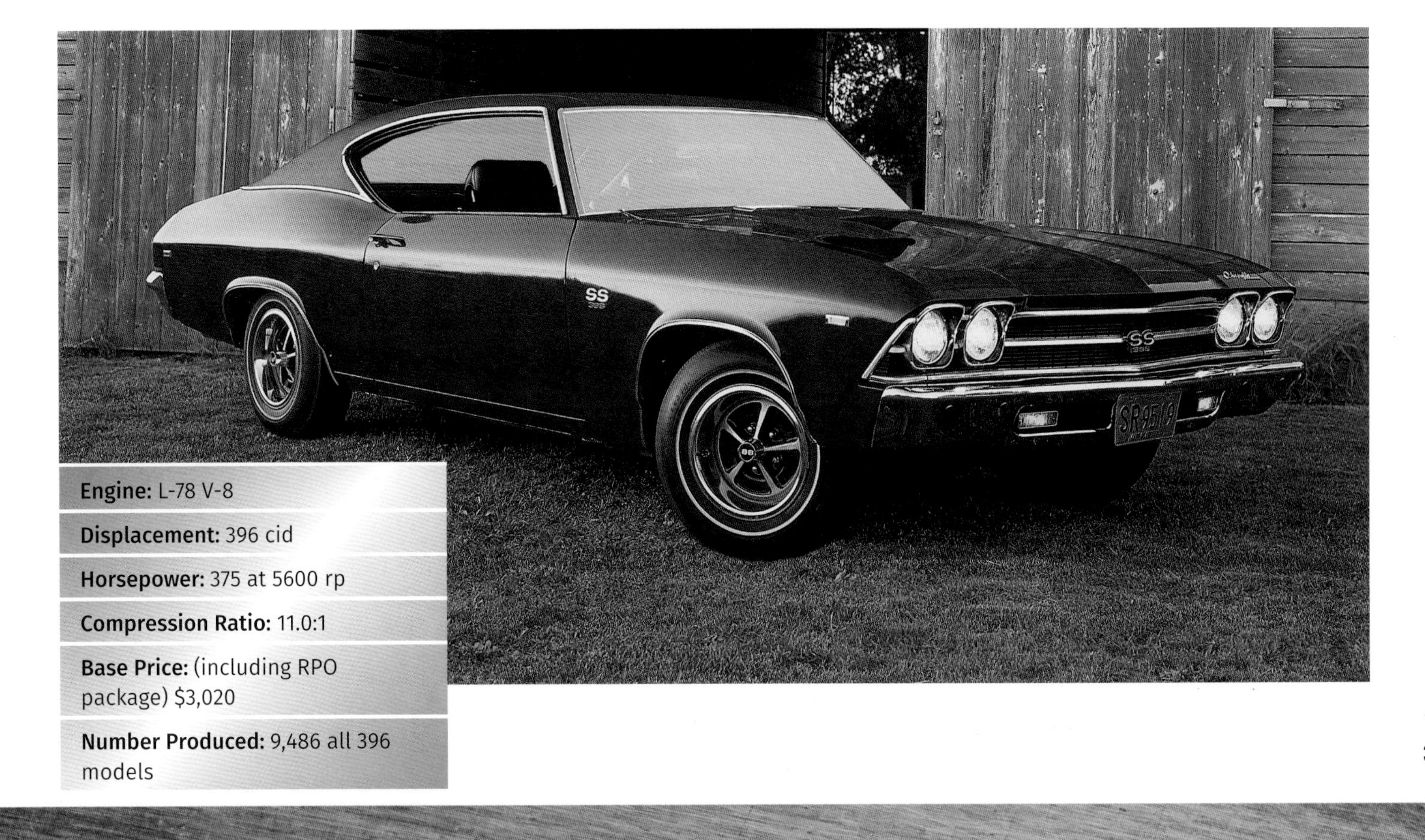

Engine: L-78 V-8

Displacement: 396 cid

Horsepower: 375 at 5600 rp

Compression Ratio: 11.0:1

Base Price: (including RPO package) $3,020

Number Produced: 9,486 all 396 models

1969 Chevrolet COPO/Yenko Camaro

Engine: L-72 Corvette V-8
Displacement: 427 cid
Horsepower: 450 at 5000 rpm
Body Style: Two-door Hardtop Sports Coupe
Weight: 3050 lbs
Wheelbase: 108 inches
Number Produced: 201 in total are known

The new 'F' body gave the 1969 Camaro a longer and lower appearance. The overhaul concentrated on the lower-body panels, which were now less smooth and more sculptural, with a side feature line. Chevy still offered sport coupe and convertible styles of the car, with their traditionally massive options list. This was quite fundamental and offered buyers the chance to customise every aspect of the car, including its exterior styling. Concealed headlights were available on the Rally Sport option, and a functional super scoop hood on the Z/28 or Super Sport.

The Z/28 was now more available, and clocked up sales of 19,014, but the engineers at Camaro, and elsewhere, continued to develop more esoteric models that were essentially race-ready cars that happened to be available to the public. Probably the most exclusive of these was the drag-ready ZL-1, equipped with an aluminium-block 430-horsepower engine, available from a few selected vehicles for a stunning $7300. Only 69 ZL-1s were built. A '69 Camaro was also the Indy Pace Car for the second year running. To celebrate, Chevrolet offered a series of replica hardtops and convertibles in the same trim style – white, with racing stripes and interiors in juicy 'Hugger Orange'.

Two engine options were available for the model, the hot 396 cid V-8, or the 350-cid block.

Corvette racer Don Yenko (who had won four SCCA national titles) had begun to customize his own hot Camaro range from his dealership in Canonsburg, Pennsylvania. Essentially, he did this by fitting bigger bore engine options than GM's head office edicts allowed the company itself to do. By '69, Chevy Performance Chief Vince Piggins was co-operating with Yenko to produce a limited range of Camaro Super Cars, circumventing the limitations imposed by the top brass by using the central office production order system (hence COPO). Yenko made the technical swaps, and then gave the cars their distinctive trim. This resulted in 201 known 'SYC' (Yenko Super Car) badged Camaros, reputedly capable of 450 horsepower, truly a classic muscle car. Yenko continued with his Chevy conversions until 1970, when GM dropped its displacement limit and started to make super cars for themselves.

Above: Yenko Super Car Camaro in Daytona Yellow.

Left: Special Don Yenko-tuned 427-cid power unit producing 450 horsepower.

1969 Dodge Charger

Engine: Overhead valve V-8, cast iron block

Displacement: 273 cid

Horsepower: 180 at 4200 rpm

Compression Ratio: 8.6:1

Base Price: $3126

Number Produced: (All Chargers) 89,704

The 1969 Charger continued to use the beautifully styled 'B' body introduced by Dodge in 1968. The styling of this classic hardtop shape is often compared to the coke bottle, being both slim and rounded. The main change in the second year was the new divided grille for this model year, and a new taillight treatment. Even though a mere 500 were built, the base engine for the model was the 225-cid Slant Six. The 318-cid V-8 was a much more popular base powerplant for these performance-orientated cars. The R/T was the sportiest of the lot; the most high-performance model in the range. This option featured all the standard Charger trim plus the Magnum 440 cid, 375 horsepower, four-barrel carburettor V-8. The car had dual exhausts with chrome tips, TorqueFlite automatic transmission and heavyduty brakes. Like our car, the model was fitted with F70-14 Red Line tires. The Charger SE was the sports/luxury model and was fitted with leather and vinyl front bucket seats, woodgrain steering wheel and instrument panel.

Coupled with the midyear-introduced R/T version of the Coronet Super Bee, the new models provided the Dodge division with a stable of fresh and exciting groundpounders. They were also members of the Scat Pack.

The Daytona Charger was an even scarier beast, a massive roof-high tail spoiler and projectile-shaped front end, gave the car shattering track performance. The latter was made from fibreglass and was fitted over the standard grille opening. Dodge was learning about aerodynamics on the job, anything to give them an advantage over the super-powerful Fords. The car could travel the super speedways at nearly 200 mph with their big 'Hemi' engines. The car was the direct ancestor of the 1970 Plymouth Superbird. The Daytona Charger car won twenty-two out of fifty-four NASCAR races for 1969, and Bobby Isaac rode the car to victory in the Daytona 500.

Ironically, the most famous Dodge Charger manufactured in 1969, is the 'General Lee' driven by the Duke brothers in the Dukes of Hazard TV series. The car was orange-painted and decorated with the famous '01' decal. In fact, the producers wrecked over 1500 Chargers filming stunts for the series. But fifteen to twenty cars survived, and are now distributed around the country.

Dodge had had its fifth record sales year in 1968, but overall car sales were down in the US market for '69, and Charger numbers fell by over 6,000 units to 89,704. R.B. McCurry remained as General Manager at Dodge.

Right: This two-tone Charger was produced in the second best year in Dodge history so far.

Below: The 440 Magnum and 426 Hemi power units were painted Street Hemi Orange.

1969 Dodge Coronet R/T

Engine: 440 Magnum V-8	
Displacement: 440 cid	
Horsepower: 375 at 4400 rpm	
Compression Ratio: 10.0:1	
Wheelbase: 117 inches	
Base Price: $3442	
Number Produced: 7328	

The Coronet R/T was one of Dodge's hot mid-size models. Powertrains were unchanged, but the new 'delta-theme' nose and tail treatments were interesting, and new grilles and taillights were included. The taillights were an unusual oval-shaped lenses. Essentially, however, the body was the same popular shape that had been introduced in '68, when the R/T version was introduced. Just 7,328 R/Ts were built, including the convertible version. As well as the base level Deluxe, there was also high-performance Super Bee model Coronet with 'Scat Pack' stripes highlighting the restyled rear. As the advertising slogan went, the Scat Pack models were 'Super Cars with the bumblebee stripes'. Both cars continued the honorable Mopar muscle tradition of offering the best in style and grunt for a modest price tag.

The big performance news for the '69 Cornets was the Magnum 440-cid 'six pack' V-8 engine. This engine was created by taking an existing 440 block, and replacing the single four-barrel carburetor with three two-barrel carburetors. The revised engine was also fitted with a fiberglass performance hood, and Ramcharger fresh-air induction was an option (this was standard on Hemi engines, that were available for a further $418).

The R/T continued as the highest-performance Coronet and the most expensive model in the range. It included all the features of the Coronet 500 but could also be fitted with a huge performance pack that included the Magnum 440-cid and TorqueFlite automatic transmission.

Coronets were also available with 426-cid Hemi V-8s, but these were more rare (around 166 two-door coupes and 92 door-door hardtops). This latter 'big block' engine option had been introduced by Chrysler in 1964 and proved that the company understood horsepower and all its implications as a production car seller. Outside, the cars were distinguished by bright sill moldings, R/T nomenclature (now in the rear black Bumblebee stripe rather than on the front fender, across the trunk lid and down the fendersides), twin simulated air scoops were also located on the rear fenders, just ahead of the rear wheelwell openings on each side of the car, and two intakes on the 'Power Bulge' hood.

The car's best dragstrip performance was 12.25 seconds for the quarter mile, at 112 mph. No wonder 'it looks guilty'.

1969 Dodge Dart Swinger

Engine: V-8	
Displacement: 340-cid	
Horsepower: 275 at 5000 rpm	
Compression Ratio: 10.5:1	
Body Style: Two-door Coupe	
Base Price: $2879	
Number Produced: 20,000	

Dodge was at seventh in the auto producer's league for 1969, producing 611,645 cars. Despite a small fall in sales, the general upturn for the company was hailed as 'Dodge Fever'. Dart continued as the compact Dodge model, and the Swinger was the sports option for the base level Dart. The car was proudly introduced as the 'Newest Member of Dodge's Scat Pack'.

Taking a cue, no doubt from the infamous hipsters Martin, David, Sinatra, Lawford and Bishop – collectively known as the Rat Pack – Dodge planners decided that their stable of hot cars fit the good times, bad boy image of the original Packers quite well. With a reverent nod to the Vegas crew, Dodge fired up the coals and branded its performance screamers the Scat Pack. Like all good branding, the program needed a recognisable image to hold it all together. A rascally bumblebee replete with crash helmet, goggles and racing tires fit the bill nicely. The cars were also decaled with dual bumblebee stripes in black, defying the traditional industry trend to run racing stripes the length of the car body. The marketing initiative heralded the Dodge muscelcars, including the Swinger, as a specific group. Effectively, the Swinger 340 was a modestly priced sports car, and proved to be a popular option for the economy-mind performance enthusiast. The car could be fitted with a whole gamut of Dodge powerplants, from the basic six-cylinder to the 383 Magnum V-8. The 340 was devoid of any creature comforts, like carpets, but was fitted with a performance option package that included the 340-cid 'high-winding, 4-barrel' V-8, Rallye Suspension, Firm Ride Shocks and a 'Power Bulge' hood with die-cast louvers. The car was also loaded with a manual four-speed gearbox with a Hurst shifter and bumblebee stripes. In fact, this all conspired to make the car a red-hot performance option, and one of the high points of the sixties Super Car Era. It could cover the quarter mile in fourteen seconds. Despite the performance heritage that went into the car, the styling had let the model down until the de-chroming of the '67 model year, and seemed disparate with the youth movement the car was aimed at, but the whole package looked great in the '69 restyle. As the slogan went: 'If you can find a hotter car (for the money) buy it!'

1969-1970 Ford Boss 302 Mustang

Like its 'big brother' the Boss 429, the Boss 302 was developed for racing – in this case for the SCCA Trans Am series. In developing the car, Ford probably ended up with the best handling, ultimate muscle car. The Ford engineers made sure that the car could corner, Bunkie Knudsen had demanded a car that was 'absolutely the best-handling street car available on the American market', and had plenty of power to call upon. The car was designed to be light, and fitted with an effective powerplant, it gave an excellent power-to-weight ratio. The car not only handled well, it could plough a straight furrow, fast. Boss 302s could cover the quarter mile in 14.62 seconds at 97.7 mph and reach 60 mph in 6.5 seconds. Top speed was 118 mph. The power was delivered by a 302-cid V-8, rated at 290 horsepower, and the car could deliver 290 pounds per feet of torque at 4300 rpm.

The Boss 302 listed at a base price of $3,'720, almost $1000 more that the base price Mustang. Some would argue that this was cheap at the price, as the car on offer was virtually ready to race out the showroom, and included all kinds of race-fit options such as quick ratio steering, competitions suspension (designed by engineer Matt Donner), E70x15 fiberglass belted tires, front discbrakes and four-speed manual transmission.

The car owed its inception to the broad-minded Bunkie Knudsen, who had been GM's youngest ever General Manager at Pontiac. He oft-quoted maxim was 'You can sell a young man's car to an old man, but you'll never sell an old man's car to a young man'. He believed that young men hankered after hot cars, the hotter the better. He view was that the market was full of ponycars, but that what distinguished the successful cars was power and performance. The Mustang was in his sights. Knudsen had poached plenty of talent from his old bosses at GM, and one of them was designer Larry Shinoda, who had created the sporty image of the Z/28 Camaro, and had worked on the Corvette Sting Ray. Knudsen wanted him to out-Camaro the car over at Ford, and the car was ultimately named after him – the 'Boss'. Shinoda styled and named the car, but Kar Kraft in Brighton, Michigan developed the prototypes, and delivered the first in August 1968.

On the track, honors were fairly even between the Boss and Z/28, who shared similar-sized engines. Camaro took the SCCA championship in 1969, and Parnelli Jones, driving the Ford, won in 1970.

Engine: Boss V-8

Displacement: 302-cid

Horsepower: 290 at 5800 rpm

Compression Ratio: 10.5:1

Body Style: Two-door Fastback Coupe

Number Produced: 7,013

1969-1970 Ford Boss Mustang 429

As part on its on-going commitment to motor racing, the Boss 429 car was designed specifically to homologate Ford's 'Shotgun' 429-cid engine for NASCAR racing, whose rules insisted that at least 500 production cars should be sold for each racing model. Ford had introduced the big-block engine family back in 1968 for the benefit of its big, luxury models. The engine was revised for racing by having its block recast as a reinforced cylinder block, topped off with cast-iron heads featuring hemispherical combustion chambers. Sound familiar? Yes, it was also known as the 'Semi-Hemi'.

The limited production run of the Boss 429s was built at Kar Kraft in Brighton, Michigan. The production car was based on the Sportsroof Mustang, and Ford delivered several of these to Kar Kraft for extensive modifications that would enable the car to accept the larger engine block, and would beef up the suspension. Kar Kraft launched the first Bosses in January 1969, three months ahead of the 'little brother' Boss 302s that Ford themselves were working on at the same time. The cars retained the Mach 1 interior styling, and basic outside appearance of the Sportsroof Mustang (including the rear roof pillar medallions and fake rear quarter scoops), but any resemblance to these other Fords was purely coincidental. The cars also had flared front fenders to supply extra clearance for the standard F60 Wide-Oval rubber mounted 15x7 Magnum 500 wheels.

As well as the bigger engine, the Boss 429s were also fitted with an upgraded engine oil cooler, larger sway bars, a rear-mounted battery (to make more room up front) and a close-ratio Manual transmission. Other go-goodies included forged steel connecting rods, power front disc brakes, power steering and staggered Gabriel shocks in the back. The cars retailed for $1,208 over the base price of a Mustang Sportsroof, and could achieve a top speed of 118 mph. They could also cover the quarter mile in 14 seconds, and get from 0-60 mph in 7.1 seconds.

Engine: V-8 with aluminum cylinder heads

Displacement: 429-cid

Horsepower: 375 at 5200 rpm

Transmission: Close-ratio four-speed manual

Compression Ratio: 10.5:1

Fuel System: Single 735-cfm Holley four-barrel carburetor

Body Style: Two-door Fastback Coupe

Number of Seats: 5

Weight: 3870 lbs

Wheelbase: 108 inches

Base Price: $4868

Number Produced: 857

1969 Ford Galaxie XL 6T

Engine: Six-cylinder, overhead valve	
Displacement: 250-cid	
Horsepower: 150 at 4000 rpm	
Compression Ratio: 9.2:1	
Base Price: $3052	

The 1969 Ford lines were introduced to the public in September '68. For this model year, FoMoCo was America's number-two automaker. Semon E. Knudsen was President of the company, and was determined to continue promoting Ford's high performance image. The company actively competed in the '69 NASCAR races, and won the series with a total of 26 Grand National victories. FoMoCo also won the Riverside 500 Grand National Race with Richard Petty driving. This sportiness pervaded models in all Ford '69 series.

The Galaxie 500 was the intermediate Ford model for 1969, having all the trim of the Custom series, plus stamped aluminium lower bodyside moldings. The cars were advertised as being 'Bigger, Wider, longer and Quieter' than the '68 version, and they were. The new Ford range grew in size and was even more luxury-orientated than in previous years. They were built with passenger comfort in mind. The Galaxie models were still available, as were the LTDs, in a range of body styles spanning the attractive Sportsroof fastback, convertible, hardtop and station wagon models.

For this model year, the Ford power team range consisted of the 240-cid six-cylinder engine, the 302-cid V-8with 220 horsepower, the 390-cid V-8 with 265 horsepower, the 429-cid Thunder Jet V-8 with 320 horsepower, the 390- cid V-8 with 265 horsepower, the 429-cid Thunder Jet V-8 with 320 horsepower and the big 429-cid Thunder Jet 4V with a massive 360 horsepower.

Transmission types for the full-sized Fords were three-fold. They included the three-speed manual, the 'four-on-the-floor' manual up-grade and the Select Shift Cruise-O-Matic three-speed automatic.

Our car is a Galaxie 500 XL, which was the sport trim version of the '69 Galaxie, coming in both 'sportsroof' (two-door fastback coupe) and convertible styles. Standard equipment included bucket seats (although bench seats were also available), wheel covers, die-cast grille and retractable headlights, pleated all-vinyl interior trim and five vertical 'hash marks' at the forward part of the front fenders. Naturally enough, the 'six' 500 XL subseries came complete with the 240-cid six-cylinder engine option.

When it came to the exterior trim, the buyer had their choice of fifteen single-tone and twenty-four two-tone color schemes. A wide range of interior trim and optional extras was also available. Some of the most popular were SelectAire air conditioning, tinted glass, six-way power front seats, power windows, a tilt steering wheel and rear window defogger.

1969 Ford/Shelby Mustang GT 350

FoMoCo hitched up with ex-racer Carroll Shelby in late 1964 to build an allnew sports car, based on the Mustang fastback. The goal was to create a car with the grunt to compete directly with the Chevy Corvette. The first Shelby GT350 was launched on January 27 1965. At first glance, the car looked just like an ordinary fastback Mustang, but that was the only similarity. The car's drive train and technical equipment was all seriously beefed-up, and the engine power raised considerably. Effectively, the GT350 was a full-bred race car that could be purchased direct from a Ford dealer. The car underwent some minor changes in '66, which Ford directed in an attempt the make line financially viable. This meant making the specifications of the car closer to those of the production Mustangs. Hertz ordered 1,000 GT350s in '66, which were known as Hertz Shelby GT350s, and had gold-stripe decals.

In 1967, the Shelby body styling was quite a radical departure from that of '66, with the extensive use of fiberglass to modify its appearance. The car was given a big rear spoiler and dual hood-mounted air scoops. The driving lights were moved to the outside of the grille. '67 also saw the introduction of the Shelby GT500s, equipped with a big-block 427-cid Medium Riser engine. The range was virtually unchanged for '68 but the cars were now known as the Shelby Cobra GT350 and GT500. A production convertible was now added to the model range, and the Cobra Jet 428-cid engine was introduced.

1969 saw the introduction of the final Shelby Mustangs (though some of the cars were held over until 1970). But, effectively, this was the last year of this fine heritage line-up. This was due to a change of thinking at Ford, who now believed that sales were less relative to high-performance marketing. Engine availability was changed from '68. The 302-cid V-8 was still fitted as standard, but the GT350s were also offered with a 351-cid four valve Ram Air cooled engine. The GT500s were offered with a single engine option, the Ram Air cooled 428. This was also the year of 'scoops' – the NASA styled hood was equipped with five, there was one on both front fenders, and one on both rear quarter panels. The Lucas driving lights were now mounted below the front bumper and the dual exhausts now exited through a massive aluminum outlet mounted in the center of the rear valance panel. The cars were fitted with five-spoke Shelby mag wheels. Technically, the car was equipped with power disc brakes and steering. Safety was accommodated with an integral rollbar and shoulder harness seat belts.

Engine: V-8
Displacement: 351 cid
Horsepower: 290 at 4800rpm
Compression Ratio: 10.7:1
Wheelbase: 108 inches

1969 Ford Torino Talladega

The 1968-69 Fairlanes were a new body style altogether. Depending on the model, the new cars were designated as either Torinos or Fairlanes... or Fairlane Torinos. Torinos were at the higher end of the range, and the performance options. Torino Talladegas were made by Ford for the 1969 model year to compete in NASCAR racing. Like Chrysler with its winged cars, Ford had turned it attention to the potential of aerodynamics. The car was based on the Torino SportsRoof model, but the Talladegas had fared-in headlights and a more aerodynamic grille design.

So that the car could qualify for the NASCAR series, FoMoCo was obliged to build at least five hundred examples of the car and to put them on general sale to the public. Around 754 cars of this type were actually built, making the Talladega one of the most rare and exotic performance cars ever made. The Mercury division of FoMoCo built a similar model based on the Mercury Montego model that was called the Cyclone Spoiler II.

The original racing version of the cars was fitted with the 427-cid racing engine at the start of the NASCAR season, but this was later switched to the semi-hemi 429-cid racing engine. The racing version of the car, known as the Talladga was designed to raise the profile of the SportsRoof (Fastback Coupe) Torino model and boost their sales. More than this, the Talladega was to become a racing legend.

NASCAR race-inspired models put on sale to the public had the 335 brake horsepower Cobra Jet 428 engine, this production model was sold as the Torino Cobra. These cars were fitted with some of the most fierce 'showroom stock' performance engines ever built, together with four-speed manual transmission as standard equipment, 735-cfm carburetion, dual exhaust, beefed up suspension and F70-14 wide oval tires. These cars were part of the Fairlane Torino GT series, which was the sporty version of the Fairlane 500 series and included all of the features of the less expensive cars plus larger engine options, factory bucket seats and console. They were available as two-door hardtop or car with fastback roofline. A ram-air functional air hood scoop was available, as was a 'Traction-Lok' differential.

From the beginning, the car was recognized as a 'screamer', and is a rare collector's car to the present day.

Engine: Cobra Jet V-8, overhead valve

Displacement: 428-cid

Horsepower: 335 at 5200 rpm

Transmission: Four-speed manual

Compression Ratio: 10.6:1

Base Price: $3680

Number Produced: 574

1969 Mercury Comet Cyclone Spoiler II

The Comet had been restyled for 1968. It is described as looking like a fullsize Mercury that had gone on a diet. The new car had a horizontal grille, rocker panel moldings, side market lights and chrome-encase vertical taillights. Among the standard safety features for the car were an energy absorbing steering wheel and column, padded dashboard and sun visors, plus front and rear seat belts. Two speed windshield wipers and washers were also fitted. '68 Cyclones had a mid-tire level body tape stripe. Those with the GT option had an upper body racing stripe, bucket seats, wide tread whitewalls, special wheel covers, all-vinyl interior and a special handling package.

The '69 Comet Cyclone was lightly face-lifted, and only the Fastback body style returned for this model year. The GT option was reduced to just an appearance group option. The hot new Cyclone was the CJ. The Cyclone had a special blacked-out grille, Cyclone emblem, wheel well opening moldings and a dual exhaust. The car came with the 428 Cobra Jet rated engine, which developed 335 horsepower. It was also fitted with the Competition Handling Package, and a plain bench seat interior.

But the hottest of the hot was the Cyclone Spoiler II, which was modified to improve its aerodynamics for NASCAR drag racing. A total of 519 Spoiler IIs were built.

Spoilers came in two trim versions, both named after Mercury NASCAR drivers. The Dan Gurney Spoiler had a dark blue roof and striping, with a signature decal on the white lower portion of the car. The Cale Yarborough edition featured red trim and a signature decal. These Spoiler signature editions had Windsor 351-cid engines, rated with 290 horsepower, and FMX Cruise-O-Matic.

The Cyclone was based on the Mercury Montego for 1970, and thus became bigger and heavier. It is said that this changed the car from a true contender to a full size pretender... The model was discontinued in 1971.

Engine: Windsor V-8	
Displacement: 351-cid	
Horsepower: 290 at 5200 rpm	
Transmission: FMX Cruise-O-Matic	
Fuel System: Autolite four-barrel	
Body Style: Two-door Hardtop	
Number of Seats: 6	
Weight: 3273 lbs	
Wheelbase: 116 inches	
Number Produced: 519	

1969 Oldsmobile Hurst

Engine: V-8 overhead valve, cast iron block

Displacement: 455 cid

Horsepower: 380 at 4800 rpm

Transmission: Hurst Dual-Gate Shifter

Fuel System: Rochester 4GC four-barrel

Weight: 3900 lbs

Wheelbase: 112 inches

Base Price: $4500 - $4900

Number Produced: 906

By the end of the sixties, Oldsmobile's market share was wavering slightly, they were now at sixth position in the auto manufacturers' league, and would begin the seventies at seventh position.

Continuing their muscle-car collaboration, Hurst/Oldsmobile again offered a Cutlass-based hardtop coupe. The car was a fine example of a pure-bred muscle car. Hurst was (and still is today) one of the premier U.S. gearbox manufacturers. When they teamed their expertise with that of Oldsmobile, a truly memorable product was born. The car was based on the Cutlass 4-4-2 two-door 'Holiday' hardtop, and was equipped with a massive 455-cid V-8 engine, and had a special Firefrost Gold and White paint scheme. The engine power output was a tasty 380 horsepower. Nine hundred and six cars were produced for this model year.

The Cutlass models were offered in three ranges for '69, the base Cutlass, the Cutlass Supreme and the 4-4-2s. The Cutlass was the next step up the Oldsmobile ladder from the base level F-85 models. The two-door Cutlass SS had a shorter wheelbase than the four-door models. A potent W-31 performance package was available for both the Cutlass and Cutlass Supreme models, as performance became important throughout the range. It was based on a 'Force-Air' inducted 350-cid V-8 with special equipment included hood louvers, an insulated fibreglass hood blanket, nylon blend carpeting, special molding package and Deluxe steering wheel. Upholstery was vinyl or cloth.

Cutlass Supremes were the ultimate luxury in the range. This three-model series was one of Oldsmobile's best sellers and was equipped with standard supreme equipment plus a few luxury extras.

The 4-4-2 series was now in its second year. Not only could the buyer for this series order the classic 4-4-2 option, but they could also order the superstock drag racing package, the W-30. Properly tuned, these cars were among the fastest show room stock cars available on the domestic market. The Hurst/Oldsmobile two-door hardtop was included in this four-model range, and was the most expensive option by far.

1969 Plymouth 'Cuda 440-cid

By the late 1960s, the Plymouth Barracuda was a quintessential muscle car, and the bigger engined versions were difficult in insure. But the original car was quite a different, milder creature all together. The model was launched in '64 as a fastback version of the Valiant Sedan, and aimed at young, sporty American drivers. These sound just like Mustang drivers... Handily, the car was fitted with a fold-down rear seat that offered 7 feet of fully carpeted 'anything' space.

The original model was by no means hot, available with a single 273-cid V-8, developing a modest 180 horsepower. Eve so, a four-speed manual gearbox with a Hurst shifter was available. Handicapped by this low performance rating, the car got off to a rather slow start, selling only 23,000 in their first eight months of production.

Luckily, Plymouth soon realized where they were going wrong, and revamped the model considerably. The V-8 remained at 273cid, but with a four-barrel carburetor and higher compression ratio, the car could now push out 235 horsepower, and was tested at 0-60mph in 8.2 seconds. The car was also offered with a Formula S performance package, with stiffer suspension, wide wheels/tires and rally stripes. The car now appealed to the market far more, and sold 64,000 in this incarnation.

For '67, the Barracuda became a line in its own right, and had a bigger and hotter engine – the 383-cid, 280 horsepower engine, coupled with a Carter four-barrel carburetor and front disc brakes.

By '69, the 'Cuda was available in three models (two-door hardtop, two-door fastback and two-door convertible). The cars were hardly changed from the '68s, except for slightly face-lifted hoods and grilles. The cars also had new side marker lamps that were rectangular. The standard engine V-8 engine option was the 318-cid, but the tried-and-tested 225-cid Slant Six was also on offer. However, higher performance options were also available, including the 440-cid Hemi V-8. The cars were fitted with standard government-required safety equipment (deep dish steering wheels and padding), all-vinyl interiors, bucket seats, Pit-stop gas cap, rally lights and red or white stripe tires.

Engine: GTX V-8

Displacement: 440-cid

Horsepower: 375 at 4600rpm

Compression Ratio: 10.1:1

Weight: 2987lbs

Wheelbase: 108 inches

Base Price: $2813

Number Produced: 17,788

1969 Pontiac GTO Judge

Pontiac's still at number three, and 'Here come da Judge' – a new $332 option package for Pontiac's '69 GTO, complete with 366 horsepower, Ram Air III induction, 400-cid V-8. For $390, the buyer could step up to 370 horsepower with Am Air IV. With this option, the car could go from 0-60 mph at 6.2 seconds, and cover the quarter mile in 14.5. The car was decorated with loud 'The Judge' decals, and was a 'hairy' looking car indeed. The first 2,000 Judges were finished in a true pop-art color, Carousel Red, and came complete with extrovert spoiler, and stripes. This in-your-face decorative style was becoming a muscle car tradition, and buyers were almost as interested in the 'look' as they were in the power under the hood.

The Judge model was introduced to try and halt the decline in GTO sales. The LeMans-based car was falling between two stools – it wasn't a budget muscle car, but neither was it a premium model. Other Pontiac models, such as the HO-powered Tempest developed almost as much power, and were cheaper to insure. An important consideration for the young drivers these cars were aimed at. The Judge was designed to be a top-of-the-range GTO, to mop up customers who were looking for something with more style and power than the regular models. For an extra $332 over the price of a standard GTO, the car came fully loaded with Ram Air III induction (later upgraded to Ram Air IV for $558), the '68 400-cid HO power unit, three-speed gearbox with floor shifter, (automatic was optional), Rally II wheels, heavy duty suspension and 14-inch tires. The cars were offered in hardtop and convertible forms, but the convertible sold only a meagre 108 units, whereas the hardtop scored nearly 7000 sales. Considering that the convertibles were over a thousand dollars more, this isn't too surprising. As well as the standard equipment, Pontiac also offered the cars with an enormous range of options, ranging from the handy to the downright bizarre (including a litter basket in several color choices, and a reading lamp). Inside the car, carpeting was fitted as standard, and there was a choice of bucket or notchback seats.

Although the Judge was now the best-equipped muscle car available for the money, it was only temporarily successful in halting the slide in GTO sales. The car was reasonably successful in its own right, but its image did not seem to reflect on the rest of the range, and GTO sales slid by nearly 20 percent to just over 72,000 for the '69 model year. Worse was to come. Sales really slumped as the '70s arrived, falling to fewer than 40,000 in 1970, and to just over 10,000 in '71. This was largely due to the spiralling insurance costs generated by the car's own scary performance image.

Engine: Ram Air V-8

Displacement: 400-cid

Horsepower: 370

Number Produced: 6725

1969 Pontiac Trans Am

Apart from the Corvette, The Pontiac Trans Am Firebird has the longest heritage of any American muscle car, its production running uninterruptedly from the mid-sixties right up to the new millennium, and was cancelled only recently. Even GM cousin, the Camaro, can't equal this record, as their production was stopped between 1974-77.

The original Trans Am debuted in 1969, originally as a low-production specialty vehicle with stock car racing the driving factor in their conception. They were designed to compete in the Sports Car Club of America's 'Trans Am' American Sedan Championship races. Get it?

But, as always, manufacturers had to retail a certain number of cars to make them race legal, and had maximum displacement criteria of 305-cid. The Mustang had been an early leader in this race series, and was ideally equipped to compete, with its 289-cid small-block V-8. The cars dominated the races, until the Mustang's performance was out-stripped by that of the Chevy Z-28 Camaro. The series effectively developed into a showcase for America's latest pony cars, and all the major manufacturers started developing and selling 'homologated' hot rod cars, basing their investment on the adage 'Win on Sunday, Sell on Monday'.

Pontiac Motor Division had introduced the muscle car to the American scene, with the Trans Am Firebird in 1968. The car was launched to the public in '69. Although the manufacture started conventionally, as a route to race legal status, the cars soon became sought after in their own right, and became a long-running success. Demand initially outstripped supply, but as soon as it caught up, the cars raced out of the doors. Ironically, the engines in the first Trans-Ams were too big (at 400-cid) to be SCCA legal in any case, but the street-racers loved the car, and it sold in big numbers right from the start. The classic 'Trans Am Performance and Appearance' cars were offered in blue-accented Cameo white paint jobs, ram air hoods, flat racing spoilers and fender-mounted air extractors. Technically, the cars were equipped with a strengthened chassis, stiffer shocks, Safe-T-Track differential and front disc brakes. The power unit was the 335 horsepower Ram Air II 400-cid V-8, and a bigger 345 horsepower Ram Air IV option was available for extra cost. Transmission options included a Muncie four-speed manual, Turbo-Hydramatic or the standard heavy-duty three-speed manual.

1970 AMC AMX

Engine: V-8	
Displacement: 390-cid	
Horsepower: 325 at 5000 rpm	
Compression Ratio: 10:1	
Number Produced: 4116	

The AMX (American Motors Experimental) was a two-seater car, based on a shortened (97 inch) version of the Javelin wheelbase. It was launched in 1968. The AMX was one of the best-handling cars of the era, and being relatively light, was also nimble, capable of quarter mile times of 10.73 seconds at 128 mph in its super stock times. American two-seater sport cars were incredibly rare during this time, the only well-known model of this kind being the Corvette. Sadly, the hot little AMX was destined for only a three-year lifespan. Its wheelbase was also slightly shorter than that of its long-lived rival.

AMC loaded the Kenosha, Wisconsin-built AMX with top performance goodies, such as heavy-duty suspension, sway bar and shocks. The highest option engine for 1970 was the 390-cid V-8, now rated with 325 horsepower. Interiors were fitted with woodgrain dashboard and woodgrain racing steering wheel, complete with AMC roundel. Having no room for a back seat, the rear area was simply carpeted.

The original AMX concept dated back to 1966 also featured a two-seater cabin, and was body-built in fiberglass. It was this prototype that inspired the steel-bodied production spin-off. When the real AMXs were revealed at the February 1968 Daytona Speed Week, they attracted rave reviews. Styling guru Richard Teague was responsible for the car's startling and yet timeless good looks, and lovingly referred to his creation as a 'hairy little brother to the Javelin'. Group Vice-President Vic Raviolo was responsible for the concept development, and referred to the car as a 'Walter Mitty Ferrari'. Like the other pony cars (the Camaro, Cougar and Firebird), the AMX followed in the venerable hoofprints of the Mustang. But the crisply defined body styling of the AMX, with its long-hood/reallyshort- deck profile almost succeeded in redefining the image of the ponycars. The cars also offered good performance and AMC were very hopeful that this style/power combination ensured good prospects for their 'runt'.

The company projected sales of 10,000 for 1968 and 20,000 for each ensuing year. But sadly, AMX sales never took off, and even a revamp for the 1970 model year (including a sexy power bulge on the hood, and simulated 'sidepipe' rocker trim) could not reverse the sagging sales figures for the model.

1970 AMC Rebel Machine

AMC were at tenth position in the auto producers' league in 1970, on an output of 242,664 cars. The company was determined to continue its muscle car legacy, and launched a new, gaudy muscle car, the 'Rebel Machine' out of nowhere. Their first muscle car offering was the 1966 Rambler Rogue, fitted with the Typhoon V-8. This engine was fitted in the Marlin the following year, but it wasn't really until the AMX was launched in '68 that AMC joined the exclusive muscle car club.

Through the development of these models, AMC had acquired a reputation for the ability to create eye-catching, high-performance machines at bargain prices. The Rebel Machine was a classic example of this combination. The car was developed with the input of the Hurst Products Company, and was based on the relatively ordinary Rebel model. But at least 1,000 Machines boasted a flamboyant red, white and blue paint scheme, and a massive shoebox-like hood. Later editions of the model were finished in a choice of solid colors and featured a blackout hood treatment with silver pin striping, plus optional red, white and blue graphics that could be added on to the grille and body.

AMC launched the car at the National Hot Rod Association's World Championship Drag Race in Dallas, Texas, during October 1969.

What AMC had saved in trim and appearance goodies for the Rebel Machine (though they were fitted with cool wheels) were spent to good effect on performance extras. Customers got a Ram Air induction set-up, heavy-duty shocks, and springs, front and sway rear sway bars, four-speed close ratio manual transmission complete with four-speed Hurst shifter and power front disc brakes. Power was supplied by AMC biggest ever publicly offered power plant, the 390-cid V-8, rated at 340 horsepower and capable of a quarter mile time of 14.4 seconds at 98 mph. Rather bizarrely, the car was also equipped with a hood-mounted tachometer. The interior was fitted with high-back bucket seats.

However, Rebel Machine sales proved disappointing, so this proved to the one and only year that the car was offered as part of the AMC line-up.

Engine: V-8

Displacement: 390-cid

Horsepower: 340 at 3600 rpm

Compression Ratio: 10.0:1

Body Style: Two-door Hardtop Coupe

Weight: 3650 lbs

Wheelbase: 114 inches

Base Price: $3475

Number Produced: 1936

1970 Buick GSX Stage 1

Engine: Stage 1 V-8	
Displacement: 455-cid	
Horsepower: 360 at 4600 rpm	
Transmission: Turbo Hydramatic automatic	
Compression Ratio: 10.5:1	
Weight: 3920lbs	
Wheelbase: 112 inches	
Base Price: $3920	
Number Produced: 400	

Opposite

Top: This Saturn Yellow GSX looks like a real muscle car, despite its gentlemanly Buick background.

Below left: The two air intakes that received air ducted through the hood scoops, feeding the Rochester four-barrel carbs.

Below right: The tachometer was mounted on the hood!

By the end of the '60s, Buick had loyal followers in many sectors of the American car market. Never a huge part of the Buick sales picture, but still catered to with great care, were Buick's performance customers. The hi-po market at this time was dominated by GM, whose Pontiac GTOs and Oldsmobile 4-4-2s had started the whole supercar race. Buick followed in their footsteps with the launch of the first of their long-running Gran Sports models, based on the mid-sized Skylark. The Gran Sports cars were built in Flint, Michigan, where Buick constructed their classier models. But the gentlemanly image of these 'better' cars didn't mean that Buick weren't serious about tapping into the muscle market. This was particularly evident in their Stage 1 GSX engine package.

Stage 1 GSXs were blessed with a great combination of good muscle car looks and zinging performance. Only coming as hardtops, the package started out as a 1970-71 option for the Gran Sport 455 coupe and was developed to offer what is now considered to be one of the muscle car era's highest profile image treatment packages. This featured two special paint colors (Apollo White and Saturn Yellow), color-coordinated headlight bezels, wild hood and bodyside stripes in black and red, front and rear spoilers, hood-mounted tachometer, dual racing mirrors and G60-15 Wide-Oval tires. Interiors were only available in all black, with bucket seats, consolette, Rallye steering wheel, gauges and a Rallye clock.

The standard power plant for the GSX was a 315 horsepower 400-cid V-8. But the Stage 1 version of the car was equipped with a modified Rochester Quadra-Jet four-barrel carburettor (topped with a ram-air hood) and a hotter camshaft so that the car developed 360 horsepower. This equipment enabled the car to cover the quarter mile in 13.38 seconds, and caused Motor Trend Magazine's Bill Sanders to remark that the car was like an 'old man's car inbred with a going street bomb... performance verges on a precipitous mechanical hysteria'.

In fact, the car was loaded with performance features. Go-goodies included the Buick Rally Ride Control Package, Positraction rear axle, four-speed manual transmission with Hurst shifter/Turbo Hydramatic, heavy-duty suspension and power front disc brakes.

The GSX carried a hefty price tag, and was effectively a $1,199 option package on the GS455. It was a special limited edition model, and sold only 678 cars in 1970, including 400 with the Stage 1 package (only 124 in '71). Many commentators feel that the car was held back by Buick's staid image, and reflected on how much better it would have sold as a GTO. Today, however, the GTX is a rare classic, and one of the most soughtafter seventies Buicks.

5610 U
OHIO 70
GOODYEAR
EAGLE

STAGE 1
GS 455

10 20 30 40 50 60 70 80
RPM
BUICK

1970 Chevrolet Chevelle LS-6 SS 454

Engine: LS-6 V-8	
Displacement: 454-cid	
Horsepower: 450 at 5200 rpm	
Transmission: Muncie 'Rock Crusher' M-22 four-speed manual	
Compression Ratio: 11.25:1	
Fuel System: Single four-barrel carburetor	
Body Style: Two-door Coupe	
Number of Seats: 5	
Weight: 3885 lbs	
Wheelbase: 112 inches	
Base Price: $4475	
Number Produced (454s): 4,475	

For Chevy fans, the 1970 Chevelle SS 454 represents probably the ultimate Chevy muscle car. Many others would agree with them. By 1970, virtually all US manufacturers were scrapping it out in the cubic inch war, and Chevy decided to up the stakes by offering the ultimate big block 454- cid V-8. It was the supreme evolution of the breed.

This engine came in two versions, the LS5 was rated at 360 horsepower (and 500 pounds per feet of torque), and the LS6 was rated at 450 horsepower (which had the same torque rating). LS6s were capable of covering the quarter mile in 13.8 seconds at 97.5mph, and could go from 0-60 mph in six seconds. It was a big, beefy engine that Car Life Magazine described as 'the best supercar engine ever released by General Motors'. Direct competition to this model was limited, mainly to the 429-cid Ford Torino and the 440-cid Plymouth Road Runner. Around 3,773 SS 454s were built for this model year.

In fact, many muscle car lovers believe that the Chevy offering should wear the crown as the king of the muscle. From a sales perspective, it is difficult to argue against the Super Sport Chevelle models, as they took over the GTO's slot as the muscle car best seller in 1969. This success was partly due to the fact that the car offered good performance at a keen price in its base form – it didn't need to have a lot of expensive optional extras bolted on to get where it was going... fast. The car was praised for having the potential to race, whilst still being a perfectly practical family car (when fitted with the handling package and brakes). Indeed, it was a car that was 'kind to women and children'. GM had finally abandoned their cid-restriction for intermediate-sized models in 1969, and this had given the engineers a whole new lease of life. The LS6 engine was produced at the Chevrolet big-block plant in Tonawanda, New York, and every component was produced with super performance in mind. The crank was a tough forged steel piece, cross-drilled to ensure ample oil supply to the connecting rod bearings. The rods were also made from forged steel. Its horsepower rating was the highest ever assigned in the true muscle car era of the sixties and seventies.

Driving a 45-horsepower Chevelle is like being the guy who's in charge of triggering atom bomb tests. You have the power, you know you have the power, and you know if you use the power, bad things may happen. Things like arrest, prosecution, loss of licence, broken pieces, shredded tires, etc.' claimed a Super Stock report. Good grief.

Opposite: Deliberately muted exterior and interior styling belie the car's true performance.

Below: Subtle badge notes the cars 454 cubic inches.

Below right: Cowl induction is when the back of the hood next to the windshield is open so that the engine can be fed by cool high pressure air that builds up there when the car is moving.

70
80
90
100
110
120
60
70

1970 Chevrolet Corvette LT-1

Corvette production dropped by half in 1970, due to a prolonged workforce dispute at the St. Louis factory. This had had the knock-on effect of extending dealer delivery times to over two months, which reduced orders to only 17,316 cars. This was the lowest Corvette production volume for over a decade, and means that cars from this 'limited-production' model year are now both rare and sought after. Other market factors were also mitigating against the success of high-performance cars at this time. High gas and insurance prices, together with a general feeling of economic uncertainty made selling this kind of luxury/power auto package tricky. In real terms, the cars were becoming obsolete. But despite their problems, Corvette continued to subtly revise the car to keep it in the forefront of design and performance.

Corvette's big block V-8 engine grew even bigger in this year – to a high point of 454-cid (producing 465 horsepower). This LS-7 option was designed for competition, and could cover the quarter mile in around 13 seconds at 110 mph. The LS-7 package wasn't fitted to regular road cars. The LT-1 engine option was also introduced to the range in '70. This was a solid-lifter 370 horsepower version of the small block engine. The LT-1 package included sharper valve timing, a bigger bore exhaust, and the same carburetion as the big-block with cold air induction. The LT-1 was actually advertised as a 350-cid set-up to accommodate Federal emissions regulations. These anti-performance government restrictions were to negatively affect all the pony car runners in the 1970s, obliging them to put an increased emphasis on comfort and luxury features to sell the cars.

The basic aerodynamic styling introduced to the Corvette in 1968 effectively persisted until 1983, but it was refined and revised for 1970. Corvettes were available as sport coupes and convertibles in this model year. The car's four fender gills were replaced by an ice cube tray grille, which was echoed in the (nonfunctional) radiator intakes at the front of the car. Corvette designers were aware that the lower portion of the car was susceptible to stone-chip damage from material flicked up by the wheels, so they flared the wheel arches to offset this problem. The twin tail-pipes were squared-ff to become rectangular rather than round. Inside the car, the seats were improved, in an attempt to give better support and more headroom, and improved access to the luggage area. This was crucial, as the cars still weren't fitted with trunk lids. Inertia-reel seatbelts were now fitted as standard.

Engine: LT-1 V-8

Displacement: 350-cid

Horsepower: 370

Weight: 3153lbs

Wheelbase: 98 inches

Base Price: $5469

Number Produced: 10,668

1970 Chevrolet Caprice 454

Engine:	V-8
Displacement:	454-cid
Horsepower:	345
Transmission:	Turbo-Hydramatic
Compression Ratio:	10.45:1
Body Style:	Two-door Hardtop
Number of Seats:	6
Weight:	3621 lbs
Wheelbase:	119 inches
Base Price:	$3474
Number Produced (All Caprice V-8s):	92,000

Chevrolet was the number two US auto-producer for 1970, on a production of nearly one and a half million cars. John Z. DeLorean continued as General Manager of the division.

Although GM was fairly convinced that the big car muscle market wasn't going to take off, they still launched the 454 Caprice. Although the car didn't boast the 'SS' assignation that had graced the Chevy Impalas from 1967-69, the Caprice was still a force to be reckoned with, when equipped with the bigblock 454-cid V-8 engine option. At almost 4,000 pounds, the car was no lightweight, but could still hit 60 mph in under 10 seconds.

The full-sized Chevys, of which the Caprice was one, continued in the same dimensions for 1970, but there were styling changes front and rear. The front fenderline, hood and grille were restyled, eliminating the encircling, integrated bumper-grille look. Round dual headlights were now arranged horizontally in square bezels, which flanked a finer-textured grille. Though the grille did retain a crosshatched insert design. The gravel pan was also reshaped to round off the front body corners, and now accommodated slightly larger parking lamps together with triple-slit side markers. At the rear end of the car, taillights took on a new vertical slot shape and were recessed into the chromed bumper.

Standard power plants for the Caprice were the base 350-cid V-8, or the (regular fuel) 400-cid V-8. The cars were also equipped with transmission controlled vacuum spark advance.

The Caprice was the top-of-the-range Chevy 'big car' in 1970, and had all the standard equipment of the less senior models (Biscaynes, Bel Airs and Impalas) together with some additional items including power front disc brakes, distinctive side moldings, color-keyed wheel covers and an electric clock. The cars were fitted with G78-15/B bias-belted blackwall tires.

1970 Chevrolet El Camino SS 454

Engine: Turbo-Jet V-8

Displacement: 454-cid

Horsepower: 360 at 5400 rpm

Fuel System: Single Four-barrel carburetor

Body Style: Car-truck

Wheelbase: 116 inches

Number Produced: 47,707

Chevrolet didn't limit their muscle car talents to passenger cars, they even worked their magic on their pick-ups. The El Camino was a Chevelle/ Malibuderived truck, introduced to the market in 1958, in the wake of the '57 Ranchero. The truck shared most of the major mechanicals of the car, including the underpinnings and trim. By 1966 almost all the muscle car options offered on the Chevelle SS models were also offered on the El Camino including the hairy 366- cid Mk IV big-block V-8, complete with 375 horsepower. Interior appointments were also sports-orientated with bucket seats, a console and mag-style wheel covers. But the SS imagery was not awarded to the El Camino at this point.

This all changed in 1968 when Chevrolet finally began to offer an SS 396 El Camino complete with the revered 'SS 396' badging, blacked-out grille and bulging power hood. Production for this first Super Sport El Camino was 5,190, helping total El Camino sales shoot up by at least twenty per cent to a new high of 41,791. A similar jump in sales was achieved in '69 when the final tally soared to 48,385.

Maximum performance also leapt in 1970, as the SS 454 option debuted for both the Chevelle and El Camino. As well as sharing many of the same goodies, including a power-bulge hood and four-speed manual transmission, El Caminos also shared the SS power plants. Most SS 454 option El Caminos for '70 were equipped with the 360 horsepower LS-5 454 big-block V-8, but a few customers ordered the higher-powered 450 horsepower LS-6 454. This latter option is considered by some pundits to be the most intimidating engine ever to be produced in the entire muscle car era. By this time, work had definitely given way to play in the car-truck field, and people were buying both the El Camino and Ford Ranchero models for enjoyable driving, rather than hauling. 'People all over the country are getting on the band-wagon' wrote Lee Kelley in the December '67 issue of Motorcade magazine.

1970 Dodge Coronet Super Bee

The Dodge Coronet Super Bee was a member of Dodge's 'Scat Pack' a collection of fun cars offering high performance. The Scat Pack was marketed to the American youth car market, and the cars were a great success. They were available in a selection of eye-scorching psychedelic-inspired colors, including Hemi-Orange, Go-Mango, Plum-Crazy and Sublime.

Introduced as a Coronet option in 1968, The Super Bee launched a flurry of advertising copy, 'Scat Pack performance at a new low price. Beware of the hotcammed, four-barrelled 383 mill in the light coupe body. Beware the muscled hood, the snick of the close-coupled four-speed... Beware the Super Bee. Proof that you can't tell a runner by the size of his bankroll'. It was as though Dodge was worried that the dowdy exteriors of the regular production cars would lull buyers of the Super Bees into a false sense of security. The Coronet Super Bee continued as the high-performance intermediate-sized counterpart to the Dart Swinger 340s for 1970. Super Bees included as the Coronet Deluxe features plus a special 383-cid Magnum V-8 engine; three-speed manual transmission with floor-mounted shifter; heavy-duty, automatic adjusting drum brakes; dual horns; heavy-duty front shock absorbers; Rallye Suspension with sway bar (or extra-heavy front duty suspension); three-speed windshield wipers; carpeting; F70-14 fiberglass belted white sidewall or black sidewall tires with raised white letters; and a three-spoke steering wheel with partial horn ring.

The 1970 Super Bee was fitted with a 383-cid Magnum V-8, complete with three-speed manual transmission. A Hurst gear shifter was optional. Priced at $3074, the car was actually cheaper than the previous year's model, and was equipped with more standard features. The Hemi engine was also offered for an additional $712, and 166 Super Bee buyers chose to order cars fitted with this option. In fact, the acceleration for the Hemi was very similar, taking just 0.2 of a second off the Magnum's 0-60 mph of 6.8 seconds. But Hemi-equipped cars could cover the quarter mile in just 14 seconds – a full second faster than the Magnum. But sales for the Super Bee slumped to 15,506 in 1970, and the Coronet models were cancelled. The value-for-money Super Bee option was available only for the Dodge Charger in '71.

Engine: Magnum V-8

Displacement: 383-cid

Horsepower: 350

Body Style: Two-door Super Bee Coupe

Weight: 3425 lbs

Wheelbase: 117 inches

Base Price: $3012

Number Produced: 3,966

1970 Dodge Challenger R/T

Engine:	Big-block V-8
Displacement:	440-cid
Horsepower:	375 at 4600 rpm
Compression Ratio:	9.7:1
Wheelbase:	110 inches
Base Price:	$3266
Number Produced in this format:	916

Short for 'Road and Track', R/T was to the Dodge muscle cars what 'SS' was to Chevrolet, although the cars were far thinner on the ground, and their profile was much lower. Sadly, they were never quite as cool. From its humble beginnings in 1967, the R/T package consisted, most importantly, of the high-powered big-block Magnum V-8. A beefed-up chassis was necessary to handle the greater output of the engine.

The package was initially fitted to the Coronet in 1967 and a Charger in the following year. Both cars were fitted with either a 375 horsepower 440-cid V-8 or the optional 426-cid Hemi. A third version of the R/T based on the new Dodge Challenger 'pony car' debuted in 1970. This car was something of an answer to Plymouth's Barracuda.

To fit in as many go-goodies as possible, the Dodge designers utilized the slightly longer wheelbase to create a slightly different slant on the same longhood/ short-deck theme. The 'S' body Challenger was offered in three different forms, a hardtop, convertible and 'sports hardtop' that featured Dodge's Special Edition package, which included a vinyl-covered roof and reduced size rear window.

This new Challenger R/T package came with a whole range of engine options, starting with the 383-cid Magnum four-barrel V-8 (335 horsepower), then came the 440-cid V-8 (375 horsepower), the 440-cid Triple-carb V-8 (390 horsepower) and the 426-cid Hemi. The last three engines were only available on the R/T models. A range of gearboxes was also available in various engine combinations. A three-speed manual, four-speed manual and Torqueflite automatic came with the various packages. Other R/T extras included the Rallye suspension package, heavy-duty drum brakes and Wide-Tread F70x14 tires (Hemi models had 15 inch tires). Inside the cars, the Rallye instrument cluster was included, as were a simulated woodgrain panel, 150 mph speedometer, a tachometer, trip odometer, clock and fuel oil and temperature gauges. The outside of the cars was decorated with either a body-long stripe, or a stripe around the tail. Rallye wheels, the functional Shaker hood scoop and spoiler were all optional. The competition-style fuel filler was standard.

Dodge built 76,935 Challengers in 1970, 14,889 of which were Challengers. 3,979 R/R hardtops and 963 R/T convertibles were produced with the Special Edition package.

Below: Several engine options were available for the R/T. This one is the 383-cid Magnum four barrel V8 of 335 horsepower.

Above: This R/T Challenger is painted mouth-watering "Plum Crazy."

Left: 150mph speedo (good) and simulated woodgrain(bad) grace the dash.

1970 Ford Torino Cobra

Engine: 429 cubic-inch Cobra Jet V-8	
Horsepower: 370 at 5,400 rpm	
Fuel System: single 715-cfm Rochester Quadra-Jet four-barrel with "Shaker" hood scoop	
Transmission: four-speed manual with Hurst shifter	
Suspension: independent A-arms w/coil springs in front; live axle with leaf springs in back	
Brakes: front discs, rear drums	
Wheelbase: 117 inches	
Weight: 4,185 pounds	
Base Price: $3,249 (429 Cobra Jet option added $229 to this amount)	
Number produced: 7,675	

After paying dearly for the rights to Carroll Shelby's "Cobra" image, Ford officials wasted little time sticking Shelby's revered snake label on almost everything that moved. Make that everything that really moved.

First came the famed 428 Cobra Jet V-8 in 1968. That was then followed in 1969 by a CJ-powered mid-sized muscle machine intended to follow in the Plymouth Road Runner's tracks. This new beast's name? What else?

With low-buck investment being the goal, the '69 Cobra was based on the yeoman Fairlane (in both formal-roof and fastback forms), not the top-dog Torino as is often commonly mistaken. Forget what you saw on NASCAR tracks; David Pearson's 1969 Grand National champion Talladega may have screamed "Torino" in on its quarter-panels, but there was nothing Torino about its street-going, long-nose counterparts. And the same was true for the non-aerodynamic Fairlane Cobra.

This confusion, however, was cleared up in 1970 when Ford's Better Idea guys did what they probably should have done in the first place, that is elevate the Cobra into the upscale Torino ranks. But this time only one bodystyle was offered, the newly restyled and renamed "SportsRoof," a sleek, sloping shape that had simply been called a fastback in previous years. As for the name game, the second-edition Cobra was unmistakably a Torino, it said so right there on the hood.

Beneath that hood, perhaps as a trade-off for the Cobra's newfound top-shelf surroundings, was a little less standard venom. In place of the formidable 428 Cobra Jet was the 360-horse 429 Thunder Jet, a torquey 385-series big-block best suited for turning pulleys on air conditioning compressors and power steering pumps for the LTD/Thunderbird crowd. Called "Ford's new clean machine" by Motor Trend, the 385 big-block family was a product of Washington's increasingly more demanding mandates to reduce emissions. "Racing and research not only improve the breed," wrote MT's Dennis Shattuck, "they also clear the air."

While the typical 429 was kind to the environment, it did have an alter-ego, a dark side that better suited the jet set. To keep up with a box-stock '69 Fairlane Cobra, a '70 Torino Cobra customer had to shell out an extra $164 for the 429 Cobra Jet, a 370-horse variation on the 385-series big-block theme. Dearborn engineers took the relatively tame 429 Thunder Jet and added a 715-cfm Rochester Quadra-Jet four-barrel on a cast-iron, dual-plane intake. They also bolted on a pair of Cobra Jet cylinder heads, they with their large rounded intake ports and big valves—2.242-inch intakes, 1.722-inch exhausts. Compression was upped to a whopping 11.3:1, and a long-duration cam was stuffed inside.

The 429 Cobra Jet also could've been ordered with or without ram-air equipment. Even though that distinctive "Shaker" hood scoop surely helped whip up a few more ponies on

the top end, engineers chose not to adjust the ram-air 429 CJ's advertised output, which mattered very little anyway considering that 370 horsepower was undoubtedly an under-statement to begin with.

Equally understated was the output figure assigned to the 429 Super Cobra Jet, a truly beefy big-block that came along as part of the Drag Pack group. The Drag Pack option transformed a 429 CJ into an SCJ by adding a big 780-cfm Holley four-barrel and a long-duration, solid-lifter cam. Additional heavy-duty hardware included forged pistons, four-bolt mains and an external oil cooler. Completing the Drag Pack cast was either a 3.91:1 Traction-Lok or 4.30:1 Detroit Locker rearend. Like the "basic" CJ, the underrated 375-horsepower Super Cobra Jet could've been crowned with the Shaker scoop, which again failed to change the advertised output figure. After Super Stock's hot-foots managed a sensational 13.63-second quarter-mile blast in a '70 SCJ Cobra, it became more than clear that quite a few more horses were hiding in there somewhere.

All CJ Cobras in 1970, Super or otherwise, were fitted with the Competition Suspension package, which typically added higher rate springs. Included as well was a stiff 0.95-inch front stabilizer bar and heavy-duty Gabriel shocks. On four-speed cars, the Gabriels in back were staggered, with one mounted in front of the axle, the other behind. This was a popular Ford trick used to hopefully control rear wheel hop created by axle wind-up during hard acceleration.

Ford offered the Torino Cobra again in 1971 before giving up on high-performance to concentrate on cleaner-running, more efficient cars. It would be more than 10 years before Blue Oval muscle again made the Detroit scene.

Above: Sleek fastback styling was christened "SportsRoof."

Far left: Chrome Magnum 500 wheels were optional on the car.

1970 Mercury Cougar Eliminator

Engine: Boss V-8	
Displacement: 302-cid	
Horsepower: 290 at 5800 rpm	
Transmission: Four-speed manual	
Compression Ratio: 10.5:1	
Fuel System: Single 780 cfm Holley four-barrel	
Body Style: Two-door Coupe	
Weight: 3610 lbs	
Wheelbase: 111 inches	
Base Price: $3200	
Number Produced: 450	

Throughout the sixties, Cougars, Z28s and Javelins fought it out in the Tran Am racing series, taking turns to take the honors. The Cougar also fought it out in the pony car sales stakes, but was beaten by its archrival the Dodge Challenger in 1970. At this time, the car was effectively an upmarket version of the Mustang, having the same performance options but more luxurious interior. Like so many worthy Mercury models, the car seemed to roll out straight into the shadows.

The Eliminator was Mercury's hottest ever pony car, and had fabulous styling to boot. It was also roomier than its Mustang cousins and had loads more class. Convenience was also on offer with a long list of optional extras available. Like the Ford Boss 302, the Cougar was designed by engineer Larry Shinoda (who had also been responsible for the lovely GM Sting Ray of '63). He had been poached over to Ford by his boss Bunkie Knudsen, and when the Boss Mustang became successful, the Lincoln-Mercury bosses hoped that he could work the same magic for their baby ponycar.

The Eliminator (named for the famous dragstrip) was originally launched as a prototype in October 1968. The front and rear spoilers were very attractive, and reputedly functional. Virtually all the styling elements of the prototype, including the hood scoop, bodyside striping and blacked-out grille were given to the production model. Unfortunately, the bosses refused to give in to Larry Shinoda's longed-for sporty mag wheels. These were available as a cost plus option. Other standard equipment included F70-14 Goodyear Polyglas tires and the Competition Handling Suspension. The prototype had been painted in Flaming Orange and the six production colors were similarly eye-catching.

Power came from a range of power sources, ranging from the base 351-cid four-barrel small-block Cleveland V-8, through a 320 horsepower 390-cid big-block V-8 to the 335 horsepower 428 Cobra Jet. The latter came with a further up-rated suspension, and a functional Ram Air hood scoop. The Cobra Jet suspension package was also partnered with the Boss 302 small-block to give the car a brilliant combination of power and excellent handling.

1970 Mercury Cyclone Spoiler

Engine: Ram Air V-8

Displacement: 419-cid

Horsepower: 370 at 5400 rpm

Compression Ratio: 11.3:1

Weight: 3464 lbs

Wheelbase: 117 inches

Base Price: $3530

Number Produced: 1631

As Ford's luxury line, you might think that Mercury would have little interest in performance, racing and muscle. But you would be wrong – for the sixties and seventies at least.

Mercury morphed their Comet into a full-blown Cyclone through various stages, starting with the Comet Cyclone hybrid in 1965. The car was by no means in the Hemi road-racing class, being more of an 'econo-racer' hot sedan. Although Mercury developed a strong interest in developing a line of strip racers in the sixties, they never got into the flashy presentation common at the time – stripes, wings and spoilers just weren't their thing.

In '68 the car was restyled as the Cyclone Spoiler II, as a true fastback, or 'Sportsroof' as Ford called it. This was one smooth curving line from the top of the windshield backwards. It looked good, and cut the air like a knife. The car strongly resembled a Ford Torino Talladega, but had more luxury trim. It was built with the NASCAR racing series in mind. A special edition of the '69 model was also built, the Cyclone Cale Yarborough, complete with special striping, badging and a rear spoiler. Signature cars were great marketing tools in the sixties, and Cale Yarborough was one of the most famous drivers of the time, driving for Lincoln-Mercury in the SCCA Trans-Am series. Dan Gurney was also honoured with an eponymous vehicle. Power for the Cyclone II came from a 290 horsepower four-barrel Windsor 351. Buyers also got FMX Cruise-O-Matic transmission and 3.25:1 Traction-Lok gears.

The 1970 Cyclone Spoiler continued the quest for aerodynamic lines, and the car grew by seven inches. It also had a strange protruding center section of the grille, front and rear spoilers, exposed headlights, mid-bodyside stripes, traction belted tires, a scooped hood and dual racing mirrors. In NASCAR, the Dodge Daytona and Plymouth Superbird were the main competition, so engine power was beefed up for this Cyclone, and the cars were fitted with a competition handling package. The race-ready power plant was the 370 horsepower Ram Air 429-cid V-8. Regular Cyclones came with a choice of the 370 horsepower Cobra Jet or the 375 Super Cobra Jet. The model lasted just two years, being discontinued in 1971.

1970 Oldsmobile Cutlass SX

Olds first introduced the Cutlass in 1964 as part of the F-85 series. In 1965 it was the three-model top-of-the line F-85 option, and remained so for 1966 with four models. The car was restyled for 1966. In 1967, the Cutlass achieved series status of its own with six-cylinder and V-8 engine options spreads over ten different models. These ranged from a two-door convertible to a four-door station wagon. The Cutlass would now begin the climb that would eventually make it the most popular name-plate on a US car. Upholstery was in cloth or vinyl. Standard equipment included carpeting, courtesy lamps, a chrome molding package, foam seat cushions and a deluxe steering wheel. Their standard tire size was 7.75 x 15 inches. Cutlasses were assembled in the same factories as the F-85s.

The Cutlass Supreme had been originally introduced as a one-model series in 1967, and was the only Cutlass offered with the high-performance 4-4-2, before this option also became a series in its own right and the high-mileage Turnpike Cruising package). 24,829 4-4-2 Cutlass Supremes were sold in this year.

For 1970, Oldsmobile introduced three models in the Cutlass Supreme series, including a convertible model. The opulent SX was exclusively available as the top of the line two-door model for the Supreme series. The car was equipped with highway cruising in mind. Oldsmobile gave the car the full luxury treatment, while ensuring that it also offered excellent fuel economy and comfortable cruising. This was achieved by a specific combination of a two-barrel carburetor on the 'Rocket' 445-cid V-8 and a 2.56:1 rear axle set up. This gave the car excellent turnpike manners and relatively low fuel consumption. Pick-up wasn't great, but the car could cover the quarter mile in a respectable 15 seconds or so. Top speed for the car was estimated at 117 mph. The tires for the cars were the G70-14 Firestones, made from polyester and fiberglass, and designed to offer less rolling resistance that, in turn, improved the gas mileage of the car.

The interior was fully fitted for luxury with a simulated walnut dash, and cloth or vinyl upholstery. The car also had Flo-Thru ventilation, a deluxe steering wheel and Custom Sport seats.

Engine: V-8 Rocket

Displacement: 455-cid

Horsepower: 365

Transmission: three-speed manual floor shift standard

Fuel System: Two-barrel carburetor

Body Style: Two-door Convertible

Weight: 3614 lbs

Wheelbase: 116 inches

Base Price: $3335

Number Produced: 4867

1970 Oldsmobile 4-4-2 W-30

'Dr Oldsmobile' has the longest historical record of any US Car manufacturer. The division had moved back into the performance field with the iconic 4- 4-2 option package, first fitted to the F-85/Cutlass models, appearing later as a fully-fledged series. Originally, the numbers stood for four barrel, four-speed, twin exhaust. 4-4-2 was marketed as a separate series in 1968, and continued as such until 1971. The car was a serious hit with high-performance lovers, and its popularity continued to grow as long as it was offered to the buyers. Later in the sixties, the seriously powerful W-cars came on stream. Oldsmobile continued to develop the 4-4-2 series so that by the beginning of the seventies, the engine was used to power the whole Oldsmobile hi-po crew.

The 1970 4-4-2 is the car that epitomises all that is good about Oldsmobile in that year. The division had now been able to overcome the GM ban on engine blocks larger than 400 cid for intermediate-sized models, and the car was equipped with a 455. The 'W-30' package offered on these 455-cid 4-4-2s was an extra performance package of go-goodies including power front disc brakes, twin-scoop fibreglass hood and a variety of transmission options. This all added up to a powerful car that also handled really well. A W-31 package was offered on the Cutlass S.

The 455 cid block had been officially off limits until this model year, and had only been available as a Hurst-fitted limited production model. Cars now equipped with the W30 option of the 455 big-block could cover the quarter mile in 14 seconds, hitting 100 mph. Top speed for the car was defined as 115 mph.

4-4-2 continued for its third year as a separate Olds series for 1970, and was the performance leader for the division. A 4-4-2 equipped convertible was the pace car for the 1970 Indy 500. High performance cars continued to sell well for the company. Three models were now offered in the series, the two-door Holiday hardtop, the two-door sports coupe and two-door convertible. There had been a fourth member, the two-door Hurst/Olds in 1969. This had been the limited production 455-cid engine model, but was now superfluous. Standard equipment included everything fitted to the Cutlasses, plus foam padded seats, a special handling package, external and internal emblems, deluxe steering wheel, low-restriction exhaust system, and paint stripes. The standard tires were G70-14 with raised white letters. Interior upholstery was in vinyl or cloth.

Engine: V-8 overhead valve, cast iron block

Displacement: 455-cid

Horsepower: 370 at 5400 rpm

Compression Ratio: 10.50:1

Wheelbase: 112 inches

Base Price: $3376

Number Produced: 19330

1970 Oldsmobile F-85 Rallye 350

Olds had launched the F-85 in 1961, as an entry-level car to it's range. This was the first time that the company had broken with its three series format. Smaller than any other post-war Oldsmobile, the first F-85 models were powered by a unique aluminum 215-cid V-8. It was launched to compete with the small car offerings of Buick and Pontiac. The Cutlasses were launched as part of the F-85 series in 1962. The cars were restyled and re-sized over the years, and went from compact to midsize in 1964. The cars gained a handsome new 'A' body platform, which was shared with the Buick Special and Skylark, the Pontiac Tempest and Le Mans and Chevy's new Chevelle. F-85s were also used as the platform for several of Oldsmobile's high-performance options, including being the first body for the 4-4-2 in 1965. This was the Oldsmobile answer to the Pontiac GTO option. The full name for the option was 'Option B- 09 Police Apprehender Pursuit'. 25,003 4-4-2s were ordered in that year.

By 1970 the F-85 line was reduced to a single model sports coupe, and Cutlass was by now the next level Oldsmobile intermediate series. The F-85 was available with both a six-cylinder and V-8 engine option. At the performance end of the scale, the W-45, Rallye 350 package was installed on 3,547 Cutlass 'S'/F-85 coupes. This enabled Olds to offer a range of what might be called 'junior muscle cars'. Although cars like this were sometimes eclipsed by the 'big-block' tire burners, the Rallye 350 was actually 'big-block quick'.

Oldsmobile launched the Rallye 350 in 1970, as a kind of bargain bruiser muscle car in the tradition of Pontiac's GTO Judge. Like 'da Judge', the car also had eye-catching livery, including an eye-scorching Sebring Yellow paint job with black and orange stripes and cool black decals. You also got hood louvers and a spruced-up interior trim. Power for the car was a 310 horsepower 350-cid V-8. This engine could also be up-graded with the scary W-31 forced induction system that boosted the horsepower rating by 15 – up to 325. A typical 310 Rallye with a three-speed manual shifter could cover the quarter mile in around 15.5 seconds at speeds of just below 90 mph. It could get from 0-60 mph in just under eight seconds.

The low sales of the F-85 Rallye 350, a mere 1020, meant that the model did not return in 1971. The hallucinogenic graphic treatment had scared off too many buyers.

Engine: V-8

Displacement: 350-cid

Horsepower: 310 at 4600 rpm

Compression Ratio: 10.25:1

Wheelbase: 112 inches

Base Price: $2676

Number Produced: 1020

1970 Plymouth Road Runner 426 Hemi

Plymouth had gone to Hollywood in 1969, adopting the popular Warner Brothers Road Runner cartoon character as both the name and symbol for a new range of low-priced, medium-sized, high-performance muscle cars. The Road Runners were instantly popular. The base price for a GTX was below $3000 and this was for a car that featured a 383-cid, 335 horsepower V-8 Hemi. Performance was shattering and the cars had loads of personality, complete with their jolly 'Beep! Beep!' horn.

The Road Runner, and 'Superbird' models were developed in a wind tunnel to develop a body that would cut through the air, straight to NASCAR stock car race victory. Qualifying speeds of over 190 mph were recorded at the super speedways like Daytona and Talladega, and highly tuned cars were capable of around 200 mph. The model dominated the 1970 NASCAR season. Pete Hamilton won the 1970 Daytona 500 in a Superbird at an average speed of 150 mph, and Superbirds rode to victory in 21 Grand National races. Plymouth was obliged to manufacture in excess of 1500 cars to homologate the model for the NASCAR series. They actually produced around 1,935 of these 426-cid Hemi V-8 fitted race-going cars, but they was obliged to discount the price of the model to stimulate demand.

The cars were replete with Plymouth heavy-duty 'fuselage' styling, which included loads of vents and scoops, both dummy and functional. The car was derived from the 1969 Dodge Charger Daytona, and employed an identical, massive aerodynamic, front-end extension that covered the headlights, and a towering rear spoiler mounted on struts above the trunk.

Production models of the car were equipped with the 383-cid V-8 that developed 335 horsepower. They also had a four-barrel carburetor and TorqueFlite automatic transmission. Options included a four-speed manual transmission and a 390 horsepower, 440-cid Street Hemi with the 440 'Six pack'. A Super Trak Package designed with racers in mind was also available, this constituted of a heavy-duty manual four-speed transmission (instead of the standard three-speed) and a 9.75-inch Dana Sure Grip rear axle.

But sadly, 1970 was the third and final year for the Road Runner body style. NASCAR changed their rules to bring the Superbird's brief reign to an abrupt end. But the iconic status of the car was assured.

Engine: Superbird V-8
Displacement: 440-cid
Horsepower: 375 at 4400 rpm
Compression Ratio: 9.7:1
Wheelbase: 116 inches
Base Price: $2896
Number Produced: 15,716

1970 Plymouth Superbird

Engine: Hemi V-8
Displacement: 426-cid
Horsepower: 425 at 5000 rpm
Compression Ratio: 10.25:1
Fuel System: Two Carter four-barrel carburetors
Body Style: Two-door Superbird
Number of Seats: 5
Weight: 3840 lbs
Wheelbase: 116 inches
Base Price: $4298
Number Produced: 135

The Superbird was the race going version of Plymouth's cheerful Road Runner range. To qualify them for NASCAR racing, Plymouth had to homologate the car. Originally, homologation had involved building five hundred regular production models to be offered for public sale. This had the effect of putting some of the wildest cars ever onto the streets of America.

The Plymouth Superbird was easily one of the wildest, capable of 200 mph, the massive spoiler and positively aeronautic styling meant that the cars looked as though they could fly – and they could. Detroit had suddenly grasped the power implications of aerodynamics – wind tunnel design could increase the mph capabilities of the car by over twenty per cent. Ford had already exploited this with the sleek Fairlane and Cyclone fastbacks, and Dodge came hot on their heels with the Charger 500. the Charger was evolved by Creative Industries of Detroit. Car designers like Raymond Loewy had been aware of the 'drag' factor of excess weight and poor design for years and had applied aerodynamic principles to their work.

The Plymouth designers started their own aero-car project in 1969, but

couldn't get anything of the ground until the NASCAR season of 1970. Plymouth's massively winged street racer also had a metal front nose clip (complete with chin spoiler) that pointed itself into the air for minimal resistance. The NASCAR homologation standards had also been raised for this season, and insisting on either a thousand cars, or a number equal to half of the company's dealers, whichever was higher. In Plymouth's case, this meant creating 1500 cars.

The main rival for the Superbird was fellow winged Mopar car, the Dodge Daytona. Superbirds won eight races in the series, Daytonas won four. Another victory went to the Charger 500. Ford brought home four wins, just before Henry Ford II cancelled their competition program.

Chrysler gave up on the Superbird when racing regulations actually pulled down their power rating. This left the car as a one hit wonder...

Above: The Superbird was without doubt one of the wildest looking street cars out there in 1970.

Left: The 425 horsepower 426 hemi.

1971 Chevrolet Monte Carlo SS 454

Monte Carlos were full-sized Chevrolets, and were the top-of-the-range offerings in this series. The new 'big' cars for 1971 were introduced in September 1970. Chevrolet recorded a production of over two and a quarter million cars in this model year, under the continued guidance of John Z. DeLorean.

The Monte Carlo model had been launched in 1970 as a luxury two-door model, an extended Chevelle. From the muscle point of view, the most significant thing about the Monte Carlo was its SS 454 package, which used Chevrolet's new 454-cid version of the Turbo Jet big-block V-8. This was offered to the public in only a mild state of tune, with a 10.25:1 compression ratio, and fitted with a single Rochester four-barrel carburetor. Even so, the engine still managed to push out around 360 horsepower at 4400 rpm. But the model wasn't really marketed as a muscle car, and only 2.6 per cent of the cars were ordered with the performance package. That honor was granted in larger measure to the Chevelle. When Chevrolet fitted the performance engine to this model, the cars could be ordered in a much high state of tune – the LS6 version had a higher compression ratio of 11.25:1, and was capable of 460 horsepower.

For 1971, the Monte Carlo was given a light face-lift. A new grille with finer insert mesh was fitted to the cars. A front bumper with rectangular parking lamps was used in the model's second year out. Another change to the trim was a raised flat disc hood ornament. The original wheelbase was retained for this year, but the overall length grew by a single inch. Standard Monte Carlo features included all the Chevrolet safety equipment, power disc brakes, power ventilation system, electric clock, sideguard door beam structure, assist straps, vinyl burled elm finish instrument panel, concealed wipers, glove box light and left-hand outside rear view mirror. Size G78-15 tires were used.

In fact, 1971 marked the final year for the Monte Carlo SS-454 model, which sold less than two thousand examples in this year. This was a considerably lower sales volume than for the other Chevy muscle car offerings in '71. The Chevelle SS 454 models scored 19,292 sales, for example.

Even so, the Monte Carlos continued to compete successfully at the NASCAR short tracks, even with the lower horsepower rating of 425 (due to a less favorable compression ratio of 9.0:1).

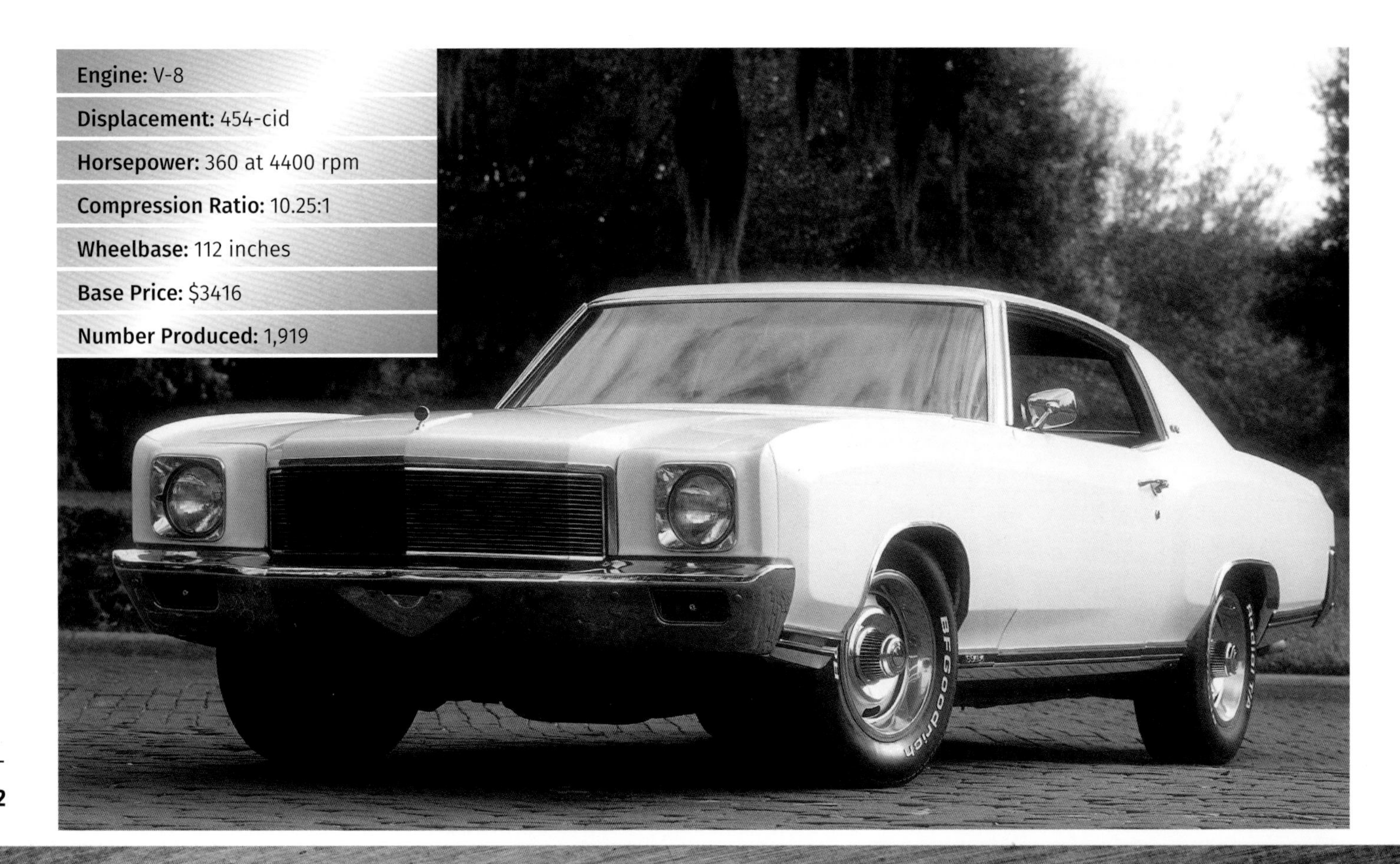

Engine: V-8
Displacement: 454-cid
Horsepower: 360 at 4400 rpm
Compression Ratio: 10.25:1
Wheelbase: 112 inches
Base Price: $3416
Number Produced: 1,919

1971 Dodge Demon

Engine: V-8 Overhead valve, cast iron block

Displacement: 340-cid

Horsepower: 275 at 5600 rpm

Compression Ratio: 10.0:1

Weight: 3165

Wheelbase: 108 inches

Base Price: $2721

Number Produced: 10,098

Dodge offered many convenience options for their 1971 models. New items ranged from slightly wider rearview mirrors to cassette tape players. Optional 'lock door' and 'low fuel' warning lights were also featured.

The 1971 Dodge Demon 340 was Dodge's compact performance offering for the year, and was a sales success, selling over 10,000 cars. Plymouth had enjoyed great success with their compact Valiant Duster muscle car, and Dodge aimed to follow this example. However, the name proved controversial, upsetting religious organizations across the United States. The Dart Demon nameplate was replaced from 1973 with Dart Sport. The 340-cid small-block V-8 engine could develop 275 horsepower, and made for a very quick package in such a relatively small body. The cars were also equipped with dual exhausts, heavy-duty suspension, E70-14 tires and three-speed manual transmission with floor shifter (four-speed manual and TorqueFlite automatic were optional). Inside the car, Rallye dashboard instrumentation was fitted together with a Tuff steering wheel. Interesting optional extras on the Demon 340 included a deluxe interior, rear spoiler and dual scoop hood with lockpins. Regular Dodge Darts were positively tame by comparison, as they were only fitted with the Dodge slant six (198-cid). The Demon also looked far more cool than the regular Dodges, coming complete with wild trim, including lots of eye-catching demon decals (complete with red pitchforks), stripes and lurid paint. A black painted hood with two gaping air scoops was optional.

Motor Trend Magazine tested a Demon 340 against a Mercury Comet GT, Chevrolet Nova SS 350 and AMC Hornet SC360. The Demon was the heaviest of the four cars, at 3,360 pounds, but it was also the quickest, with a standing quarter mile time of 14.49 seconds, and a 0-60mph time of 6.5 seconds. The compression for the car was lowered to 8.5:1, though the power drop to 240 horsepower was now measured by the SAE net rating, so wasn't as dramatic as it seems.

The essence of the Demon's success was that it was an affordable muscle car, which was also much cheaper to insure than its big-engined buddies.

1971 Dodge Charger R/T Hemi

By 1971 the muscle car era was drawing to a close, but Dodge decided to integrate both Super Bee and R/T muscle cars into their newly restyled Charger series. The cars now rode on a new 115-inch chassis, and the model range was expanded to six coupes and hardtops in three series, the Chargers, Charger 500 series and Charger R/T (one car) series. All Chargers were now semi-fastback coupes featuring rear quarter window styling that swept up from the fender to meet the sloping upper window frame. The full-width bumper-grille shell was split by a large vertical divider on all Chargers and the rear end of the car featured a small trunk lid spoiler. Six square taillights were located in the oval rear bumper.

The top-of-the-range R/T model was the most luxurious and high-performance Charger model. It included all the standard equipment fitted to the 500s, together with a 70 amp/hour battery, heavy-duty brakes and shock absorbers; pedal dress-up kit; extra heavy duty Rallye suspension; Torqueflite automatic or four-speed manual transmission. Interiors were kitted out in keeping with the contemporary trend, and were awash with white-pleated vinyl and plastic wood. The exterior of the car was dressed with a black R/T decal.

Engine options were the 440 Magnum (fitted as standard) and the Hemi was available as a $707 option. Some cars (178) were also equipped with the 440- cid 'Six Pak' V-8. The Hemi Chargers were equipped with twin Carter four-barrel carburetors and had a compression ratio of 10.25:1 producing 425 horsepower. According to Mopar authority Galen V. Govier, 63 R/Ts were fitted with the Hemi engine option (30 four-speed manuals, and 33 TorqueFlite automatics). These cars were feared at dragstrips up and down the country.

NASCAR performance limitations severely reduced Dodge's interest in this race series, and they cut back heavily on their factory backing of stock car racing. Following 22 Grand National wins in 1969, and 17 in 1970, there were only eight Dodge victories in 1971. 1971 also marked the final year when the Hemi engine was offered for sale to the public, in either the street or racing form. Insurance premiums were astronomical on these high-performance cars and beginning in 1972, all cars had to be run on regular gas. Dodge decided to retire the Hemi while it was still in front. It returned in 1992, for sale to racers and restorers via the Chrysler Performance high-performance parts division.

Engine: Street Hemi V-8, overhead valve with hemispherical combustion chambers

Displacement: 428-cid

Horsepower: 425 at 5600 rpm

Compression Ratio: 10.25:1

Number Produced: 3,118

1971 Ford Mustang Mach 1

Engine: Cleveland V-8

Displacement: 351-cid

Horsepower: 285 at 5400 rpm

Number Produced: 36,499

Ford had broken the mold to offer the first Mustang to the American public in April 1964. The car was a completely unique fusion of Detroit's power and comfort, with the styling of the classic European sportscars. The Mustang was an immediate run-away success, and for the moment, the first of the ponycars was ahead in a one horse race.

One of the main reasons for the success of the Mustang range was always the breadth of its appeal. Ford was careful to offer a selection of cars to appeal to different kinds of buyers, from the boy-racer to the more sedate middle-aged customer interested more in image than performance. The first Mach 1 was introduced in 1969 as a middle-of-the-road Mustang, with muscle car heritage in a relatively civilized road-going package. The car was styled with the new Ford Sportsroof (the latest version of the fastback). Mach 1s were immediately successful, and sold 70,00 units in their introductory year.

By 1971, Mustang production peaked at 149,678 unis, and the division was known as the Ford Marketing Corporation. The Boss Mustang 351 was now the most serious package on offer. But for buyers who wanted something slightly less hairy, but still with a sporty flavor, the Mustang Mach 1 remained the streetfriendly alternative for this model year. The car came complete with a full sports appearance package and was available with a range of power plant options. These varied from the relatively mild 210 horsepower 302-cid V-8 all the way through to the Cobra Jet 429, or even the completely wild 370 horsepower 429-cid Super Cobra Jet fitted with Ram Air induction. In fact, only five of these cars were reputedly built, and 1971 was the final year that this big-block 429 engine was on offer. The 351-cid Cleveland four-barrel V-8 was another engine on offer with the Mach 1, and developed an impressive 285 horsepower. Other sports goodies could also be included in the package, including E70 tires and Special Handling suspension.

Mach 1s were also blessed with the classic good looks you would expect from such a fine muscle car heritage. Buyers got a blacked-out hood with twin NASAstyle air scoops, a mean black spoiler and side striping, with Mach 1 decals. The package was completed with a honeycomb grille with driving lights, color-keyed mirrors and bumpers, black or 'argent' colored lower bodysides. Inside, the buyer could upgrade the interior with high-backed bucket seats and a center console. The Mach 1 perfectly reflected a market that was now tending towards appearance, rather than performance concerns. But with the range of powerplants on offer, Mach 1 buyers could still specify a seriously hot car from the Mustang stable.

1971 Ford Mustang Boss

Engine: 351 cubic-inch "Boss 351" V-8

Compression: 11:1

Horsepower: 330 at 5,400 rpm

Fuel System: single 750-cfm Autolite four-barrel carburetor

Transmission: Hurst-shifted four-speed manual

Suspension: independent A-arms w/coil springs in front; live axle with leaf springs in back

Brakes: power front discs, rear drums

Wheelbase: 109 inches

Weight: 3,625 pounds

Original Price: $3,746.90

Number produced: 1,806

Opposite top: The last, for a while anyway, of true muscle Mustangs.

Opposite bottom: The 330 horse 351 was standard Hurst shifter on a four-speed manual gearbox.

Below: Full instrumentation was standard on the car while the Mach 1 sports interior and high back bucket seats were optional.

Fans of Ford pony car performance were hit hard when both the small-block (302) and big-block (429) Boss Mustangs were unceremoniously cancelled in 1970. But not all was lost. In November that year, Dearborn officials rolled out their new Boss 351, a truly hot 1971 Mustang based on the totally restyled SportsRoof body, as fast a fastback as yet come down the pike.

The heart of the '71 Boss 351 was the 351 High Output (HO) Cleveland V-8, an able small-block that could throw its weight around like most big-blocks. Rated at 330 horsepower, the HO featured superb free-flowing heads, which were nearly identical to those used by the Boss 302 save for revised cooling passages. Those excellent canted valves carried over from Boss 302 to HO right down to their diameter, as did much of the valvetrain. Both engines also shared screw-in rocker studs, hardened pushrods, and guide plates. The Boss 351's solid-lifter cam, however, was more aggressive than the Boss 302's.

The 351 HO's lower end was also more stout. As was the case inside the Boss 302's modified Windsor block, the HO's crank was held in place by four-bolt mains, but the latter had four-bolt caps at all five main bearings, not just three. The HO crank was cast (of high nodular iron) instead of forged, and it was specially tested for hardness. Forged connecting rods were shot peened and magnafluxed and were clamped to the crank by super-strong 3/8-inch bolts. Pistons were forged-aluminum pop-up pieces. On top was a 750-cfm Autolite four-barrel on an aluminum dual-plane manifold.

The Boss 351's standard supporting cast included a ram-induction hood, a special cooling package with a flex fan, and a Hurst-shifted wide-ratio four-speed. In back was a Traction-Lok 9-inch rearend with 31-spline axles and 3.91:1 gears. Underneath was the Competition Suspension package, which featured heavier springs, staggered rear shocks, and sway bars front and rear. Power front disc brakes were standard, too, as were F60 raised-white-letter rubber tires on 15x7 steel wheels adorned with the flat hubcaps and trim rings. Ford's flashy Magnum 500 five-spoke wheels were optional.

The Boss 351's standard appearance features were all but identical to those found on the 1971 Mach 1. Included up front was Ford's functional "NASA hood, a chin spoiler and a honeycomb grille with color-keyed surround. That ram-air hood incorporated twist locks and was done in either an argent or blacked-out finish, depending on the body paint choice. Like the hood, Boss 351's standard lower-body paint accents and accent tape stripes were either black or argent, again depending on the chosen exterior finish. Black or argent treatment once more showed up at the rear. Inside or out, from nose to tail, the Boss 351 Mustang was a big winner in most critics' minds. Sure, some reviewers complained about visibility problems inherent to the 1971 SportsRoof restyle. But they couldn't deny the Boss 351's aggressive appearance and high-spirited nature.

RAM AIR
MUSTANG
BOSS
351
BFGoodrich
Radial T/A
BOSS 351
MUSTANG

1971 Mercury Cougar XR7 429

Mercury introduced its Cougar model in 1967. It was one of the handsomest cars of the year, and was an immediate marketing success for the division. The car successfully bridged the gap between the performance of the Mustang and the luxury of the Thunderbird. The two-door hardtop coupe was trimmed with a sports interior, and had the Cyclone V-8 engine. The model sold over 150,000 units in its first year.

Mercury produced an especially hot version of the Cougar in 1969 – the Cougar Eliminator. It was the Mercury version of Ford's Boss 302 and Mach 1. The base engine offered was the 290 horsepower 351-cid, but the range of powerplants went right up to the 428 Cobra Jet. This produced 335 horsepower and could cover the quarter mile in just 14 seconds. With this engine, the Eliminator also came with an oil cooler, staggered shocks, tachometer, power front-disc brakes and heavy-duty suspension with various transmission choices.

The Eliminator was a limited production model of only around 500 examples. The Eliminator made a second and final appearance in 1970, and could now be ordered with the Boss 302 and Boss 429 packages. The latter came with Ram Air induction. Styling changes were minimal, special striping, hood scoop and rear deck spoiler were the only signs to mark out the model as Cougar's performance special.

The Cougar was restyled more completely for 1971 and grew once more. It was available as a two-door hardtop and two-door convertible. The concealed headlights had disappeared, and were now exposed and recessed. The protruding center grille had vertical bars and was framed in chrome. The rear bumper was integrated into the rear deck panel, which housed the large rectangular taillights. The sports interior included high-backed bucket seats, consolette with illuminated ashtray, glove box and Flow-Thru ventilation system. Buyers also got full instrumentation with toggle switches, a cherrywood appliqué dash and leather seat facings. The Cougar XR7 was the most luxurious model in the line-up. When the 429-cid Ram Air cooled Super Cobra Jet was ordered for the car, a functional hoodscoop was fitted, joined to the air cleaner via a rubber seal around the rim.

1971 Plymouth GTX 440+6

Engine: GTX V-8	
Displacement: 440-cid	
Horsepower: 390 at 4700 rpm	
Transmission: Four-speed manual	
Compression Ratio: 10.5:1	
Fuel System: Three Holley two-barrel carburetors	
Body Style: Two-door Hardtop	
Number of Seats: 5	
Weight: 4022 lbs	
Wheelbase: 115 inches	
Base Price: $3969	
Number Produced: 135	

Plymouth's GTX had first been launched in 1967 as a repost to the Pontiac GTOs. At least the release of the model meant that Chrysler's low-budget Plymouth division was finally in the high-performance race.

The Plymouth Division's ability to fill the space under the hood was never in question. The hottest of their hot powerplants on offer for 1966 was the 426 Hemi. What they really needed was a sexy, well-styled body to hold their mean go-goodies. They used the Belvedere hardtop as the basis for the new car, and bolted on various racing equipment. This included a (non-functional) twin-scoop hood. The car was equipped with a competition flip-to fuel cap, bright exhaust tips (fixed to the standard emissions system) and optional twin racing stripes. Red-line Goodyear tires came at the corners of the car. The standard GTX engine was the gold standard RB big-block 440-cid V-8 that had been introduced into the Chrysler luxury liners. This was made from a 426-cid block, bored out to the larger displacement. Extra performance equipment was married up with the block, including better-breathing heads, a warmer cam, a free-flowing intake and single four-barrel carburetor. All of these resulted in Plymouth's Super Commando 440, a powerplant capable of 375 horsepower, the pumping heart of the GTX.

Hot Rod magazine's Dick Scritchfield wrote of the '67 GTX (equipped with the Super Commando) 'it's exciting to look at and it's exciting to drive'. The triple-carb 440 V-8 package launched in 1969 must have increased the feeling. By 1971, the basic body shell for the GTX performance machine was the Plymouth Satellite, as the Belvedere had been retired in 1970. The new car was a stripped-down race-ready rocket. So stripped down, in fact, that the cars didn't have either wheelcovers or even hood hinges – the lightweight fibreglass lids just lifted off after the release of the four corner pins. Other power options for the 1971 GTX 440 were the 'six' (three Holley two-barrel carburetors, supplied for $262), and the 426 Hemi (at an extra $750). Only thirty Hemi-equipped GTXs were built in this year.

1971 Plymouth Hemi 'Cuda

Engine: Hemi V-8
Displacement: 426-cid
Horsepower: 425 at 5000 rpm
Compression Ratio: 10.25:1
Wheelbase: 108 lbs
Base Price: $4300
Number Produced (with four-speed): 2

Along with Plymouth's new E-body platform of 1970, three distinct pony car model lines were available, the base Barracuda, the sporty 'Cuda and the top-of-the-range Gran Coupe with its leather bucket seats and overhead console.

Chrysler was looking for a car to compete head to head with those Pontiac GTOs and Chevelles that GM produced in the mid-sixties. But as they already had the fabled 426 Hemi, the powerplant at least was ready and waiting. This engine was officially rated at 425 horsepower, but independent tests put this closer to 500. Hemi-equipped '60s Plymouths had already been timed covering the quarter mile in around thirteen seconds. But for most people, the best was yet to come.

The Plymouth Barracuda was introduced in 1964, and immediately won an enthusiastic following. In 1967, the car got an all-new body and look of its own, and the big block powerhouse was fitted for the first time. A third version of the car, now called the 'Cuda (with the Hemi option fitted) hit the streets in '69. The revised 'E' body, designed by the Cliff Voss Advance Styling Studio, was a great asset to the model, as the engine bay was easily large enough to shoehorn the big Hemi into the car. The exterior appearance, credited to John Herlitz also

Below: Luxury interior for a car that cost over $4000 back in the day but now worth $3.5 million!

gave the car a thoroughly rakish appearance. The 'Cuda was dressed up with front foglamps, racing style hood pins, simulated hood scoops, a blacked-out rear panel and hockey stick bodyside stripes. Inside the car, the interior was equipped with full instrumentation, bucket seats and a console.

Despite their complete desirability, 'Cudas were rare predator-fish. The badge was only awarded to Barracuda models when the 426 Hemi was fitted. Only 652 Hemi 'Cudas were built in 1970 and this fell to 114 in 1971, now equipped with four fender gills on either side. The option also involved a selection of mandated go-goodies, including the new (fully functional) 'Shaker' hood scoop, Dana rear end axle and a choice between four-speed manual transmission or TorqueFlite automatic. A heavy-duty radiator was all included in the Hemi package, as were eleven-inch drum brakes and a beefed up chassis. This led to the car being dubbed 'the ruggedest pony car suspension in the industry'.

Thus fitted, and tuned up by Car Craft, the cars could turn in quarter mile times around 13.10 seconds. This had led some to rate the car as the greatest Hemi-fitted car ever, and rare too, as it was discontinued in the 1971 model year.

Above: The 1971 Hemi 'Cuda convertible is a very rare species indeed, only seven were built in all, just two with the four-speed transmission.

1972 Plymouth Duster 340

Dusters were members of Plymouth's entry level Valiant Group series, as were Scamp two-door hardtops and Valiant sedans. Interestingly, the mighty 'Cudas had also started out as Valiants back in '64, but by this time, the 'Cudas were configured on the larger 'E' Mopar body. The Dusters soon eclipsed the sales of the bigger car.

Dusters were engineered in 1969 for a 1970 launch. The name would ultimately outlast 'Valiant'. The 340 Duster achieved a great power-to-weight ratio that allowed the car to out-perform many larger cars fitted with more powerful engines. Car Life rated the 1970 model with a 0-60 mph time of just 6.2 seconds, and a quarter mile time of 14.7 seconds at an average 94 mph.

The 1971 Valiant models had set great sales levels, with a volume of 256,930 units in that year. Consequently, there was very little impetus to change this successful model. Only details of the model taillights and grille were changed. The front side marker was also now slimmer and longer. The base Duster model for '72 shared this pattern of light revision, but the side market lamps were moved an inch or two higher above the lower feature line. Base Duster equipment included ventless side windows, a concealed spare tire and twospeed wipers.

Outside, the car had a lower deck tape stripe, bodyside tape stripe, a unique grille, roof drip rail moldings and wide tires. Mechanically, the car was fitted with the 340-cid V-8 with dual exhaust, a three-speed manual floor shift, optional axle ratio, heavy-duty suspension and dual snorkel air cleaner. Duster interiors had been upgraded in 1971, and the car was equipped with an optional fold-down rear seat, optional electrically heated window defogger and had front and rear fender guards as standard. Lowback-style seats were a Duster option. Just as they had done with the Demon, Plymouth offered a mighty little compact in the shape of their 1972 Plymouth Duster 340. Like the earlier Demon, the car also made use of fun 'character' graphics to appeal to a young audience, and this 340 version had a wide tape strip running the full length of the car along the beltline with bold '340' lettering on the rear fender edge. Effectively, the Duster was Plymouth's entry-level muscle car and came fully loaded with both plenty of performance goodies and a wallet-friendly price. The car continued to be a great success in '72, with almost identical sales to the 1971 model. Its fun image and targeted marketing was right on the money.

Engine: V-8 overhead valve, cast iron block

Displacement: 340-cid

Transmission: Three-speed manual

Compression Ratio: 8.5:1

Base Price: $2728

Number Produced: 15,681

1972 Pontiac Trans Am

During the 1971 calendar year, Pontiac captured third place in the US auto producers' league for the tenth time in eleven years, but the image of the division was simultaneously losing interest from the buying public, and things were very difficult by 1972.

Pontiac had launched the Trans Am as a $725 Firebird option package in 1969. A 335 horsepower engine was offered as standard, with the option of a 345 horsepower Ram Air IV. Buyers got a limited slip differential and threespeed manual transmission. Despite its name and heritage, the Trans Am never actually raced in the eponymous series. Nevertheless, with 0-60 mph times of 6.5 seconds and quarter mile times of 14.3 seconds (averaging 101 mph), they were undoubtedly fully loaded in the performance department.

The Trans Am was heavily restyled in 1970, and the model was fitted with a new 'bull nose' front grille and streamlined fastback profile that made the car look longer, lower and wider. The 1972 Trans Am series reflected the same look, but was now fitted with the 300 horsepower 455-cid V-8 as standard. The Trans Am interiors had the same standard features as the Firebirds, plus a Formula steering wheel, engine-turned dash trim, rally gauge cluster with clock and a tachometer. Outside, the cars were equipped with a front air dam, wheel flares front and back, full-width rear deck spoiler, shaker hood, 15-inch Rally II rims with trim rings and black-textured grille inserts.

Sadly, a combination of circumstances almost resulted in Pontiac axing the car in 1972. Part of the problem was that buyers were deserting muscle car models in the fuel conscious seventies. But this wasn't the only problem Pontiac had to overcome in this year. The most damaging was a five-month long strike on the Firebird/Trans Am manufacturing line at the Norwood, Ohio plant. This resulted in the loss of thousands of cars, and millions of dollars. It very nearly caused the company to drop the model altogether. Further complications arose from the increased volume of Federal safety regulations. These generally resulted in a lessening of performance, and this was certainly true for the Trans Am. Compression ratios were scaled back to cope with lower octane fuel and stricter emissions limits.

Despite all its problems, the Trans Am turned out to be a real survivor, and it was a best-selling model by 1974.

Engine: V-8

Displacement: 455-cid

Horsepower: 300 at 400 rpm

Compression Ratio: 8.4:1

Wheelbase: 108 inches

Base Price: $4256

Number Produced: 1286

1973 AMC Javelin

Engine: Base V-8, overhead

Displacement: 304-cid

Horsepower: 150 at 4400 rpm

Compression Ratio: 8.4:1

Base Price: $2983

Number Produced: 25195

The Javelin had first been added to the AMC range in 1968, fitting into the slot vacated by the unsuccessful Marlin. It was AMC's attempt to follow in the 1964 hoofprints of the Ford Mustang, and like the classic ponycar, boasted four seats. The car was built on a 109-inch wheelbase platform, and had an overall length of 189 inches. Power for the car was varied, supplied by either a 232-cid six-cylinder, or a 343-cid V-8. The top Javelin option, the SST was fitted with the 390-cid V-8 that worked so well in the Rebel Machine. This option was introduced late in the '68 model year – so late that it didn't make it into that model year's brochures. At first, Javelin power was hampered by the wide-ratio three-speed manual gearbox, but the car came into its own in '69 with 'loud and proud' options aimed squarely at muscle car aficionados. This option was called the 'Go-Package'. It included heavy-duty suspension and wide wheels together with a selection of garish paint colors, including Big Bad Orange and Bid Bad Green. If even a standard Javelin 390 wasn't quick enough to fit the bill, there were tuning parts that could be dealer-fitted. AMC was undoubtedly keen to leave its sensible image in the past.

Original styling characteristics for the model included a split grille with blackout treatment and form-filling bumper, single square headlamp housings were integrated into the fenders, and round parking lamps were integrated into the bumper.

The Javelin had a clean-lined body with smooth-flowing lines and a semi-fastback roofline with wide, flat sail panels. The Javelin was given new styling in 1973, which it substantially shared with the AMX sports hardtops. Both cars featured a new, smooth roofline. A new taillight treatment was adopted, with twin-pod lamps at each side of the car. The base level Javelin now had a recessed plastic grille that incorporated rectangular Rallye lights. Bucket seats were standard in the car along with interior styling of the aircraft cockpit-type.

Race driver Mark Donohue drove a Javelin in the 1970 Trans Am race series, and lost the championship by just one point to Parnelli Jones in the Boss Mustang. The car proved itself to be both powerful and reliable, and was a great combination with this experienced and talented driver. The car went on to win seven out of ten races in 1971, to take the championship. George Follmer repeated this feat in 1972, but pulled out of the races in '73.

1973 Ford Gran Torino Sport

The full-size Fords were for '73 the only models to receive significant restyling. The new full-size LTD was honoured by Motor Trend Magazine as the 1973 'Full-size Sedan of the Year'. Road Test Magazine went further, calling it the 'Car of the Year'.

The rest of the Fords received only minor trim updating. More federally mandated safety requirements were initiated. They were reflected in the form of massive 'park bench' safety bumpers, designed to tolerate a full-on impact at five miles per hour without any damage. Pollution standards were tightened, and engines were detuned. Sadly, this had the negative effect of making some of the least fuel efficient and poorest performing engines ever built. The embargo on fuel imports from the Middle East brought fuel economy into the spotlight.

Ford also attempted to make the 1973 models thief and vandal proof. For example a new fixed-length type of radio antenna was adopted, and the hoods could not be opened from outside.

The Torino and Gran Torino series were built on a slightly shorter wheelbase than the full-size models. They were slightly modified for '73, with revised grilles having slightly more rectangular openings than those of '72. There was a larger front bumper, and the technical equipment was also up-graded. The rear brakes were enlarged, an interior hood release was fitted, and there was an optional spare tire lock. The Torino models were the base trim for the series, and were offered in three body styles. The Gran Torinos were the top trim series models for the year, and included all the Torino trim, plus manual front disc brakes, cloth and vinyl upholstery trim, carpeting, and chrome trim on the foot pedals. A deluxe two-spoke steering wheel was fitted.

The Gran Torino Sport was the sporty version of the Gran Torino line and included all the Gran Torino features, plus the 302-cid v-8 engine that developed 138 horsepower. Upholstery for this model featured pleated, all-vinyl trim, hood scoops, color-keyed dual racing mirrors and a unique grille.

Well-known names on the FoMoCo board during this year were Chairman Henry Ford II, Corporate President Lee Iacocca and Ford Marketing Corporation Vice-President and Divisional General Manager B.E. Bidwell.

Engine: Overhead valve V-8

Displacement: 302-cid

Horsepower: 135 at 4200 rpm

Compression Ratio: 8.0:1

Weight: 3670 lbs

Wheelbase: 114 inches

Base Price: $3154

Number Produced: 51,853

1973 Oldsmobile Hurst/ Olds Cutlass

Engine: V-8	
Displacement: 455-cid	
Horsepower: 230	
Compression Ratio: 8.5:1	
Wheelbase: 111 inches	
Base Price: $3323	
Number Produced: 1097	

The big story for the 1973 Oldsmobile Cutlasses was the same as for the other two-door hardtops in the GM division, a completely new 'colonnade' styling package. Sadly, the new body was introduced one year later than planned, due to a strike at General Motors. The Cutlass model was a great success for Oldsmobile throughout the early '70s and helped boost overall Oldsmobile sales. Cutlasses were still available with the 4-4-2 option, but as the decade wore on, this became more of an appearance package on regular cars, rather than an independent line.

The all-new Cutlass Supreme was at the top of the Olds intermediate line. It was a hands-down favorite with the intermediate buying public, who loved its formal roof styling. Standard equipment for the Cutlass Supreme included armrests, ashtrays, front disc brakes, carpeting, an interior hood release, dome light, molding package, windshield radio antenna, seat belts with shoulder harnesses, a deluxe steering wheel and chromed hubcaps. Upholstery was in vinyl or cloth.

Oldsmobile and the gearbox specialist, Hurst, had been working together for several years by this time, and their combined efforts had resulted in some legendary autos. Their coupled names were a performance institution. But like the iconic 4-4-2 package, the penultimate Hurst/ Olds package of '73 (available at an extra cost of $635) depended more on its appearance than being a serious performance option. A sports console and swivelling sports bucket seats were fitted as standard in the car. Even so, the cars were powered by the mighty 455- cid Oldsmobile Rocket V-8, and could go from 0-60 mph in 6.5 seconds.

A Hurst/Olds special model (known as option W-45) was also offered in the following year, for the very last time. This car was chosen as the pace car for the '74 Indy 500, and Oldsmobile celebrated this honor by manufacturing a handful of special replica cars. Although the smaller 350 engine was available on the car, the standard powerplant was the big-block 455. The W-30 package was also fitted. Sadly, the power output of the car was now rated at a meagre 230 horsepower (though its 0-60 time remained 6.5 seconds).

Despite the powerful performance package fitted to this final Hurst/Olds, its major selling features were the padded rood and special decals, which says a lot about the priorities of the car-buying public of this time.

1973 Plymouth Road Runner

Chrysler came to the muscle car arena rather late, but soon got into the swing of things. The fact that they had the Hemi and Max Wedge engines available to them was immediately helpful. By the late sixties, muscle cars were getting increasingly expensive, and moving away from their target audience because of this. Mopar now led the field with the first budget muscle cars. They evolved a simple but successful formula, take the lightest, cheapest, two-door body available to you, strip it of all the options and stick in the most powerful off-the-shelf V-8. The new Road Runner outdid even this brilliant idea, by adding all the desirable options back in, and then offering the whole package at a very attractive price.

Although the car didn't look anything like a typical muscle car, with an almost complete absence of chrome, bulges and hood scoops, it was an instant success, selling well over 40,000 cars in its first year. That made up one in five of the Plymouth intermediate sales. The car interiors were seriously basic, with rubber mats instead of carpet, a plain bench seat. Only the defining Road Runner badges hinted that there was a special kind of power under the hood, and a special kind of fun to be had from the car.

The standard power unit for the original Road Runners was a special 335 horsepower 383-cid V-8, but the Hemi was also available for $700. Road Runner sales nearly doubled to 80,000 in 1969, and it was named as 'Car of the Year' by Motor Trend Magazine. Only a very few of these cars were Hemipowered, but this option was tested against five other econo-racers, and outpaced them all completely. It could get from 0-60 mph in 5.1 seconds. By 1971, however, only the GTX and the restyled Road Runner were equipped with the Hemi in a changing performance climate.

Styling was revised for the 1973 Road Runners, which were now part of the Satellite Group. The cars were now wider but not as long as the 1972 models. They now had a combined Satellite/Sebring grille, the hood had a wide center power bulge and rectangular taillights were fitted.

Engine: V-8, overhead valve, cast iron block

Displacement: 318-cid

Transmission: Three-speed manual

Compression Ratio: 8.6:1

Body Style: Two-door Hardtop Coupe

Weight: 3525 lbs

Wheelbase: 115 inches

Base Price: $3115

Number Produced: 19056

1974 Pontiac Trans Am 455 Super Duty

By this time, the muscle car era had truly drawn to an end. But, as the manufacturer of the first muscle car, Pontiac bravely continued to fly the performance flag for the American car industry. The gas crisis, Arab oil embargo and federal emission standards had thrown the US car industry into a serious tailspin, and the manufacturers had been obliged to adjust their priorities. Gas-guzzlers were now almost impossible to sell, and buyers were looking for more economic cars with smaller, 'sippier' engines – which meant that they often turned to imported models.

Pontiac were really bucking the trend by continuing their development of the Trans Am, which was completely restyled for 1974, losing the aggressive bull nose front end styling of 1970, and gaining round headlights and a much smoother grille. They had effectively decided to by-pass the search for better fuel economy, in the greater search for excitement. The rise of appearance packages over performance criteria seemed relentless across the industry as a whole, and this was certainly true for this Trans Am model – the Shaker hood was now non-functional and was about as useful as the screaming chicken painted on the hood.

Having said this, Pontiac continued to invest in technological improvements, and the package now included a heavy-duty oil pump, forged aluminum pistons, a Rochester Quadra- Jet four-barrel carburetor, a limited slip differential and four-speed manual transmission. Outside, the cars were equipped with dual exhausts, rear ducktail spoiler (standard since the 1970 restyle), fendermounted air extractors and F60-15 tires. The standard powerplant for the car was now the 225 horsepower 400. Optional engines were the 215 horsepower 455-cid, or the mighty 290 horsepower Super Duty 455-cid. The Super Duty option marked returned an interesting return to performance at Pontiac, and cars thus equipped were capable of covering the quarter mile in around thirteen seconds, at an average speed of 103 mph. 'Super Duty' was first used to identify race ready cars in 1961 and '62.

Inside the '74 model, the Trans Am buyer got a pretty cool interior, with a turned aluminium dash, rallye gauges with tachometer, and a Formula steering wheel.

The motoring press was completely shocked by the new Trans Am. They had believed that all the powerful, fast cars had been relegated to museums.

Engine: Super Duty V-8

Displacement: 455-cid

Horsepower: 290 at 4000 rpm

Transmission: Turbo 400 automatic

Compression Ratio: 8.4:1

Number Produced: 943

1975 Chevrolet Corvette

Engine: V-8, overhead valve, cast iron block	
Displacement: 350-cid	
Horsepower: 165 at 3800 rpm	
Compression Ratio: 8.5:1	
Wheelbase: 98 inches	
Base Price: $6857	
Number Produced: 4829	

Most of the changes to the Corvette for the 1975 model year were hidden from view, and the design remained unchanged. 1974 had seen the introduction of the restyled body-color rear end to the car, which perfectly match the new nose introduced in 1973. The bumpers were improved from a safety point of view for '75, but looked exactly the same. Less positive changes were made to the power train.

This was also the final year for the Corvette convertible, which wouldn't return to the market for another twelve model years. Convertible sales had been on the slide for a few years now, despite the fact that the roadster had embodied the spirit of the range since its inception. The last ones came off the line in July '75. Out of a total Corvette sale of 38,465 cars, only 4629 were ragtops. The Corvette convertible would not return to the range for another twelve years. In 1975, it seemed like this day would never come, and that the marque was sliding into a graceful middle-age.

These concerns were reinforced by the engine revisions for '75. The 1974 Corvette turned out to be the last fitted with the optional 454-cid big-block V-8. After a full decade of use, these engines were finally killed off by emission controls and economy concerns. Things would only get worse in this department. The Energy Policy and Conservation Act became law in '75, but its fuel-economy standards would not be enforced until 1978.

The standard engine now fitted to the range was the 350-cid, de-tuned to offer a modest 165 horsepower. A catalytic converter was fitted under the hood as an attempt to avoid even further de-tuning, whilst still conforming to emission controls. The performance with this set up was perfectly acceptable (0-60 in 7.5 seconds, with a top speed of 125 mph) but it wasn't what Corvette owners were used to. Positively, the car was now equipped with servo-assisted power brakes. Inside the car, there was a feeling of increased luxury. Options included leather upholstery, electric windows, air-conditioning and a stereo. The speedometer registered kilometers-per-hour for the first time.

The new attitude to performance meant that possibly the biggest change of the year at Corvette was in the marketing of the cars. Gone were the adverts offering blood-and-guts road burners, the cars were now styled as luxury grand touring cars.

1975 Ford Thunderbird

Engine: Thunderbird V-8

Displacement: 460-cid

Horsepower: 218 at 4000 rpm

Transmission: Automatic

Compression Ratio: 8.0:1

Fuel System: Motorcraft four-barrel

Body Style: Two-door Hardtop Coupe

Weight: 4893 lbs

Wheelbase: 121 inches

Base Price: $7701

Number Produced: 42,685

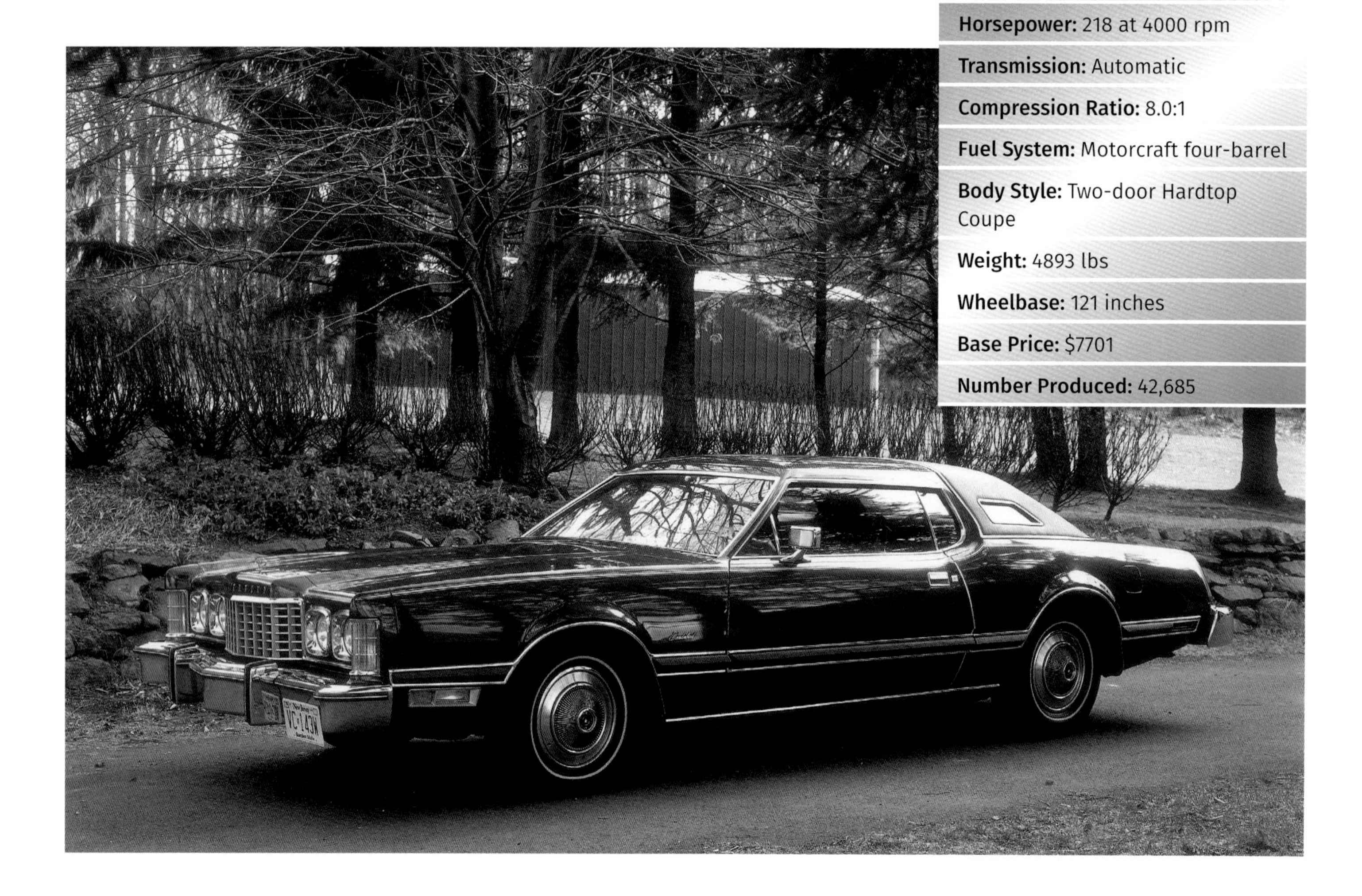

With the exception of the full-size models, the 1975 Fords remained as they were for 1974. Ford had introduced their Thunderbird model to the market in 1955, almost simultaneous to the launch of the Chevy Corvette. It was an exceptionally handsome-looking car and one that met with far greater commercial success than the original Corvette models. The Thunderbird was originally a two-seater personal car, but grew to a fourseater in 1958 and shrugged off its original pretensions to being a sportscar. The Thunderbirds were now squarely aimed at the lucrative personal luxury sector of the market. Even so, in the pre-Mustang days, Thunderbirds were easily Fords sportiest models.

Despite a brief flurry of Supercharged Thunderbirds in the sixties, by 1975, the model had come along way since its exciting origins. It was now FoMoCo's top–of–the–line personal/luxury car. A highly promoted new feature was the Sure-Trac rear brake anti-skid device. There were also other chassis refinements, such as four-wheel power disc brakes and the Hydroboost hydraulic brake boosting system. Standard features to the car included automatic transmission, power steering, power brakes, power windows, air conditioning and an AM/FM radio. Optional collectors' cars came with either the Silver or Copper Luxury Groups packages. These came with special heavy-grained half-vinyl roof, velour of leather seats, deluxe trunk lining and specific wheel covers.

Various options were on offer, including six-way power seats, Cruise Control, Climate-control air conditioning, electric rear window defroster, power moon roof and the Turnpike Convenience Group (which for $138 included cruise-control, a trip odometer and manual passenger seat recliner).

A single two-door hardtop coupe Thunderbird was offered for 1975. They were manufactured at either the Los Angeles or Wixom, Michigan plants.

1978 Dodge Aspen Super Coupe

The year 1978 wasn't a good one at Dodge, who ranked at number eight in the auto production league, with model sales down by 5 per cent. Production was further depressed, by as much as 21 per cent. Manual gearboxes for this year's models were manufactured by Volkswagen, whilst automatic gearboxes were made by Chrysler. Dodge advertising from this period featured a Sherlock Holmes character. Ho hum...

Similar in overall appearance to the 1977 version, this year's compact Aspen sported a revised front end. The new narrower one-piece grille was made up of horizontal bars, but without the former recessed side portions. Instead, those twin areas between the grille and headlamps were occupied by square amber parking lights, each with two horizontal trim strips. Round headlamps were again recessed in bright square housings. Rear end appearance changed too, with revised taillamps. Each rectangular unit was about one-third the car width (with backup lamps to the center of the car) was split by two horizontal trim strips that continued across the center trim panel. That panel displayed 'Dodge' block letters in the center row. As before, the standard model had no bodyside moldings. This year's Aspen weighed less, as a result of the changes in the windshield, inner body panels, headlining materials, and brakes. Ten new body colours were on offer, mostly metallics, and five colors were carried over. Only one trim level was now offered, but a long list of options was available for buyers to personalzse their cars.

The car was offered in three body styles. The three Aspen sporty models were also on offer, the Super Coupe, R/T and the R/T Sport Pak. The Super Coupe arrived later in the season, and included dark brown body paint with black-finish hood, front fender tops, headlamps, wiper arms, front and rear bumpers and remote-control racing mirrors.

Bodyside and roof tape striping was included, along with wheel flares, front/rear spoilers, quarter-window louvers, GR60 x 15 Aramid fiber-belted raised white letter radial tires on eight-inch GT wheels, heavy-duty suspension with rear anti-sway bar, and the 360-cid four-barrel V-8. Standard gearboxes were three-speed manual, but four-speed manual and Torqueflite automatic were available as options.

Engine: 90-degree, overhead valve V-8

Displacement: 360-cid

Horsepower: 175 at 4000 rpm

Compression Ratio: 8.0:1

Weight: 3255 lbs

Base Price: $3917

Number Produced: 48,311

1978 Ford Mustang King Cobra

Some people have viewed the Mustang as the ultimate ponycar, throughout the seven generations of this model's design and technological innovation. It certainly is an American icon. Sadly, the King Cobra II is by no means one of the most sought after incarnations of this legend. Some said that the car was all face, no muscle. Others said it was as 'slow as molasses'. But no one could reasonably argue that the car isn't eye-catching.

By the end of 1973, the once-mighty Mustang was on the ropes, the ponycar market was crowded with any number of also-ran cars, and the demand for big, powerful, clumsy dragsters was shrinking drastically as the oil crisis loomed. The original Mustangs had been based on the Falcons, but Lee Iacocca wanted to scale up the modest, bread-and-butter Pinto II into the new Mustang model in order to fill the lucrative sporty sedan niche that made so much money for Detroit. These cars became the Mustang IIs. The Pinto was beefed up to use the heavy-duty parts necessary to control the vibration of the larger engines. The emphasis at Mustang was now on flash rather than performance, and oldtime Mustang fans were soon crying out for a return to the good old days.

The striped and spoilered Cobra II hit the streets in 1976, and 1978 saw the introduction of the King Cobra Mustang II. Mustang hoped it would be the answer to their prayers. The package was offered with four-speed manual transmission and featured a better suspension with front and rear stabilizer bars, together with power front discs and power steering. The top power train was the 302-cid V-8 with a Variable Venturi Carburetor. This option empowered the car to cover the quarter mile in 16.59 seconds at 82.41 mph, and did 0-60 mph in 11.2 seconds. Goodyear P195/70R13 radial tires were fitted, and the cars were adorned with a completely wild paint and graphics appearance package. Most King Cobras were built in Bright Red, but technically, any Mustang color was available. The King Cobra package (at a cost of $1,277) pushed the price of a Mustang II from a base of $4088 to over $6800. To add to the marketing furore, Ford teamed up with Munroe Shocks to create a total of eight show cars, collectively known as the 'Munroe Handler'. These cars featured custom bodywork and numerous handling upgrades and were great ambassadors for the production models.

Sadly, this would be the last year for the Mustang II models. An all-new Mustang was to debut in 1979, modelled on the Ford Fox platform.

Engine: V-8, 90-degree, overhead valve

Displacement: 302-cid

Horsepower: 139 at 3600 rpm

Compression Ratio: 8.8:1

Base Price: $6800

Number Produced: 5000 King Cobras

1979 Ford Indy Pace Car Mustang

The year 1979 started with Jimmy Carter telling the nation about his six-point plan to combat America's oil crisis. Inevitably, this led to a search for better gas economy.

Mustang had already decided to end their run of Mustang IIs, and introduced their third generation – Mustang III - models in 1979. In fact, the basic idea for these dated back to 1974, even before the second generation Mustangs were launched. These were based on the new 'Fox' Ford global platform, and it was hoped that it would be possible to manufacture the car in Ford factories worldwide. Unfortunately, this 'world car' concept proved impractical. Ford commissioned four different design teams to style the car, including Ghia in Italy. The chosen design came from Jack Telnack's group at Ford's own North American Light Car and Truck Design department. The new cars had a longer wheelbase and roomier cabin, yet they weighed 200 pounds less than the previous models, and were far more aerodynamic. Mustang IIIs were offered as a two-door booted model, or a three-door hatchback. A convertible wasn't added to the range until 1983. The car still had the live rear axle, but the front suspension was a new MacPherson strut layout. Three levels of ordinary suspension were offered, and a Special Suspension package was optional, complete with cast aluminium wheels. The same engine range was initially retained, but even so, the improved weigh-to-power ratio and wind tunnel tested lines imparted far better performance to the car.

The new shape Mustang proved to be an immediate success, boosting model sales by over 80%, and was credible enough to act as a pace car for the 1979 Indianapolis 500 Race. The true 'pace' cars used a highly modified 302-cid engine.

Ford built around 6,000 Indy Pace Car replicas to celebrate this honor. These were fitted with either a turbo four-cylinder, or 5-litre V-8. A Front air dam with integral fog lamps, non-functional rear-facing hood scoop, rear spoiler and sunroof were standard features on Pace Car models, together with a special silver, black orange and red paint scheme. But the factory left the lettering decals to be fitted, or not, by the dealer or customer.

Almost 2.5 million Mustangs built on the IIIrd generation platform between 1979 and 1992, with only minor changes to the front and rear of the body.

Engine: V-8 Cast iron block

Displacement: 302-cid

Horsepower: 140 at 3600 rpm

Compression Ratio: 8.4:1

Base Price: $5,000 plus

Number Produced: Approximately 6,000

1981 Chevrolet El Camino

Engine: V-8

Displacement: 305-cid

Horsepower: 145

Compression Ratio: 8.6:1

Induction: 4Bbl. Rochester 4ME

Body Style: Car-derived truck

Wheelbase: 108.1 inches

Number Produced: 41,091

The El Camino resulted from a mating of car with truck, to produce a wonderful hybrid that resulted in a model war with the Ford Ranchero that lasted for several decades. El Camino means 'the road' in Spanish, and it turned out to be a long and winding one for this model, lasting for more than thiry years of production.

Ford introduced the Ranchero back in 1957, Chevrolet brought out their contender late in the following year. The El Camino may have been slightly derivative, but it outlasted the Ranchero, which was discontinued in 1979. The Ford's sales had faded after 1973, largely due to the oil crisis.

Both vehicles strayed from their roots in the seventies, when they were absolutely loaded with every kind of comfort and performance option. Rather than being rugged haulers, both models became the forerunners of the huge American SUV market. In this decade, El Caminos were fitted with vinyl roofs, air conditioners and 8-track players, and a lot of hot performance equipment was also offered. It was the performance that really stimulated sales. Both Rancheros and El Caminos were advertised as being great second cars that offered fun and flexibility, as well as being available for hauling those larger purchases home. The El camino remained very popular into the late seventies, especially with the 'SS' (Super Sports) performance package.

The Chevrolet designers kept coming up with more and more outrageous styling packages. The epitome of El Camino flashiness was achieved by the Royal Knight special edition. Gold-accented black paint, and a fabulous long sinewy design style were highlighted by a gold 'dragon' hood treatment. The Royal Knight option also came with a large front air dam, painted sports mirrors and rally wheels. Power came from regular production option Z16. The fifth and final generation of Chevelle/Malibu-based El Caminos sold very well in the eighties, with its annual tallies remaining above 20,000 until 1986, before falling to 13,743 in 1987. But the increasing popularity of mini vans and SUVs meant that there were other options to distract the natural El Camino buyer. GM bosses could see that the writing was on the wall, and finally discontinued the model in early 1988, over thirty years from its inception.

1983 Chevrolet Cavalier

Engine: Four-Cylinder	
Displacement: 121-cid	
Horsepower: 88 at 4800 rpm	
Compression Ratio: 9.3:1	
Number Produced: 627	

Chevrolet introduced the Cavalier in May 1982. It was Chevy's next-to-base model, just senior to the subcompact Chevette. This new front-wheel drive, four-passenger Chevy was described as subcompact on the outside, with compact roominess inside. Chevy described the car as a 'high-content vehicle' which meant that it was fully-loaded with a good array of standard equipment, including a radio, power brakes, reclining bucket seats, stabilizer bar and remote trunk/hatch/tailgate release. Four models were offered in this opening line-up, two- and four-door sedans, a two-door hatchback and a four-door wagon. These models were arranged in three tiers of trim, the lowbudget Cadet, interim base and high-end CL. Only a single engine was offered in the first place, a transverse-mounted 112-cid, four-valve engine, rated at 88 horsepower equipped with four-speed (overdrive) manual transmission. The Environmental Protection Agency fuel rating was 30 mpg for city driving. A two-liter version was announced but did not arrive for the moment.

The cars were designed with low, horizontal-style grilles, single recessed rectangular headlamps, and taillamps split into two sections by a horizontal divider. In 1983, the basic appearance of the model line was similar to that of '82, but a modified standard equipment list allowed significant price cuts to the subcompact Cavalier. This was due to General Manager Robert C. Stempel's new 'pricing strategy' that reduced the sticker price of over half the 1983 Chevrolets. These new base prices were as much as $1868 lower than in the previous year. But quite a few items that had been standard on the original cars were now classed as options. The range now included seven models, on three trim levels. The top line-up was the CS while the CL now became an option package. A new convertible model was available in the CS range. The convertibles were built by American Sunroof in Lansing, Michigan. Produced in limited numbers, it was the first Chevrolet ragtop since 1975. The car was equipped with tinted glass, power steering, power windows and twin sport mirrors (the left-had mirror was controlled remotely). The budget-priced Cadets were dropped completely, along with the base hatchback coupe.

The US Economy picked up in 1983, but the position at Chevrolet remained precarious. Model year sales improved by nearly six per cent, but other domestic manufacturers improved theirs by nearer seventeen per cent. Ironically, most of the improved sales came from mid- and full-size models, rather than the subcompact and compact models like the Cavalier.

1985 Avanti GT

Engine: V-8 OHV

Displacement: 305.4-cid

Horsepower: 180 at 4000 rpm

Raymond Loewy was approached by Sherwood Egbert, the President of Studebaker, in March 1961, with a brief to design a completely new car for the company. Loewy was probably the most important industrial designer of the twentieth-century. When the car was introduced to the world the following April, the Avanti inspired immediate acclaim, even though it was launched in what was possibly the most imaginative auto design era in history. 'Avanti' was derived from the Italian for 'go forward' and the car was aptly named. It represented a great leap forward both in design and technological advances, and was a brilliant hybridisation of American know-how and European styling. As the Avanti was built from fiberglass, it was possible to use completely unique sinuous styling. The material also offered safety and practicality.

Unfortunately, Studebaker was already in a dire economic position, and was unable to give the Avanti and its other two new models (the Lark and the GT Hawk) long enough to float the company out of trouble.

The last true Studebaker was produced on March 17 1966. But the Avanti model remains in production, still highly recognizable, to the present day.

In 1965, Nathan Altman and Leo Newman set up The Avanti Motor Corporation in Studebaker's old factory in South Bend, Indiana, to manufacture the Avanti II. This car was in not a replica of the original, but a continuation of the original Studebaker production. They used the original tooling. As Studebaker no longer manufactured engines at this time, Avanti decided to use Corvette powerplants, but the frame, suspension and body panels remained unchanged. The company was sold to Stephen Blake in 1982. Blake made some changes to the car, including dropping the 'II' from its name, and introducing a convertible version in 1985. But quality control problems and an erratic management style resulted in bankruptcy. Avanti's remaining assets were purchased by Mike Kelly in 1986. Kelly then formed the New Avanti Motor Corporation, with financial backing from John J. Cafaro. He believed that he was 'handcrafting the finest motorcar in all the world', but he and Cafaro were unable to agree as to whether the company should pursue the performance or luxury markets.

1985 Chevrolet Monte Carlo SS

Chevrolet made a big claim in 1985, that they were offering 'America's highest-mileage car, America's fastest car, America's most popular full-size car and America's most popular car, regardless of size'. That just about covers it. But internally, 1985 was a year of refinement and evolution rather than revolution. Several interesting models and engine options were announced, including the IROC-Z Camaro, and high performance Z24 Cavalier.

The first Monte Carlo was introduced by Chevrolet back in 1970, at the height of GM muscle car power. The car was designed to compete with the Pontiac Grand Prix and Buick Riviera. But just as soon as it was introduced to the market, performance fell out of favor, and the model's brief life as a muscle car was over almost before it began. However, this first generation of Monte Carlo's was a great hit with the buying public, who snapped up 145,975 of the cars in their launch year.

The top model in the range was the SS 454, bought by 3823 performance-inspired souls. The car was equipped with the mighty Chevy 454 big block, and was came fully-loaded with go-goodies, including heavy-duty suspension, a heavy-duty battery, automatic level control suspension, air shocks and 15x7 inch rally wheels. Ten Monte Carlos sneaked off the line equipped with the LS-6 454 engine – rated at a massive 450 horsepower.

The car evolved over the years, as did the SS model. By 1985, the standard Monte Carlos were equipped with a 130 horsepower powerplant, which was aimed at V-8 fans looking for improved gas mileage. Chevrolet claimed that it was almost as 'frisky'. The SS model 'cousin to the Grand National stock cars' racing on the NASCAR circuit was very strongly marketed. Formerly offered only in a white or blue body, this year's car came in silver, maroon, white or black with gray or maroon interior trims, and either bench or bucket seats. This street version of the NASCAR model had special instruments, sport suspension, rear spoiler, sport mirrors and steering wheel and special rally wheels. The car had very plain styling, and only discreet badging. The blackout grille was simple crosshatch pattern, flanked by recessed quad rectangular headlamps with deeply inset park/signal lamps below the forward crease line. The car was fitted with the RPO L69 up-rated version of the familiar 305-cid V-8, now with four-speed overdrive automatic transmission. The car now had beefier equipment, SS mufflers grew in capacity and axles got larger ring gears. Removable black roof panels were to become available on the SS later in the model year.

Engine: V-8

Displacement: 305-cid

Horsepower: 180 at 4800 rpm

Compression Ratio: 9.5:1

Number Produced: 35,484

1986 Ford Taurus

Engine: 4-cylinder transverse or Vulcan V6	
Displacement: 2.5/3.0 litre	
Horsepower: 90/140	
Transmission: 5-speed manual or 4-speed automatic	
Body style: 4-door sedan	
Wheelbase: 106 inches	

The first-generation Ford Taurus was produced by Ford as the first of six generations of the new concept. Introduced in December 1985 as a 1986 model, the front-wheel drive Taurus was a very influential design that is credited with saving Ford from bankruptcy, bringing many innovations to the marketplace and starting the trend towards aerodynamic design for American automakers in the North American market. Ford of Europe had launched the 1980s move to aerodynamic design for the company with the 1982 Ford Sierra. Development for the first-generation Taurus started in the early 1980s to replace the Ford LTD, at the cost of billions of dollars, with a team led by vice president in charge of car development Lewis Veraldi dubbed "Team Taurus." Ford was suffering from a lackluster product line from the late 1970s to the early 1980s and then-chairman Philip Caldwell staked much of the finances and future of the company on Veraldi and his team's success, giving them unprecedented leeway in developing what would become the Taurus. The Taurus' development employed a strategy of teamwork and customer communication that would prove very influential for the automotive industry as it consolidated all of Ford's designers, engineers, and marketing staff into a group who worked on the car collectively. When production ended in 1991, more than 2,000,000 First-generation Tauruses had been sold.

1986 Ford Mustang SVO

Engine: Four-cylinder	
Displacement: 140-cid	
Horsepower: 200 at 5000 rpm	
Compression Ratio: 8.0:1	
Base Price: $15,272	

Ford launched the second edition of the Mustang, Mustang II back in 1974, in part as a reaction to the prevailing oil crisis. They described it as 'the right car at the right time'. But the car was far removed from the inspiration at the inception of the pony car concept, and many aficionados were disappointed by the relatively lack lustre performance. But sales held up, and buyers were eager for Mustang III, which arrived with all new sheet metal in 1979.

Performance was also back. The Boss 302 V-8 came back on the scene in 1982, even stronger than before. The revised model line-up now contained a GT instead of the old Cobra option, but sales declined for the third year in a row.

Things looked up in 1983, when a convertible re-joined the line-up. Mustangs also received a restyled front and rear end, and the power of the V-8 was boosted to 175 horsepower. The turbo model also rejoined the Mustang family, now with multi-port fuel injection. Borg-Warner's close-ratio five-speed gearbox could help the GTs hit 60 mph in around seven seconds.

SVO (Special Vehicle Operations) department had been established in 1981, to oversee the company's renewed involvement in motorsports and limited edition production. Its first fruit -the SVO- appeared in 1984, with an air-to-air intercooler on its turbocharged four-cylinder engine. The car also had Hurst linkage, four disc brakes and big 16-inch tires. The car was immediately heralded as 'the best driving street Mustang the factory has ever produced' by Motor Trend. Road & Track magazine claimed that the SVO 'outruns the Datsun 280ZX, outhandles the Ferrari 308 and Porsche 944... and it's affordable'. Praise indeed. By 1986, SVO turbos were capable of 200 horsepower, ready for the Hurst-shifted five-speed. The cars retained their single-slot grille and luxury interior. In fact, high performance was a serious theme at Mustang between '84 and '85 for both hatchbacks and convertibles, and the cars were taken to heart by the V-8 brigade.

The Mustang line-up for 1986 was made up of five models, with the hatchback SVO Turbo Four the ultimate Mustang of the year. The car now carried a computer-controlled 200-horsepower 2.3 liter four cylinder engine with inter-cooled turbocharger.

1986 Dodge Daytona Turbo Z

Engine: Chrysler K 4-cylinder
Displacement: 2.2 liter
Horsepower: 146
Transmission: 5-speed manual/4 speed automatic
Body style: 3-door hatchback

In America, the traditional front engine rear drive pony car was being re imagined for the 80's with Ford's Mustang in 1979 followed by GM's Camaro/ Firebird in 1982. Last out of the bag was Chrysler's attempt, the Daytona/ Laser in late 1983.

The Dodge Daytona was produced from 1984 to 1993. This modern take on a pony car was based on the Chrysler G platform, which was derived from the Chrysler K platform with a transverse power unit and front-wheel drive. The Daytona originally used the 2.2 L Chrysler K engine in normally aspirated (93 hp) or turbocharged (142 hp) form. The 96 hp 2.5 L K engine was added for 1986. In 1985, the 2.2 L Turbo I engine's horsepower was increased to 146 hp. In 1986 the Daytona was available in two trim lines: standard and Turbo Z.Aiming high had the benefit of turning out a traditional American pony car concept with front wheel drive and a turbo charged four cylinder power unit which challenged the V8s cars of the day.

A new T-roof package was added to the option list, but just 5,984 Daytona owners chose this option. Total production this year would be 44,366.

The Daytona derived its name from the famous 1969-70 Daytona Chargers with their aero noses. Aside from the name, little linked the 1982 Daytona to its legendary namesake. The Daytona had the distinction of being the most aerodynamic Chrysler product ever, with a cd of 0.378.

The Daytona Turbo was on Car and Driver magazine's Ten Best list for 1984. A performance oriented "Shelby" version of the Daytona was introduced in 1986.

1988 Ford Mustang GT

Engine: V-8	
Displacement: 5-liter	
Horsepower: 225 at 4000 rpm	
Compression Ratio: 9.0:1	
Base Price: $16,610	
Number Produced: 18,174	

When Ford decided to re-introduce the Mustang GT in 1982, it was the first GT pony car model since 1969. 'The Boss is back!' The car was built on the Fox-based Mustang platform, and debuted with the 302-cid HO engine as standard equipment. The engine had also undergone an engineering restyle, and could now produce a further 17 additional horsepower, bringing its rating up to 157 at 4200 rpm. This was achieved with the addition of a 2V Motorcraft carburetor and the equipment to feed it with clean air. A big part of the new GT's allure was its specially designed exterior, complete with non-functional hood scoop. The racy cockpit styling, with special blackout trim and bucket seats was also attractive to buyers. In fact, the whole package was reminiscent of the classic American muscle cars, and was rated as the quickest domestic car produced in the year.

In 1987, the GT was extensively restyled inside and out, and more wild ponies were pushed under the hood. The competition was hot on the heels of the car, and Ford was forced to turn up the heat. Which it did, with the introduction of the '87 5.0 liter HO. Engine power was upgraded by a massive 25 horsepower with a hotter 302-cid block fitted with cylinder heads direct from the Ford Truck division. These heads featured better breathing valves, opened up exhaust ports and dished-top forged pistons replacing the previous year's flat tops. A stainless steel exhaust system improved airflow. As in past years, handling and suspension upgrades also showed gradual improvements, new strut housings were fitted, plus braking and cornering were improved. This and the new exterior design set the tone for the next 6 years of GT production.

Everything had gone so well in 1987, Ford decided not to change a thing, resuming GT production with no major changes for '88. Sales bore out this wisdom of this decision, hitting an all-time high for both hatchback and convertible models, with a joint total of 68,468. Externally, the cars also retained their 1987 'aero' look, with flush-fitting headlamps, aerodynamic nose, rear wing, round fog lights, louvred taillights, flush rear quarter window, ground effects package and Turbine wheels. Ergonomic GT design continued inside the car, with a new dash and instrument panel. But optional leather articulated sport seats were newly on offer for the convertible. The '87-'88 models were so successful, that Ford decided to continue their production until at least 1993. In fact, they simply run out of parts in '87, leaving over 10,000 orders unfulfilled.

1988 Ford Thunderbird

The Thunderbird was Ford's answer to the Chevy Corvette. But whereas the Corvette never strayed far from its original two-seat sports car concept, the Thunderbird would soon add a hardtop coupe version, a back seat, and even two more doors...Ford made a serious attempt to offer a do-everything car in the Thunderbird, but it always provided performance, along with a dash of style.

The massive Thunderbirds of the 1970s seemed fully forgotten as the new generation Thunderbirds arrived for 1983. This one was loaded with curves – and before long, a turbocharged, fuel-injected four-cylinder engine, equipped with close-ratio five-speed and 'quadra-shock' rear suspension.

The Thunderbird Turbo Coupe became quite an attraction, helping sales of the line to more than double. A convertible Thunderbird reappeared in 1983, and a five-liter Thunderbird Sport model was added to the line in 1987. All the sheet metal was completely restyled for this year, although the personal luxury coupe's basic profile remained virtually untouched, and the underpinnings were unchanged. The Turbo Coupe was the senior Thunderbird for '87, and its body featured a grille-less nose and functional hood scoops. Its top speed was 131 mph.

An unchanged management guided Ford through 1988, with corporate sales and earnings breaking records for the third year in a row. World sales for the corporation were $92.4 billion, topping General Motors.

The base engine for 1988 Thunderbirds was the V-6, but a further 20 horsepower were wrung from the engine by replacing the former single-point system with a multi-point fuel injector. Inside the engine, a new balanced shaft produced smoother running. Dual exhausts were now standard with the five liter V-8 engine, which was fitted to the Sport model and optional on the base/LX editions. As before, the Turbo Coupe carried a turbocharged 2.3-liter four-cylinder engine. Other standard Turbo equipment included a five-speed gearbox, anti-locking barking, electronic ride control and 16-inch tires. Sports models switched from a standard electronic instrument cluster to analog gauges, and came with articulated sport seats.

The Mustangs of 1987 and '88 were so successful, that Ford scrapped their plans to launch the Mazda-based Probe model as the 'Mustang IV'.

Engine: 6-V 90-degree overhead valve

Displacement: 232-cid

Horsepower: 140 at 3800 rpm

Compression Ratio: 9.0:1

Body Style: Two-door Hardtop Coupe

Wheelbase: 104.2 inches

Base Price: $14,320

Number Produced: 147,243

1990 Chevrolet Corvette ZR-1

Launched in 1953, the Corvette had had to adapt to survive. Although the hefty rise in the price of gas, Federal emissions regulations and the 55 mph speed limit had conspired to make the car redundant, the Corvette retained and developed a loyal customer base that survives to this day. The half-millionth Corvette came off the St. Louis production line in 1977. The Silver Anniversary car of 1978 was effectively more of a luxury tourer than a sports model. Manufacture of the cars was subsequently switched to Bowling Green, Kentucky to a specialist Corvette plant.

Jerry Palmer, the Head of Chevrolet Three Studio and Dave McLellan, the Corvette design engineer, were now developing a new model, that would combine aerodynamic styling, nostalgia and modernity to create a new, sixth generation Corvette that would launch in 1984. The car was so successful in everything it set out to do, that it remained the basis of the car until its fortieth birthday in 1993, even its fiftieth in 2003.

The '84 car cleverly combined the great heritage of the model, while looking new and exciting. Styling was restrained, but all the signature Corvette features were still there to be seen – fold-away headlights, four round recessed taillights, wraparound rear window and fender vents harking back to the indented coves of the '56 model. Top speed was measured at 140 mph, and the car could get from 0-60 mph in 6.7 seconds. The car was, in fact 'one of the half-dozen fastest production automobiles in the world.' However, it soon became evident that the brilliant new 'backbone' chassis and tough suspension could handle even more power. This resulted in the decision to rework the thirty-year-old 427-cid V-8 engine. The result would be known as the ZR-1. It was Corvette's intention to rival the performance of the hot Europeans – Ferrari, Aston Martin and Porsche at a fraction of the price. Tuned port fuel injection, gas-pressurized shocks, an oil cooler, ABS from Bosch, and six-speed manual gearbox were all added to the Corvette during the eighties.

When it arrived in 1990, the super-performance ZR-1 was effectively a seventh series Corvette, which had cost around 24 million dollars to develop. It was worth it, the car is one of the all-time greats. It was fitted wit the fantastic Texas-built LT5 engine, complete with huge acceleration, high top speed, masses of torque, and fuel efficiency. It all added up to making the car one of the most potent machines ever to hit the streets, that held the driving experience as its paramount raison d'etre.

1992 Pontiac Firehawk Firebird

The third generation Firebird made use of many unseen build improvements in its final model year of 1992. These included the use of improved structural adhesives, improved body panel fit and finish, reinforced structure for rigidity and silicone impregnated weather striping. Even so, GM were still concerned about the forces generated by their big 5.7 liter Firebird engines, and wouldn't allow this option to be ordered with either T-tops or a manual transmission.

1992 was the dawn of a new era of serious performance with the introduction of the Formula Firehawk. This was the fastest, most extreme, hard edged, Firebird ever. 250 cars were slated for production (in street and competition versions), but only 25 actually made it into production, including a single convertible.

This Firehawk, Option code B4U, was the fastest street Pontiac ever produced. It arrived as a new model to the Firebird line-up as a special kind of Formula model. The cars were actually normal Pontiac-produced Firebirds that were then sent to development partner SLP (Street Legal Performance) for the Firehawk package to be installed – for a hefty sum. The SLP package included a number of engine and suspension upgrades (including heavy duty 4-bolt main engine blocks with forged crank), Brembo brakes, a tire and rim package – and the Firehawk graphics. The upshot of this was sparkling performance: 0-60 mph in 4.6 seconds, quarter mile times of 13.2 seconds and a top speed of 160 mph. Amazingly, the car still gave good gas mileage at 25 mpg, and was street legal in fifty states.

'Mandatory Red' was the standard color option for the 1992 cars, but single Artic White, Medium Quasar Blue Metallic, Medium Green Metallic and Dark Jade Gray Metallic examples were built. Two of the cars were fitted with the Super Duty 366-cid engine that was rated at 375 horsepower. SLP had built a prototype Firehawk to develop the package. Early Firehawks also got a special interior package, including Recaro racing seats, and a full roll bar. After the full SLP treatment, the cars retailed for around $40,000, easily the most expensive Pontiac ever to hit the streets. It was Pontiac's hot ticket for performance, designed to compete with the Mustang 5.0 LX Notchback coupe. The Firehawk helped the company to regain its position on both the street and the track with an infusion of pure performance. SLP began to install a similar package to Camaros in the late '90s.

Engine: V-8, 90 Degree

Displacement: 350-cid

Horsepower: 350 at 5500 rpm

Compression Ratio: 9.9:1

Wheelbase: 101 inches

Base Price: $39,995

Number Produced: 25

1992 Saturn

Engine: Inline four-cylinder
Displacement: 116-cid
Horsepower: 85 at 5000 rpm
Base Price: $9195

It was GM President Richard G. 'Skip' LeFauve who sanctioned the $3.5 billion investment that was needed to establish Saturn, in the belief that Americans would buy domestic cars 'as good as the Japanese imports', he wanted to 'out- Japan the Japanese'. The aim was to sell American cars to Honda and Toyota buyers. Initial discussions began in 1982 when a 'new, innovative, 'small car project'' was mooted. The company was actually launched in 1985 as a whollyowned subsidiary of GM, and the first new division of the company for sixty-six years.

'Saturn' was the original codename for the company, named for the small rocket that carried American astronauts to the moon. The new firm was based in Troy, Michigan and had a brand new 4.1 million manufacturing and assembly complex in Spring Hill. The first Saturn to roll off the line in 1990, was a red metallic four-door sedan. Teething troubles necessitated some recalls, but the company managed to handle these so effectively, that they made them look almost beneficial.

Saturn's slogan was 'A different kind of company. A different kind of car'. This also extended to its sales practices, and the cars were offered with nonnegotiable set prices in no-hassle dealer environments. But by the late '90s, the small-car market was shrinking, and Saturn introduced a mid-sized car to the range. Saturn's basic principle was to concentrate on a few core vehicles, and steer clear of the 'instant collectible' market. Their design was attractive and uncontroversial, but never earth-shattering. But there were some firsts in the company history, a three-door coupe was introduced in 1999, intended to breathe new life into the coupe market. It was said to be the first three-door American production model.

But the company's position became gradually more and more difficult. GM began to demand a return on their investment, and inter-company rivalry developed with the product developers in other divisions. Ultimately, GM decided to bring the company in-house, and abolish its separate status, that it felt it could no longer afford. However, Bob Lutz, GM Vice Chairman maintained that 'Saturn's going to be fine'. Rumours abound that GM intend to double the size of the Saturn product line by the end of 2006, adding a sports roadster, mid-size car and large SUV.

Saturn offered four models in 1992, all with very similar styling. A single shorter-wheelbase two-door coupe was offered, and three four-door sedans, in different tiers of trim.

1993 Corvette: 40th Anniversary Model

On July 2 1992, the one millionth Corvette was driven off the Bowling Green, Kentucky assembly line. Like the first Corvette to be manufactured in the original Flint, Michigan plant, it was pure white convertible. An iconic car, it just travelled right across the street to the National Corvette Museum at 350 Corvette Drive.

1991 saw a subtle re-styling of the Corvettes, including the 'King of the Hill' ZR-1. To the annoyance of ZR-1 drivers, the entire range was now fitted with the same rear panel and square taillights. On a brighter note, the 1991 Daytona 500 was won by Jeff Gordon, driving a Corvette.

Dave Hill took over as Corvette chief engineer in 1992, and oversaw the introduction of the LT-1 version of the 250 small-block, this took output to 300 horsepower, and performance of the standard 'Vette to only a shade lower than that of the ZR-1. All the cars were still equipped with ABS II-S anti-locking braking and driver's side airbag, as well as an anti-theft system. The ZR-1 itself was still powered by the 32-valve DOHC 5.7-liter V-8 matched with a six-speed transaxle. Regular Corvetes used the 5.7-liter V-8 fitted with the four-speed overdrive automatic or optional six-speed manual transmission.

Corvette marked its 40th anniversary I 1993 with a special appearance package that was available on all Corvettes. This consisted of a metallic Ruby Red commemorative edition, and it accounted for about a third of the sales for this year. The leather interior was color-coded to the paintwork, and adorned with embroidered headrests, and the bodywork was embellished with various badges. Under the hood, the motor was supplied with sound-deadening polyester rocker. The ZR-1 horsepower rating climbed from 375 to 405, in response to the launch of the Dodge Viper, one of the very few cars in its class. The car was still fitted with the six-speed transaxle. Regular Corvettes were equipped with the 5.7-liter V-8. 1993 Corvettes were also the first cars to be fitted with GM's Passive Keyless Entry System.

A convertible Corvette have reappeared in 1986, having been absent from the range since 1975. Ragtops had been offered in every subsequent year. The stiffened chassis of the sixth generation had made it possible for the car to appear in roadster form once more.

Engine: V-8

Displacement: 350-cid

Horsepower: 300 at 5000 rpm

Compression Ratio: 10.5:1

Wheelbase: 96.2 inches

Base Price: $41,195

Number Produced: 5,712

1993 Oldsmobile Eighty-Eight LSS

Oldsmobile has always been able to present a good variety of cars, with both creature comforts and performance for a fairly wide range of buyers. Historically, the company also has a reputation for introducing many technological advances, including automatic transmission in 1940, the short-stroke overhead valve V-8 in 1949 and air conditioning in 1953. In the final decade of the twentieth century, the division attempted to incorporate recognisably 'Oldsmobile' design elements into their new models. This was an attempt to capitalize on the brand loyalty that has always been associated with this GM company.

In the full-size family sedan market, comfort, reliability and value are the key motivators. Oldsmobile had always known exactly how to tap into this market, by keeping their cars fresh and desirable with a rolling program of continuous improvement.

Even though Oldsmobile had made several changes to the Eighty-Eight series for 1992, the 1993 models in this sedan-only series featured several further refinements. In the previous model year, improvements were concentrated on significantly upgrading the interiors and safety related features, but the 1993 developments were mostly concentrated under the hood. The standard 3.8-liter V- 6 engine for the range was now tuned with a higher compression ration, new fuel injectors and revised emission controls. This resulted in a powerplant with more torque (225 pounds per foot, up from 220 in the previous year). The car also had better fuel economy, cleaner exhaust and improved acceleration. The engine was linked to an electronic shift four-speed automatic transaxle. Anti-lock brakes and a driver's side airbag were now standard across the Eighty-Eight model range.

The Eighty-Eights were still offered in both base and LS trim levels. They now came with parking lamps in the front bumper, and back up lamps were again positioned in the back bumper. The cars also had fixed-vent front windows; flush, rectangular body-colored door handles and rear windows that roll down just half way. An LSS package was offered on the LS sedan that transformed the big comfortable car into a sport sedan, equipped with bucket seats, full instrumentation, touring suspension, variable effort steering and 16-inch aluminium wheels.

Satisfied customers remarked on the good gas mileage of the cars, extremely comfortable and spacious interior, enjoyable ride and general reliability. On the downside, some buyers commented that the cars had trouble getting up hills.

Engine: 90-degree OVH valve V-6

Displacement: 231-cid

Horsepower: 170 at 4800 rpm

Compression Ratio: 9.0:1

Base Price: $21,949

Number Produced: 47,428

1994 Chevrolet Camaro Z28

Engine: Z28 V-8

Displacement: 350-cid

Horsepower: 275 at 5000 rpm

Compression Ratio: 10.5:1

Base Price: $18,745

Number Produced: (Base and Z28) 7260

GM launched the Camaro in 1966, and the GM copywriters went into full swing, 'Meet the masked marvel. Meet Camaro'. The car was very well received, despite the difficulty that GM had had in naming the car. They were looking for a European-sounding name that began with a 'c', in the Chevy tradition. Evidently, what they turned up actually means 'shrimp-like creature' in Spanish.

But it wasn't until the Z28 models were first introduced that GM really had a car that could run head-to-head with the Mustang. Z-28s were built specifically for racing in the Trans Am series, and necessitated a substantially new car. You can always tell when a concept really works – it keeps coming back over and over again. The tag actually remained in use right up until the end of the twentieth century.

Like the Corvette, the Camaro was redesigned over several generations. By the early 1990s, the wedge-shaped Camaro, that had looked so radical when it was first introduced in 1982, looked old and stale. In the meantime, a whole new generation of two- and four-seater Japanese sports cars had entered the American car market, with a degree of technical sophistication that somewhat overcame their small engine sizes. The fourth generation Camaro, first launched in 1993, was a real product of the 1990s, with soft, smoothed over lines. The car was a totally new product from nose to tail. But the Z28 performance option remained a true American under the bodywork, complete with traditional V-8 power and rear-wheel drive. The new Camaro was introduced as a single model hatchback coupe, but a convertible was added to the range for the following year after a year's absence. It was offered in both the base V-6 and Z28 series, as were the coupes. The new convertible had a sleek new body of composite panels fixed to a steel structure. The front suspension and rack-and-pinion steering were also new.

When the Z28 package was fitted to the convertible, it meant a V-8, rather than V-6 powered the car, and raised the Camaro to the muscle car league. The associated go-goodies, including aluminum cylinder heads, Borg-Warner six-speed manual gearbox, ABS brakes and eight-inch wheels with Goodyear 235/55R16 tires meant that the car actually had to be speed limited – to 108 mph. With Z-rated tires, the real top speed was over 150 mph. This meant that this Z28 was the fastest road-going factory-built Camaro to date.

1994 Chevrolet Impala SS

Engine: LT1 Corvette V-8
Displacement: 350-cid
Horsepower: 260 at 5000 rpm
Compression Ratio: 10.5:1
Weight: 4218 lbs
Wheelbase: 115.9 inches
Base Price: $22,920

Back in the sixties, the Impala was a car with classic muscle car potential, a prime candidate to be beefed up for the performance brigade. The car had been downsized in the early '60s, and the lowered weight meant greater performance potential. This had been a reaction to the compacts that hit the market at this time, including the Ford Falcon and Chevy's own Nova. Coupled with the size issue was the range of powerful Chevy V-8s available. For example, there was the meaty 348-cid Turbo Thrust available with either 340 or 360 horsepower, or the 409 Turbo-Fire that inspired the Beach Boys song. This could offer up to 409 horsepower, equipped with twin four-barrel carburetion. An SS (Super Sports) handling package could also be fitted, complete with power steering and brakes, heavy-duty springs and shocks, sintered metal brake linings and a tachometer. But although the Impala could offer a truly hot package, and was a serious contender in the NASCARs of the early 1960s, only a moderate number of buyers were convinced.

Over the years, SS status was sometimes compromised, badges were affixed to cars with no performance equipment at all, but the legend persisted.

Chevrolet has often been inspired to resurrect old and evocative names from its rich and varied history. The Impala SS of 1994 came out of this impetus. The original SS had been capable of generating 400 horsepower, quarter mile times of around twelve seconds, and fierce admiration on the streets. Thirty years later, kids who had been too young to drive when the first cars were available had their dreams brought to life by Chevrolet. The new Impala SS was based on an ordinary Caprice Classic Sedan, but the engine was uprated to the Law Enforcement 260 horsepower V-8 and suspension, complete with sequential fuel injection and all-wheel disc brakes. This propelled the car to fantastic performance. The Caprice was already a popular model, and its big smooth shape look really good when lowered. The all-black body of the SS, contrasted with polished aluminum wheels looked great straight from the factory. The interior was slightly dull, in grey cloth, but performance was the inspiration behind the car. The model was offered for only four years, but sold in excess of 40,000 units. It was living proof that the original muscle car concept of four-seater cars still had a loyal following.

1994 Chrysler LHS

Engine: V-6 overhead valve
Displacement: 214-cid
Horsepower: 214 at 5800 rpm
Compression Ratio: 10.5:1
Induction: Sequential fuel injection
Body Style: Four-door Sedan
Wheelbase: 113 inches
Base Price: $30,283
Number Produced: 44,739

There were big changes to Chrysler's line-up for 1994. In fact, it was vastly different to that of the previous year, with the 'new' New York LH sedan and LHS sedan joining the remaining LeBaron models and the Concorde Sedan. The LeBaron ranks were thinned considerably. Gone were all the coupe models and the base and LX convertibles. The LE and Landau sedans and the GTC convertible were offered again, but it was the final year for sedan availability.

The New Yorker LH series was introduced in May 1993, and featured the cabforward design sported by most other Chrysler models in this year. Running changes, from one model year to the next were planned as an intrinsic aspect of the model range. Standard features of the car included dual air-bags, anti-lock four-wheel disc brakes, air conditioning (with non-CFC R-134A refrigerant), stainless steel dual exhaust, solar control window glass, windshield wipers with vertical speed sensitive operation, and courtesy lamps with fade off time. The 'Fall' model of the car also had variable-assist speed-proportional power steering and aluminised exhaust system coating, 'Simplex' wheel covers and touring wheels and tires.

The LHS series was the new top-of-the-line Chrysler, and senior member of the New Yorker series. The cars were identical to the New Yorker LH cars, and it too was subject to a program of rolling model changes. The power train was the same for both cars, and they shared most of the standard equipment. The difference was that variants that were optional on the LH were standard on the LHS. These included low-speed traction control, automatic temperature control, vehicle theft security alarm system, eight-way power driver and passenger seats, automatic headlights with delay feature, automatic day/night rearview mirror and power moonroof.

Both cars also shared the same wheelbase and chassis, and were offered in the same range of colors, Black Cherry, Radiant Red, Char-Coal, Emerald Green, Black, Bright White and Bright Metallic Silver. Interiors were available in Agate, Medium Driftwood, Medium Quartz and Slate Blue.

1994 Ford Taurus

The second-generation Taurus shared all of its mechanical parts with the first-generation Ford Taurus, yet its exterior and interior were nearly completely redesigned. However, its exterior still strongly resembled that of the first-generation Taurus, leading many to falsely believe that the second-generation was simply a facelift of the first-generation Taurus. This is partially true, however, as the wagon model, from the B-pillar to the rear of the car, was a carryover from the first generation.

A high performance version the Ford Taurus SHO (Super High Output) was produced by Ford from 1989 until 1999.The SHO was redesigned in 1992, although it continued with the same powertrain as before: The Yamaha Built V-6 engine and 5-speed manual transmission. The second generation SHO borrowed from the Mercury Sable's front fenders, hood, and headlights, but used a different bumper, fog lamps, and no middle lightbar. The SHO also got unique seats, side cladding, dual exhaust, as well as a unique rear bumper. The 1992 version mildly up dated for 1994 can be visually identified by not containing a rear trunklid spoiler, having downturned exhaust tips, and only a driver's side airbag (later models have both driver's and passenger airbags).

The second generation of Taurus proved to be very popular, selling 410,000 units in its first year, becoming the best selling car in the United States

Engine: V6 SF Vulcan or Essex V6

Displacement: 3.0/3.8 liter

Power: 140

Transmission: 4-speed automatic

Body style: 4-door sedan

Wheelbase: 106 inches

1994 Ford Mustang GT

Engine: V-8

Displacement: 302-cid

Horsepower: 215 at 4200 rpm

Compression Ratio: 9.0:1

Wheelbase: 101.3 inches

Base Price: $21,960

Number Produced: (all coupes) 78,480

Ford first introduced the Mustang in 1964 and it is the progenitor of the entire ponycar herd. The car was designed and marketed to be a low cost sports car, and did not have any serious intent to be a performance model. But as the series developed, all kinds of high performance options were offered in the line, beginning with a 'Hi-Po' 289-cid 271 horsepower engine only a few months after the car was first launched. The first generation of Mustang was offered between 1964-1973, first as a two-door sports car and later as a four-door. The Mustang II's rolled out in 1974. Initially they weren't very well received, but were soon selling well, and several special models were added to the line with extra power options. The cars were restyled again in 1982 onto the Ford Fox platform. This model persisted until 1993, and was the base for all kinds of performance models including GTs, Cobras and several SVO (Special Vehicle Operations department) cars.

In 1994, in its thirtieth anniversary year, the long-overdue next-generation Mustang was unveiled in two body styles – notchback coupe and convertible – and three separate series: the base level Mustang, GT and limited edition Cobra. The sheetmetal was completely new, but the exterior and interior dimensions were virtually unchanged. The three-door hatchback, four-cylinder engine and LX series were now all things of the past. Engine options were now the 3.8 liter V-6 fitted to the base model, or the 5.0 liter V-8 fitted to the GTs and Cobras. The Cobra's sequentially fuel-injected engine was modified through the use of GT-40 cylinder heads, roller rockers, stronger valve springs and a tunedlength cast-aluminum upper intake manifold. Other changes included an oilcooler and a lightened flywheel. The cars were still rear wheel drive, but allwheel disc brakes were now fitted as standard. The gears in the Cobra's T5OD five-speed manual transmission were coated to better handle the engine's 240 horsepower output. Standard equipment in all Mustangs included dual airbags.

The 1994 Mustang convertible was now assembled in house. These now included power tops and glass rear windows with optional rear defogger.

The new 1994 Mustang was named as Motor Trend Magazine's 'Car of the Year', and a Mustang Cobra paced the 1994 Indianapolis 500. Mustang production totals for the year were almost double that of 1993.

1994 Pontiac Trans Am 25th Anniversary Model

Pontiac launched their fourth generation Firebird/Trans Am in 1993. The cars were completely reengineered and redesigned into a well-balanced package that was both in tune with stringent government regulations and the high expectations of the performance-buying public.

The twenty-fifth anniversary model of 1994 was the epitome of this early fourth generation car. It followed a Pontiac tradition of issuing anniversary models that had begin in 1979 with a ten-year anniversary model, and was continued in 1989 with a twenty-year commemorative car. The most outstanding feature of the 1994 special edition is its unique styling. The car was white with a blue 'screaming chicken' decal on the nose, and blue stripes running up the hood of the car, down the deck lid and over the rear spoiler. These were borrowed from the 1970-72 Trans Am model years. The door bottoms also had blue twenty-fifth anniversary Trans Am script. The car also had special lightweight rim, five spoke wheels that were painted white to match the car. There were also ornamented with the blue bird, and twenty-fifth anniversary Trans Am logos.

The cars interior was white and black, with white leather Prado seats decorated with blue embroidered twenty-fifth anniversary Trans Am script, and white door panels.

All the 1994 anniversary models were powered by the LT1 350 'Corporate' V-8 and equipped with a four-speed automatic transmission. They could be ordered as hardtops, T-tops or convertibles. In fact, for the whole 1994 Trans Am line-up, convertibles were added to each series, in addition to the coupes offered in '93.

Sadly, Firebird production was discontinued in 2003, in the thirty-fifth anniversary years of the car. Pontiac issued a 'Collectors' Edition' car with special 'Collector Yellow' color scheme, detailed interior appointments and special edition emblems. A Gold Rush Trans Am show car was also manufactured to be displayed at car shows around the country.

Engine: LT1 V-8

Displacement: 350-cid

Horsepower: 275 at 5000 rpm

Transmission: Four-speed automatic

Compression Ratio: 10.5:1

Weight: 4200 lbs

Wheelbase: 101.1 inches

Base Price: $21,395

Number Produced: (total Firebirds) 51,523

1995 Ford Bronco

Engine: OHV V-8	
Displacement: 302-cid	
Horsepower: 205 at 4000	
Wheelbase: 104.7 inches	

Ford had first introduced the Bronco in 1966 to compete against Jeep's CJ-5 and International Harvester's Scout in the burgeoning recreational four wheel drive vehicle market. The first Broncos were very Spartan, devoid of any performance options or creature comforts. The debut models were only available with 105 horsepower, 170-cid six-cylinder engines, derived from the Falcon line-up. The only available transmission was the 3.03 three-speed manual with column-mounted shifter. The simple sturdy, construction of the first Broncos, and their great performance and manoeuvrability that ensured great fun on and off the road mean that this classic model is sought after to this day.

Responding to public demand, Ford decided to enlarge the Bronco in 1978. Its design closely followed the F-150 of that year, and many parts are interchangeable. This full-size Bronco II was available until 1996, with every kind of model from mild to wild. Ford's full size four-wheel utility vehicle could hold more than 100 cubic feet of cargo or seat six people in a roomy interior. It could also haul as much as 7800 pounds. Originally, Broncos came only in a two-door body style, with a two-way tailgate, but four-door versions were added later. Antilock rear braking was standard by the late 1980s, but operated only in two-wheel drive mode. A four-speed automatic transmission, first introduced in 1990, gradually edged aside the three-speed unit. A twenty-fifth anniversary model marked the Bronco's 1991 season. In 1992, a new grille and more rounded sheetmetal were introduced, together with new mirrors. ABS was fitted to the Bronco in 1993, and a driver-side airbag was added for '94. Things got cooler outside for 1995, with a new sport trim package for the XLT that included a body-colored grille and bumpers, new running board steps and deep-dish aluminum wheels. Engine power in the optional 5.8 liter V-8 dropped from 210 to 200 horsepower. Until 1993, the standard engine was a 4.9 liter inline six-cylinder, packing 150 horsepower, but a second, smaller (5 liter) V-8 became standard at this time, which initially developed 185 horsepower. This had gone up to 205 horsepower by 1996. Bronco drive was standard ondemand part-time four-wheel drive, not for use on dry pavement, and a conventional transfer-case shift lever was fixed to the floor.

1995 Ford Ranger Splash

When Ford first offered the Model T in pickup form back in 1925, as part of the 'Improved Ford Package' they really started something. Drive past a parking lot in any mid-American town and pickups will outnumber sedans by two to one. The combination of comfortable transportation and loadcarrying ability make the pickup a great transport choice. Now that extended cab options are available, you can also move the whole family or labor team.

In Henry's day the pickup was the cheapest option but today it is usual for the truck to be loaded with luxury features. In 1993 Ford's Model T equivalent, the Ranger, received a facelift. Our featured model is the 1995 Ranger Splash 4x4 with extended cab and flareside rear fenders. This model series featured a 125.2-inch wheelbase, which had been increased by 4.5 inches. Coupled with a 1.4-inch wider track, this gave the vehicle much more stability. The old V-6 was replaced by three engine options: an overhead cam four-cylinder inline unit of 140 cubic inches giving 112 horsepower, and two overhead valve V-6 units of 182 and 245 cubic inches respectively, developing 145 or 160 horsepower. The larger of the two six-cylinder units is the favorite as its fuel ratio, town to highway, is only a mile or two more to each gallon. This engine was also the reviewers' choice for hauling power and overtaking. The vehicle was judged to 'handle well - for a truck that is - and ride nicely'. Like many Fords of the past the suspension is conventional, with front coil and rear leaf springs, which do the job, nonetheless.

For 1995 the interior was restyled with a drivers side airbag and a new grille was added. Four-wheel antilock brakes, center High-Mount Stop lamps and 15-inch wheels were standard fittings.

Engine: Four-cylinder inline OHC/V-6 OHV.

Displacement: 140/182/245-cid

Horsepower: 112/145/160

Transmission: Five-speed manual or optional four-speed auto, 2/4 wheel drive

Induction: Fuel injection

Body Style: Two-door Pickup with extended cab

Number of Seats: 5

Weight: 3300 lbs

Wheelbase: 125.2 inches

Base Price: $1458

1995 Lincoln Continental

Engine: V-8 DOHC
Displacement: 281-cid
Horsepower: 260 horsepower
Compression Ratio: 9.0:1
Wheelbase: 109 inches
Base Price: $40,750
Number Produced: 32,851

Until 1994, the long-established Continental was a six-cylinder (170 horsepower) giant economy car, but although the moniker didn't change, the hefty revises of 1994 transformed the car into a small luxury model with a trimmer body design and far more exciting styling. The new inheritor of the long-established Continental heritage now featured a new transverse V-8 engine with dual overhead camshafts. The new car had more contemporary styling with a curved character line that ran the length of the car in an eye-pleasing update of the slab-sided appearance of the previous Continental. Lincoln openly acknowledged that they hoped to make the new car a hybrid sports sedan whilst retaining its traditionally luxurious feel. With its new four-speed automatic transmission, the new V-8 could now generate a sparky 260 horsepower. The body shell of the car was also a great deal stronger than that of the old model. Standard equipment on the car included dual airbags, antilock braking, automatic climate control and an air-filtration system to trap dust and pollen. A redesigned instrument panel displayed virtual-image graphics. These were strikingly bright at night, but difficult to read on a sunny day. The driver could also adjust the ride quality and the steering assist with dashboard buttons.

The Continental was now much quicker, and a bit more agile, although it retained its original front-wheel drive and was thus given to serious understeer. It was loaded with desirable electronic gadgetry. In acceleration, the newly energetic car could match a Cadillac Seville SLS, going from 0-60 mph in about eight seconds. However, the gas mileage was unimproved at 16.3 mpg, and premium fuel was recommended. Interior space in the car was very good, with plenty of leg space for the occupants in the front and rear. Storage was also capacious with a wide, deep and long trunk.

The new Continental was positioned in an extremely competitive market sector, positioned against more powerful Cadillac Seville, the superb Lexus LS 400, plus BMW and Audi models. Although none of these other cars came close to the Lincoln for electronic wizardry, the new Continental had only modest success, and after some changes and add-ons in the intervening years, the distinguished model was discontinued in 2002.

1996 Chevrolet Camaro Z28 SS

Chevrolet's Super Sport legacy has a long heritage, dating back to the 1961 Impala. It has been awarded to several models over the years, including Novas, Chevelles, El Caminos and Camaros. Previous to the '96 version, the SS badge last appeared on a Camaro in 1972. Latterly, some select Chevy pick-ups have also been equipped with this venerable package.

With the rebirth of the American muscle car in the 1980s, came a great wave of nostalgia for the original era of power and true muscle. Some of the same buyers were interested in the new inheritors of the genre. More and more muscle came on stream in the '90s for a while variety of cars, as performance became more and more of an issue with the buyers. A typical hot V-8 of 1992 made around 205 horsepower, but within a mere four years, this had been pumped up to more than 300.

Chevrolet had revivified its Super Sport image in 1994, by awarding it to a four-door Caprice, the Impala SS (q.v.). But just as these cars were discontinued, the Camaro Z28 SS came on stream. In fact, this model didn't actually originate with Chevrolet, but an outside contractor - SLP (Street Legal Performance) Engineering Inc. of Troy, Michigan. This firm had made similar conversions to a 1992 Pontiac, which resulted in the Firehawk Firebird. Effectively, SLP became an out-of-house department of the Chevrolet division, and were much like Ford's Special Vehicle Team, in that their brief was to design performance versions of the regular production models.

The 1996 Camaros were manufactured at GM's F-body factory in St. Therese, Quebec, and converted into SSs at the nearby SLP Boisbriand plant. The trick was to ensure that the cars remained emissions legal for the US. SLP President, Ed Hamburger was insistent that the SS Camaros were engineered to the very highest standards, whilst offering world-class performance with a simple ownership process. This meant that you could order the car at your local Chevy dealer, just asking for an order form. Nothing too onerous in that, and the $4000 extra for the juiced up car wasn't bad either.

The made-over SSs had bulging forced-air hood, which helped to boost the Z28's LT1 small-block from 285 horsepower to 310. The suspension was enhanced, and the car was fitted with 17x9 castaluminum wheels with Z-rated Goodrich Comp T/A tires.

Engine: LT1 V-8

Displacement: 350-cid

Horsepower: 310 at 5500 rpm

Transmission: Borg-Warner T56 six-speed manual Compression Ratio: 10.4:1

Induction: Electronic Sequential-port Fuel Injection

Body Style: Two-door Coupe

Number of Seats: 4

Weight: 4365 lbs

Wheelbase: 101.1 inches

Base Price: $24,500

Number Produced: 2,410

1996 Dodge Viper GTS

Engine: V-10 overhead valve	
Displacement: 488-cid	
Horsepower: 450 at 5200 rpm	
Transmission: Six-speed manual	
Compression Ratio: 9.6:1	
Induction: Electronic Sequential Multi-port Fuel Injection	
Body Style: Two-door coupe	
Number of Seats: 2	
Weight: 3445 lbs	
Wheelbase: 96.2 lbs	
Base Price: $66,045	

Opposite top: The slicked back hardtop gave the Viper a huge improvement in its aerodynamics.

Opposite bottom: Full exhausts were extended all the way to the GTS coupe's tail.

Below: The Viper's 8.0 liter 450 horsepower V10 power unit was borrowed from a Dodge truck.

The Corvette is widely known as 'America's sports car', and the fiberglass two-seater has reigned supreme for over fifty years. Of course there have been contenders for the title, such as the Thunderbirds, the Avanti, Cobras and the AMX. But the Corvette outlasted them all, and sold in increasingly large numbers.

But if ever there was a serious threat, it came from the Viper. Dodge developed a prototype for the car by 1988, and showed it to the public for the first time on January 4 1989. A pre-production version of the car went on to pace the 1991 Indianapolis 500, and a production version was finally introduced to the marketplace in 1992. But the open-air cockpit and very basic nature of the comforts on offer didn't really go down well at all. Even the fantastic V-10 engine caused offence, it was truly loud. AutoWeek Magazine's writer Matt DeLorenzo said that the Viper RT/10 roadster sounded like a 'UPS truck at idle'. Hardly surprising, as the beast under the hood was actually borrowed from Dodge's truck line. Dodge saw the obvious solution on offer. In 1995, they offered a closed version of the roadster, which they hoped would make the car a little more civilised and less of a handful. The slicked back hardtop also offered huge improvement to the aerodynamics of the car. The car was launched as a 1996 model. It looked rather similar to the Shelby American Dayton coupe, which was no surprise as Carroll Shelby was one of the inspirations behind the car. The Viper Blue and white striped livery was a combination directly copied from Shelby's victorious car of 1964.

Dodge managed to keep the weight of the GTS coupe down to that of the RT/10 roadster, and this was achieved by cutting down the weight of other components – like the engine. The block was now eighty pounds lighter. Elsewhere, the car had a little more sophistication, with a softer suspension, electric windows, adjustable foot pedals, dual airbags and more convenient instrumentation. However, the roof made the car totally unsuitable for the claustrophobic.

But the power under the hood was fantastic. The V-10 was revised to make a huge leap in power. As well as being lighter, it had a stronger camshaft, better-breathing heads and revised exhausts. Everything conspired to jump from 415 horsepower all the way to 450. This was the most powerful engine offered by Detroit in this year, and for sheer muscle, nothing American could beat the Viper in 1996.

For the second time, the Viper was chosen as the pace car for the Indianapolis 500. This time it was the GTS coupe. The car was hailed as instantly collectible, one of the world's greatest touring cars.

69M522
MANUFACTURER

1996 Ford Cobra Mystic

Engine: V-8 90-degree, dual overhead cam

Displacement: 282-cid

Horsepower: 305 at 5800 rpm

Compression Ratio: 9.85:1

Induction: Sequential Fuel Injection

Wheelbase: 101.3 inches

Base Price: $24,810

Number Produced: 5496

The base, GT and Cobra Mustangs all returned for the 1996 model year. Ford's SVO team (Special Performance Operations) had developed a convertible to add to the range, but this did not go into production until the spring of 1996, whilst the coupe was available at the beginning of the year. The Cobras were just slightly longer than the other Mustangs, at 182.5 inches. Under the hood was where the major changes for this season occurred. The 5.0-liter V-8 previously fitted to the GT and Cobra Mustangs was discontinued, replaced by a modular 4.6 liter V-8. Teksid cast the aluminum cylinder heads. They are an Italian company that also works for Ferrari Road and Formula One cars. In fact, the engine had such great performance that it was voted one of the ten best by Ward's Auto World for 1996. The Cobra also now used a new T45 five-speed manual transmission, which replaced its predecessor, the T5.

The already stiff chassis was furthered stiffened and rigidized by the addition of a redesigned front crossmember. This also provided clearance for the deep oil sump fitted to the engine. Also new for the Cobra was a 3.27:1 rear axle ratio. Standard equipment on all V-8 Mustangs now included the Passive Antitheft System.

In this model year, Mustang production totalled 124,698. 2,000 'Mystic' Cobra coupes were finished in ultra-technological light-reflecting Mystic Clearcoat paint (produced by BASF), as an $815 option. These cars were the first to wear a paint developed with this color-shifting technology. The paint contains special 'light-interference' pigments, which allow the paint to show drastic color transformations as the observer's angle to the car changes, or the angle that light hits the car alters. The microscopic paint layers act like tiny prisms, and the color produced is controlled by precise variances, measured in atoms, in these prismatic layers. As Todd Hayes wrote for the North American Auto Writers Syndicate, 'The car literally changes from emerald green to violet blue, magenta, burgundy, amber, and a dark brown. It is truly fascinating and the object of many open-mouthed stares on the freeway'.

A second line of 1000 'Mystichrome' painted Cobras was launched in 2004. This special optical paint was manufactured by DuPont.

1997 Ford Thunderbird

Engine:	V-8
Displacement:	281-cid
Horsepower:	205 at 4250 rpm
Compression Ratio:	9.0:1
Body Style:	Two-door coupe
Wheelbase:	113 inches
Base Price:	$19,015
Number Produced:	73,814

Ford introduced the Thunderbird in 1955 as their answer to Chevrolet's Corvette. But whereas the Corvette remains true to its two-seater sports car origins to this day, the Thunderbird was more of a do-everything car, and has been through many changes, restyles and rethinks over its existence of more than four decades.

The Thunderbird has always been about style, refined power and creature comforts and the combination of these elements made it a good seller right from the start. The original car came only as a two-seater convertible, but it was soon available as a hardtop coupe and with a back seat and even two more doors... It wasn't really until the 1957 car that Thunderbird got serious about performance and new versions of the 312-cid 'Y' block V-8 were offered.

Several Thunderbird restyles were initiated by Ford over the years, the final makeover took place in 1989. The car became trimmer in its overall dimensions, but the wheelbase was actually nine inches longer than that of the previous model. Fully independent suspension was introduced with this model, which came in two versions, the base and LX. Power came from a choice of V-6s, the base being a 3.8 liter with four-speed overdrive automatic, all the way through to a 210 horsepower V-6 fitted to the Super Coupe. V-8 power was re-introduced to the line for 1991, with a five liter engine rated at 200 horsepower.

Thunderbird launched four models for 1992, the base Thunderbird, Sport, LX and SC. There was also a light facelift for this year. The line-up was trimmed back to two models in the next season. Just the LX and Super Coupe were offered in 1993. Performance was back under the spotlight in 1994, with the introduction of an overhead cam 4.6 liter V-8. By 1996, the line was tapered further to just a single car, the Thunderbird LX, and this remained the situation in 1997. Several revisions were made to various aspects of the car for 1997, including a new instrument cluster, decklid spoiler with integrated stoplight, and three new exterior colors and a single new interior color, Light Praire Tan. New optional equipment included 16-inch wheels and an upgraded suspension.

At the end of the year, Ford announced that it was going to discontinue Thunderbird production after forty-two continuous years. But watch this space...

1998 Corvette C5

Engine: V-8
Displacement: 346-cid
Horsepower: 345 at 5600 rpm
Compression Ratio: 10.1:1
Base Price: $44,425
Number Produced: 12,090

The eagerly awaited C5 fifth series Corvette was launched in January 1997 at both the Detroit and the Los Angeles Auto Shows. A new line was to have been introduced in 1993, but due to the dreadful financial problems at GM, the development of the new car had been delayed. The body shape was now reworked, and completely engine and chassis were introduced simultaneously. In fact, there was almost no carry-over from the previous series. The new chassis was based on a pair of immensely strong hydro-formed rails, stiff enough to support the fiberglass body. Ride quality was improved by the lengthening of the wheelbase to 104.5 inches, and this also gave the cars a more spacious interior. Five-spoke alloy wheels were fitted as standard, recalling the ZR-1.

In its 45th anniversary year, the Corvette returned to offering convertible and coupe models with the debut of a topless version. The convertible's glass window was heated and the top had an Express-Down feature that released the tonneau cover and automatically lowered windows part way at the touch of a button. But essentially, it remained manually operated. One of the most-welcome innovations of the new convertible was the provision of a trunk lid! This convenience hadn't been seen on a ragtop 'Vette since the early sixties and was greeted with great enthusiasm. It even featured remote opening via a control button the key fob.

New for this model year was the magnesium wheel option featuring lightweight wheels with a unique bronze tone. Two new exterior colors were offered and a new light oak leather interior color was available. Standard features included stainless steel exhaust system, Extended Mobility Tires capable of running for 200 miles with no air pressure, dual heated electric remote breakaway outside rearview mirrors, Daytime Running Lamps and five mph front and rear bumpers. The LS1 V-8 and four-speed automatic transmission were again offered as standard, with the T56 six-speed manual transmission available as an option.

Corvette was also able to celebrate its fourth selection as the pace car for the Indianapolis 500, with Indy 500 veteran Parnelli Jones driving the wild purple and yellow car (with black and yellow leather interior). Corvette had first been given the honor in 1978, then in '86 and '95. Corvette also made a longawaited return to Trans-Am racing in '98, achieving a first place in the season's opening event – a street race around Long Beach, California. The winning car was the Number 8 Autolink Corvette driven by Paul Gentilozzi.

1998 Ford Contour SVT

Ford launched the Contour in the spring of 1994. It was their new 'World Car', the replacement for the Tempo. It was designed to run against the Honda Accord and the other imported compact and mid-size sedans. Mercury offered a similar four-door sedan, the Mystique. Both cars were based on the European Ford Mondeo model. Contours were offered in a three-tier range – the GL, LX and sporty SE. Ford took a huge step forward in performance, refinement and overall execution with this compact sedan, and many felt that the car was a formidable rival to both the Japanese imports and domestic cars such as the Chrysler Cirrus and Dodge Stratus. There were only two criticisms levelled at the car, that the exterior styling was bland, and the legroom in the back of the car was limited. Dual airbags were standard on the cars, as were impact resistant bumpers that could absorb an impact of five mph. Antilock brakes and traction control were optional on all models. The base engine option for the range was the four-cylinder with a 170 horsepower six optional. Both engines were available with a five-speed manual gearbox or an optional electronic four-speed automatic transmission.

Ford added a high-performance car to the range in 1998, in the shape of the SVT (for Ford Special Vehicle Team). The car was equipped with the same Duratec 2.5 liter (155-cid) V-6 available on the other Contours, but this one was tuned to deliver 195 horsepower. This was a full 13 per cent more power. Developing 76.6 horsepower per liter of displacement in a non-supercharged engine, and a top speed of 143 mph gives the SVT some of the best performance figures in the business. The car also came with a five-speed transmission as standard. Antilock brakes, sport suspension, larger tires and other go-goodies were optional. Road-testers felt that the SVT felt responsive, but that the driver really had to work the transmission to develop all the available power. They also felt that the car felt more German than American to drive, with its precise steering, sporty handling and firm ride.

The SVT designers showed a great deal of restraint in the exterior restyling of the car, the only differential between them and the rest of the Contours were sports rocker panel extensions, a modestly restyled front fascia with round fog lamps and 16-inch alloys. They managed to avoid the customary spoiler. The SVT interior was distinguished with black-on-white gauges, SVT logos and dark blue leather upholstery. The SVT was retained for the 1999 model year, but was only available through selected Ford dealers by 2000.

Engine: DOHC six-cylinder

Displacement: 155-cid

Horsepower: 195

Wheelbase: 106.5 inches

Base Price: $22,405

Number Produced: 5,000

1998 Ford Escort ZX-2

The 1981 introduction of the front-wheel-drive subcompact Escort heralded a whole new era for Ford. Ford described it as 'The World Car' due to its international presence. The car was equipped with a CHV (hemi) engine, and immediately became Ford's best-selling model. Within a year, the car was also the best-selling car in America. The GT model was offered with turbocharging in 1984, and a Mazda-built diesel could also be ordered. The Escort line-up was revised for the 1986 model year, with the GT model fitted with a high-output engine. The car continued to sell well, although Ford's overall sales fell, leaving them with less than 12 per cent of the market. But Escort's continued sales successes helped Ford to bounce back to the top in 1987. The company broke all records, netting $4.6 billion in income (out-performing GM) and $71.6 billion in worldwide sales. Escort continued as America's favorite for 1988, and by 1989 the car was approaching its three millionth sale. Ford continued to maintain a high level of product development for the line, and launched an all-new Escort design in both 1990 and 1997. Fewer body styles were now on offer and both the two- and four-door hatchback models were discontinued. A new SE series was introduced to the line. The wheelbase remained unchanged, but the overall length of the new model went up by four inches. The new body design featured completely new body panels and frameless door glass. It weighed 120 pounds more than the previous version so even increased horsepower meant performance was by no means sparkling.

The ZX2 two-door coupe was launched in January 1997 for the 1998 season, and was available in 'cool' and 'hot' trim versions. The 'cool' package had limited options and a manual five-speed gearbox. The 'hot' option had more options available, including a choice of wheels, plus rear defogger and air conditioning fitted as standard. ZX2 exterior and interior styling was unique, and the car had a sportier suspension setting. It was aimed at younger buyers with an average age of 32, who got peppy performance from the Zetex 2-liter, four-cylinder, that developed 130 horsepower. Optional equipment included anti-lock brakes, a premium sound system, power windows, power moonroof, remote entry system and an automatic transaxle. Three packages were offered with the car, 'Comfort Group', 'Power Group' and 'Sport Group'. The latter featured 15-inch alloys, a rear spoiler, foglamps, upgraded bucket seats and special badging. The performance coupe sold well, but some auto reviewers considered the ZX2 to be unacceptably noisy. Sadly, Ford discontinued the stalwart Escorts in 2001, and the ZX2's final model was launched in 2003.

Engine: In-line Four-cylinder

Displacement: 2-liter

Horsepower: 130 at 5750 rpm

Compression Ratio: 9.6:1

1998 Ford Taurus

Engine: V-8 dual overhead cam

Displacement: 207-cid

Horsepower: 235 at 6100 rpm

Compression Ratio: 9.8:1

Wheelbase: 108.5 inches

Base Price: $28920

Number Produced: (all sedans) 85,364

Ford first introduced the Taurus in 1986, to general acclaim. There was a mild refreshment of the model in 1992, and a major restyle in 1996. Until that time, the car was a perennial best seller. The line also got a new engine for 1996, the Ford designed DOHC 300 horsepower Duratec. A 3.4 liter V-8 was optional, and fitted as standard to the SHO. Only an automatic gearbox remained.

The SHO (Super High Output) models changed the dull image of the Taurus. SHOs debuted in 1989 with a sticker price of around $20,000. Its great performance reflected on the entire range. It had mid-six second 0-60mph times, and a top speed of 145 mph (out-pacing even the Ford GT). But the car wasn't the sales success that Ford was hoping for. Although the SHO was a great drive, and had a distinctive interior, it looked just like the other boring Taurus models on the outside.

The full overhaul of 1996 was an attempt to reintroduce some of the original thunder to the series. But although, technologically, the new Taurus was a better car than its predecessor, the stunning new shape was completely out of step with any other car coming out of Detroit. The car was now larger and much more rounded. It was not at all well received by the press or the public, and the line lost its top sales spot. The new model was also exported (as a right hand drive car) to Japan, Australia and New Zealand, but wasn't taken up with much enthusiasm in these places either.

For 1998, Taurus gained a more assertive front end, and other subtle styling changes were made. The line was reduced to a smaller number of models, and a dizzying array of option packages was dropped to simplify things for the buyer. There were now four Taurus models, the LX sedan, SE sedan and station wagon and the four-door SHO sedan. The SHO was equipped with a 235 horsepower V- 8, which turned the mild mannered family car into a racy touring car. All Taurus engines now worked through a four-speed automatic. These were recalibrated to provide smoother shifting.

The car was mildly restyled again in 2000, away from the controversial oval design theme, to a less startling and more angular look, but the Taurus never reclaimed the top spot in car sales again. Taurus remains part of the Ford lineup, having been 'refreshed' for 2004. The new car is advertised as having 'clean, sculptured lines and a stylish look. With exceptional value and comfort, Taurus is a symbol of luxury for the real world'.

1998 Lincoln Navigator

Engine: SOHC V-8	
Displacement: 5.4-liter	
Horsepower: 230 at 4250	
Transmission: Four-speed automatic	
Body Style: Five-door SUV	
Weight: 5150 lbs	
Wheelbase: 119 inches	

In the full-size SUV category, the Lincoln Navigator is the only vehicle with 'Made In America' stamped on its rocker panel. This is even more surprising in the light of the fact that this Yankee challenger comes from a marque with no history of truck development. Lincoln launched the Navigator in 1998, a full-size Lincoln sport-utility vehicle, based on the Ford Expedition model. The difference was in the styling. The Lincoln had unique (and much more lavish) front and rear styling, extra sound deadening, softer suspension tuning, plus standard leather upholstery and walnut trim. Of course, Lincoln is no stranger to luxury on a grand scale, its current Town Car sedan, a perennial favorite with limousine services, is the biggest passenger car sold in the US.

The main rivals to the Navigator included the GMC Yukon/Denali, Land Rover Range Rover and Toyota Land Cruiser. Quality control on the model was good with a special post-assembly inspection and a twenty-five mile pre-shipment road test.

These first Navigators were available with either rear-wheel drive or Ford's Control Trac four-wheel-drive system, which could be left engaged on dry pavement. The 4x4 option came with automatic self-levelling air-ride shock absorbers at front and rear. On rear-wheel 2WD Navigators, these was fitted at the rear only. Antilock four-wheel disc brakes were standard either way. A three-seat third-row bench seat was installed, with a three-seat split middle row seat available at no extra charge. The Navigator could carry a total of eight people in great comfort. Towing capacity for the truck was 8000 pounds – or four tons to you and me.

Navigators were equipped with a single engine, the Ford Expedition's larger V-8, the 5.4 liter, rated at 230 horsepower. Although Lincoln's SUV is too big and heavy to drive like a car, it does have reasonable acceleration for such a large vehicle. Although the model is not truly agile, it handles reasonably well on twisty roads.

1998 Mercury Mountaineer

Ford's classy cousin Mercury decided to go after a piece of the increasingly popular SUV pie in 1997 by introducing the Mountaineer. Aimed at the upscale customer, it was effectively a loaded four-door Explorer, complete with V-8 and All Wheel Drive, lavishly garnished with a chromed-out grille and luxury trim.

Only eighteen months after its introduction, the Mercury Mountaineer was awarded a host of changes, including a new SOHC V-6 engine and five-speed automatic overdrive transmission, Control Trac permanently engaged four-wheel drive and a major freshening of the exterior. The new front-end appearance included a new grille, complex reflector headlamps, bumper fascia and a lower valance, complete with fog lamps. From the side view, there were new standard cast aluminium wheels and a lower accent color, Medium Platinum. The vehicle glass had a deeper tint to reduce glare and sun load. At the back, the Mountaineer now had a new liftgate with new tail-lamps and high mount stop lamp, plus larger glass and a revised wiper arm from improved visibility under all conditions. A new one-piece rear bumper improved impact performance and had pre-drilled mounting holes for a trailer hitch. A sole base engine was now on offer, the overhead cam V-6. The engine's cast-iron block was reinforced for increased durability, and a balance shaft has been added to help reduce noise, vibration and harshness. The Mountaineer's original five-liter V-8 was now offered as an option, complete with four-speed automatic transmission and permanent four-wheel drive.

Inside the cars, there were refinements in comfort, convenience and utility. Seats provided increased support and a more attractive appearance. Prairie Tan was introduced as a new interior color choice. New and more sophisticated audio equipment was offered for 1998, and a one-touch power moonroof was new for this season.

Like nearly all the vehicles in this market sector, the downsides of the Mountaineer were the meagre gas mileage on offer (at around 16 mpg) and the rather bouncy and truck-like ride it offered. On the upside, the Mountaineer offered a roomy and comfortable interior, with good visibility in all directions. The permanent all-wheel drive option was very useful, and easier to use than many other systems offered by the model's competitors. The vehicle also handled reasonably well, with confident cornering and good acceleration.

Even so, Ford's Explorer retained its number one sales position in this year, having achieved sales of nearly three million units by this time.

Engine: V-6 SOVC

Displacement: 4-liter

Horsepower: 205 at 5000 rpm

Transmission: Five-speed automatic

1999 Mercury Cougar

Engine: V-6
Displacement: 2.5-liter
Horsepower: 170 at 6250 rpm
Compression Ratio: 9.7:1
Weight: 2941 lbs
Wheelbase: 106.4
Base Price: $17,095

Very different from the overweight Thunderbird-related Cougars of the past, the revived 1999 Cougar was a compact, front-drive sports hatchback coupe with seating for four. Its stated goal was to liven up Mercury's staid image, whilst being true to the heritage of the 1967 Cougar by combining style with performance. Although the styling of the car was sharp edge, the basis for the car were the Ford Cougar/Mercury Mystique sedans. The Cougar was an inch longer than both of these and was designed with a more luxurious feel. The three cars shared engine options, but the Cougars were set up to feel sportier, with firmer suspension and tighter steering.

The power options were a base four-cylinder 16-valve Zetec equipped with a manual shift, rated at 125 horsepower. For around $600 more, the buyer could order a 24-valve, 2.5-liter V-6, rated at 170 horsepower. This engine came with a four-speed automatic as standard. Five-speed manual and traction control were both V-6 options. Although the four-cylinder engine had better fuel economy, road-testers rated it to be noisy and to have excessive vibration, preferring the smoother ride of the V-6. They also preferred the five-speed manual transmission for better pick-up times and a sportier feel. Some felt that both engines were outperformed by those fitted to their Japanese rivals (particularly the Mitsubishi Eclipse and Honda Prelude). The car handled reasonably well, but the long wheelbase diminished its twisty-road agility.

The strongest selling point of the car was its revised 'high tech' design, specifically designed to attract trendy young buyers. These would be different from the usual Mercury customers, better-off middle-aged people. The New Edge Ford styling, of sculpted surfaces and sharp intersections, also incorporated some strong European references. These included the roofmounted radio antenna and many other subtle details that were attractive to the image-conscious market. As well as the detail, the design fundamentals were right in the car, such as the basic body styling and exemplary driving dynamics.

A sport packages was available for $775 on theV-6, which included fog lamps, four-wheel disc brakes, a leather-wrapped steering wheel, leather shift knob, spoiler, 16-inch aluminium wheels, a center armrest and more distinctive bucket seats. The interior featured a blend of good styling and ergonomics, and commentators were reminded of Ford's European Puma model.

1999 Plymouth Prowler

The Plymouth Prowler, later the Chrysler Prowler, is a retro-styled production car manufactured and marketed in 1997 and 1999-2002 by Daimler Chrysler, based on the 1993 concept car of the same name. The car was front-engined rear wheel drive layout and in fact the first rear-wheel drive Plymouth since the 1989 Plymouth Gran Fury and would stand as the last Plymouth model with that layout. Indeed the last Plymouth model period. Its overall production was 11,700 units including those badged as Chrysler.

The car was inspired by the vibrant hot rod market and a return to retro styling values at the end of the 20th century.

The Prowler utilized a powertrain from Chrysler's LH-cars, a 24-valve, 3.5 L Chrysler aluminum-block 253 hp V6 engine coupled to a four-speed Autostick automatic transmission. The transmission was located at the rear of the vehicle and joined to the engine by a torque tube that rotated at engine speed, an arrangement similar to that used by the C5 Corvette giving a desirable 50-50 front-rear weight distribution. Some critics bemoaned the lack of a V8 option but in fact the V6 unit had a horsepower rating comparable to the company's Magnum V8s of that era. While not making nearly as much torque as a V8, Prowler's light weight translated into impressive off-the-line acceleration.

The car prominently featured aluminum construction, in many cases adhesively bonded, chiefly in the chassis. The body was produced in Shadyside, Ohio, and the car was assembled by hand at the Conner Avenue Assembly Plant (CAAP) in Detroit, Michigan.

Engine: V6 aluminum block 24 valve

Displacement: 3.5 liter

Transmission: four-speed automatic

Horsepower: 253@6,400

Wheelbase: 113.3in

Length: 165.3 in

Weight: 2800 lbs

Price: $39,300

Production year: 3.921

1999 Pontiac Trans Am – 30th Anniversary Edition

Engine: V-8, aluminum block and heads	
Displacement: 346-cid	
Horsepower: 320 at 5200 rpm	
Transmission: Four-speed automatic	
Compression Ratio: 10.5:1	
Induction: Sequential-port fuel injection	
Body Style: Two-door convertible	
Number of Seats: 5	
Weight: 3474 lbs	
Wheelbase: 101.1 inches	
Base Price: $35,495	
Number Produced: 535	

The Trans Am turned thirty in 1999, and Pontiac celebrated by launching a special, limited run of a Cameo White Trans Ams trimmed in Lucerne Blue stripes... and a set of rather bizarre blue alloys. Coupe and convertible models were offered. The livery is pretty much the same as that given to the car on its twenty fifth birthday model, all of which helps to underline the fact that the Trans Am is the longest-running, continuous, performance nameplate in America – as there was no 1983 Corvette, and Camaro Z28 production was discontinued between 1975 and mid-77.

The special paint scheme pays tribute to the WS4 option package of the very first Trans Am of mid-year 1969. At the time, Ford Mustang and Chevy Camaro seemed unassailable in the SCCA Trans Am series, and to be taking it in turns to take the honors. Pontiac's basic problem was that even its smallest engine was too big to be race legal for the series – Trans Am stipulated that engines could be no larger than 305-cid, theirs was 350. The street Trans Am was launched in this year, and Pontiac were obliged to pay a royalty to the SCCA for every Trans Am sold. The single-choice paint scheme didn't help sales, and they built just 697 of this landmark model. The most popular model of the year was the Ram Air III with a four-speed gearbox.

The 1999 car follows a more 1990s aesthetic, with a monochromatic paint scheme, white inside and out. Inside, white leather bucket seats carried special '30th Anniversary' embroidery on their headrests. But the 17-inch five-spoke 'medium-blue tinted clear-coat' metallic rims were either loved or loathed. But part of the design success of the new Trans Am is the many nods it gives to

Below: Pontiac celebrated 30 years of the Trans Am in 1999. Here are the two models 30 years apart.

its own past. The 'screaming chicken' hood emblems (though they were quite discreet by now) and the twin hood scoop, for example.

The car also came with the customary serious muscle power under the hood. It is equipped with a slightly detuned version of the C5 Corvette's 5.7- liter LS1 V-8. The car went straight to the top of the contemporary muscle class. The fully functional WS6 Ram-Air hood punched up the power even further. For transmission, there was a choice between four-speed automatic of six-speed manual. The Pontiac performance and handling package was fitted as standard, including power-assisted four-wheel ABS disc brakes.

Sadly, GM has recently cancelled its F-body line, and the Pontiac Firebird Trans AM has been consigned to history. This will probably translate into soaring collector-car values for the anniversary models.

Above: The new car retains many styling cues of the original Trans Am.

Below: At this time the Trans Am was America's longest running muscle car.

2000 Ford Focus Sony Edition

Just as Ford's old stalwart, the Escort was fading, the company brought out the replacement Focus. The first Focus hit US dealerships in October 1999, and quickly became one of the ten best selling cars in America, and yet another winner from the Ford stable. In fact, it outsold every other economy model except the Honda Civic. It went on to win the title North American Car of the Year for 2000. The Focus combined a tall cabin and upright seating to create massive interior room, especially for the rear passengers. It was offered as a hatchback, sedan and wagon. This compact model was designed to appeal to the highly influential Generation X (20-25 year olds) and the Echo Boomers (aged 19 and under). To understand their market, Ford executives immersed themselves in the youth culture of major cities such as Los Angeles, New York and Miami. The execs soon realized that this age group is passionately interested in music, fashion and the internet. Ford instigated a huge 'live' advertising campaign to promote the car, the biggest since the '50s. This involved all kinds of media plugs for the car, that were aimed at the target age group, such as the 'Dawson's Creek National Sweepstakes'.

This strategy also led to the first hitch up with an electronics brand in automotive history, with the launch of the Sony Edition Focus in 2000. Responding to the popularity of Sony products with their target market, FoMoCo teamed up with Sony Electronics to create a car featuring custom-designed Xplod red and black speakers and amplifiers. Sony had launched these Xplod products in the previous year, and they were already a big hit with 16-24 year olds across America. The incorporation of this advanced equipment into the car effectively turned the Focus into a kind of mobile sound system.

Seven thousand Sony Limited Edition Ford Focuses were manufactured. The cars displayed special decals on each front door panel, and were equipped as four-door SE Sport Focus models with a 60/40 split rear seat, air conditioning, power mirrors, all-door remote entry and powerlocks, 15-inch painted aluminum five-spoke wheels, five-passenger seating and black cloth or optional black leather seats. The car was offered in three standard Ford colors, Rainforest Green, Infra-Red and Pitch Black plus one new color made exclusively for the Sony Limited Edition models – Going Platinum.

Ford also made a special edition Kona Focus, supplied complete with its own 'Out of Bounds' Kona mountain bike and bike rack.

2001 Chevrolet Corvette Z06

Millenium celebrations over, American car enthusiasts found something new to celebrate in the shape of the new hardtop Corvette, the Z06, resurrecting the heritage racing model designation. The car shared the revised aluminum engine block design of the mighty LS6 V-8, which raised the power output to 385 horsepower, making it even more powerful, and half-a-second quicker than the ZR1 of 1992. The engine block was fitted with red covers. The car was exclusively fitted with the M12 six-speed manual transmission.

In effect, the car was a more sporting development of the hardtop Corvettes, and reminded Corvette aficionados of Duntov's first Z06 production racers of 1963. The car arrived as Corvette's highest priced variant at $47,500. Corvette brand manager Jim Campbell said that the car was 'aimed directly at the diehard performance enthusiast at the upper end of the high-performance market'.

The 2001 chassis was rigidized with a stiffer rear leaf monospring, a larger diameter front stabilizer bar and revised front camber settings to improve cornering. The Chevrolet secondgeneration active handling package was standard on all 2001 Corvettes.

The exterior of the car was accented with stainless steel highlights, and the model was available in five colors: Quicksilver Metallic, Speedway White, black, Torch Red and Millenium Yellow. It also sported fully operated brake vents and unique wheels that were even wider than those fitted to the coupe and convertible, fitted with Goodyear Eagle Supercar tires, supplied complete with a sophisticated puncture repair kit. The interior of the car was slightly less refined that those of the coupes or convertibles. It's nature was slanted towards performance at the expense of touring. The interiors were available with only two color choices, black, and black with Torch Red accents.

In 2002, standard Corvettes were rated at 350 horsepower, and the Z06 was rated at 405 horsepower. This increase in power was achieved by the removal of two of the pre-catalytic converters from the exhaust system: a well-known NASCAR engine-builder's trick. The Corvette went on to pace the Indy 500 for the fifth time in 2002, but unlike previous years, Chevrolet produced only three pace car replicas that didn't go on sale.

Engine: LS6 V-8

Displacement: 346-cid

Horsepower: 385 at 6000 rpm

Transmission: Six-speed Manual

Compression Ratio: 10.5:1

Induction: Electronic Sequential Fuel Injection

Body Style: Two-door hardtop coupe

Number of Seats: 2

Weight: 3115 lbs

Wheelbase: 104.5 inches

Base Price: $47,500

Number Produced: 5,773

2002 Ford Thunderbird

Engine: DOHC V-8	
Displacement: 3.9-liter	
Horsepower: 252 at 6100 rpm	
Weight: 3775 lbs	
Wheelbase: 107.2 inches	
Base Price: $35,390	

Ford restored a famous name to its line-up with its 2002 Thunderbird. The car had reverted to a two-seat convertible, and was designed with a retro panache that recalled the classic model of '55. Like the Thunderbirds of the mid-fifties, the new car has reverted to the 'Personal Luxury Car' formula. The new T'bird was built on a shortened version of the rear-wheel drive platform employed by the Lincoln LS and Jaguar S-Type sedan, and was equipped with a power-fold soft top with a heated glass rear window. A removable hardtop was also available, with the marque's trademark porthole windows.

The Thunderbird 2000 is the production version of the concept car that was launched to great acclaim in US auto shows of 1999 and 2000. Although the car's design is very evocative of the car's heritage, the new Thunderbird uses the latest engineering technologies to install confidence in the driver. It is a balanced rearwheel drive car that utilizes a rigid, computer-engineered chassis and a fine-tuned four-wheel independent suspension system employing lightweight materials to improve the car's responsiveness. The car also has rack-and-pinion steering assembly, four-wheel anti-lock disc brakes, electronic brake force distribution, seventeen-inch cast aluminium wheels and all-season tires. The car was also equipped with an optional all-speed traction control system, which gave the car excellent start-up on slippery surfaces and improved its cornering stability. Although the car was successful in achieving the feel of a relaxed cruiser, it is too big and heavy to be sports-car agile, especially in tight manoeuvres.

The standard engine for the car was an all-aluminum 3.9-liter dual overhead cam V-8, producing an estimated 252 horsepower at 6100 rpm coupled with a specially engineered close-ratio, five-speed automatic transmission. The entire assembly gave great performance and efficiency along with minimal levels of noise, vibration and harshness. The car had good acceleration, with 0-60 mph times of seven seconds. But the acceleration between 35 and 50 mph was soft, due to a failure to downshift promptly. Fuel economy wasn't sparkling either, at around 17.4 mpg. The car's interior felt luxurious with leather seats and leather-wrapped steering wheel.

The car was launched into a hotly competitive sector of the market, running against the Acura TL, Lexus ES 300 and BMW 3-Series. Although originally successful, there has been talk of discontinuing Thunderbird production, but for the moment it remains in the Ford line-up.

2003 Ford Mustang Mach 1

Engine: V-8	
Displacement: 4.6-liter	
Horsepower: 305 at 5800 rpm	
Transmission: Tremec TR3650 five-speed manual	
Compression Ratio: 10.1:1	
Induction: Electronic Sequential Fuel Injection	
Body Style: Two-door hardtop	
Weight: 3465 lbs	
Wheelbase: 101.3 inches	
Base Price: $28,805	
Number Produced: 9,652	

Traditionally, Ford has not been as keen as Chevy or Pontiac to commemorate the anniversaries of its heritage models. No special edition commemorated Mustang's 30th anniversary in 1994, although the car was completely restyled. Neither was there the slightest celebration of the cars 35th birthday in 1999. But when Ford's centennial came 'round in 2003, the company started to review its family jewels with greater interest.

So despite their reluctance towards general nostalgia, Ford decided to issue a special-edition Mustang, the Mach 1, for 2003. The car celebrated the original 'Mach 1' Mustang fastback of 1969. This model had been designated the 'first great Mustang' by Car Life magazine. The new Mach 1 'offered performance to match its looks, handling to send imported-cars fans home mumbling to themselves, and an interior as elegant, and liveable, as a gentleman's club'.

The car had a great many styling clues to the original car. In effect, it was a 'modern interpretation of an American icon'. The Mach 1 was offered with a 'shaker' hood scoop, stripes, the pony medallion and a crisp beltline molding. The five-spoke Heritage wheels were unique to the 2003 Mach 1 and were part of its retro appeal, as were the blacked-out front rear spoilers, old-fashioned rocker panel stripes and fog lamps. The interior carried through these styling clues, with a bright aluminium shift ball, foot pedals and 'nostalgic instrument cluster' all as seen in the 'Bullitt' Mustang. Six color choices were on offer, Torch Red, Azure Blue, Dark Shadow Gray, Zinc Yellow and Oxford White.

Performance was supplied by a 2003 revision of the 2001 Cobra's 4.6- liter dual overhead cam, four-valve-per-cylinder V-8. This powerplant had an output of 305 horsepower, making the car even faster than its venerable ancestor. The new Mach 1 could also out-handle its forerunners, fitted as it was with a stiffened suspension including stiffer springs, Tokico struts and Goodyear Eagle ZR45 tires.

For 2004, the Mach 1 finally achieved commemorative identification, marking the car's 40th anniversary.

2004 Avanti

Engine: LS1 V-8
Displacement: 5.7 liter
Transmission: Four-speed automatic
Induction: Fuel injection
Weight: 3640 lbs
Wheelbase: 101.1 inches
Base Price: $83,000

Studebaker commissioned Raymond Loewy and his design team to breathe live into the Avanti concept back in 1961. Although the original version of the car was launched in an extremely fertile era for automotive originality, it was immediately lauded for its aerodynamic beauty and became an instant classic. Loewy himself drove several examples of the car. But the Avanti fell victim to the financial problems of its parent, and was almost immediately discontinued in 1963. However, the model was such a fresh and original concept that, over the years, several attempts were made to salvage the Avanti from the Studebaker wreckage. In essence, although these were fragmented, and production was stopped for several years, the car survived. The Avanti just refused to die, and its flowing fiberglass body appears startlingly modern to this day. The car celebrated its 40th anniversary in 2002.

The current owner of the model, the Avanti Motor Corporation, is the largest independent car manufacturer in the US, and is owned by Michael E. Kelly. Kelly hails from Studebaker's hometown of South Bend, Indiana, and has been involved with the Avanti for a number of years, since he first bought the marque in 1986. After selling his shares in 1988, he bought back Avanti in 1999, and describes the company as being 'committed to handcrafting the finest motorcar in the all the world'.

Although the slightly revised styling of the contemporary Avanti is recognizably similar to that of Raymond Loewy's original masterpiece, the car is technologically much more advanced. It is now powered with a 5.7 liter fuel-injected V-8 LS1, with optional Vortec supercharging that gives 0-60 mph times of five seconds. Four-speed automatic or six-speed manual transmissions can be specified. The car is available as either a T-Top sports coupe or convertible. The cars are equipped with plush leather interiors with Burlwood accent trim, air conditioning, cruise control, CD player, rear window defogger, dual airbags, 17-inch polished Avanti wheels, power locking and antenna, keyless entry and full instrumentation.

Only around 150 Avantis are manufactured each year at the company's Georgia production facility. Unsurprisingly, this exclusivity comes at a high price, with the coupe and convertible available for $79,000 and $83,000 respectively, with a supplement of $10,000 for a supercharged GT engine option.

2004 Pontiac GTO

Pontiac built their original GTO 'Goats' between 1964 and 1974, and to many these were not only the original American muscle cars, but also the 'best of breed'. They produced over half a million during this period, and they were the best selling muscle car between 1964 and 1968. If you couldn't afford a Corvette, this was the car for you. Unfortunately, GM had discontinued their F-body in 2002, so no rear-wheel V-8 platform was available to Pontiac. Pontiac got around this by using the Australian Holden Monaro GM model as the basis for the new car. This was a two-door coupe first seen at the Sydney motor show in 1998.

When the new car hit the streets in 2004, it was met with rave reviews. In fact, the car was universally well received both by aficionados of the original muscle cars, and newcomers to the scene alike. The car was introduced to fill the gap left by the retiring Trans Am, and was put together in just 17 months from conception to delivery. Pontiac put this speed down to their 'focus on customer enthusiasm' and their consummate desire to 're-energize the car market with vehicles that command attention and excite the customer's senses'.

The main source of the model's undeniable excitement was under the hood. The LS1 V-8 (Corvette's standard powerhouse since the 1997 debut of the C5) was fitted to the car. It was rated at 350 horsepower, and was mated to a standard 4L60-E four-speed automatic transmission. A close-ratio Tremec six-speed manual (as seen on the Z06) was optional.

The car was also equipped with power-assisted four-wheel ventilated disc brakes (complete with four-channel anti-lock system), four-wheel independent suspension, power rack-and-pinion steering, and 17x8 cast aluminum wheels with BF Goodrich 245/45ZR-17 tires. The car was adorned with a low profile spoiler, GTO identification, full instrumentation and comfortable bucket seats. Compared to the Z06, the car was still reasonably priced at $33,000, and could milk the available power to achieve a 0-60 mph time of 5.3 seconds, or a quarter mile time of 13.62.

Engine: LS1 pushrod V-8

Displacement: 346-cid

Horsepower: 340 at 5200 rpm

Transmission: Six-speed manual

Compression Ratio: 10.1:1

Induction: Electronic Sequential-Port Fuel Injection

Body Style: Two-door coupe

Number of Seats: 5

Weight: 3,821 lbs

Wheelbase: 109.7 inches

Base Price: $33,000

Number Produced: (projected) 18,000

2005 Chevrolet SSR

Engine: Aluminum V8 OHV	
Displacement: 5300 cc	
Horsepower: 300	
Transmission: Four-speed Automatic	
Body Style: Two-door Sports Roadster	
Number of Seats: 2	
Wheelbase: 116 inches	
Base Price: $42,990	

Is it a Roadster? Is it a pickup? Is it a sportscar? Is it a hotrod? Well its actually all four. The SSR is pure Chevy and a celebration of everything that has gone before. Its sleek retro lines reflect classic body shapes of yesteryear – from the 30s onward – coupled with state of the art technology. Powered by the latest version of the legendary small block Chevy unit, with aluminum heads and block, the Vortec 5300 produces 300 horsepower.

On the practical side a press of the remote keyless entry system flips up a tonneau to reveal the rear cargo area with its durable bedliner. Activate the power-retractable hardtop and the car becomes a convertible. This must be the ideal beach vehicle.

Contact with the asphalt is taken care of by Goodyear Eagle RS-A Touring tires mounted on cast aluminum wheels, 19x8 at the front and 20x10 inches at the rear. Stopping is by courtesy of four-wheel ventilated discs and ABS. The ZQ 8 sports suspension features a five-link rear-end.

Externally the styling is real cool with the chunky wheels hunkering down under funky flared fenders which are formed by a new process- Inverted Toggle Draw – eliminating weak spots in the metal. And hey....running boards are back!

Inside the twin-cockpit interior are body hugging Ebony leather seats with color-keyed console and satin chrome details.

2005 Ford Mustang GT

Engine: V-8
Displacement: 281-cid
Horsepower: 300
Transmission: Close-Ratio 5R55S Automatic
Induction: Electronic Throttle Control
Body Style: Two-door Sports Coupe
Number of Seats: 4

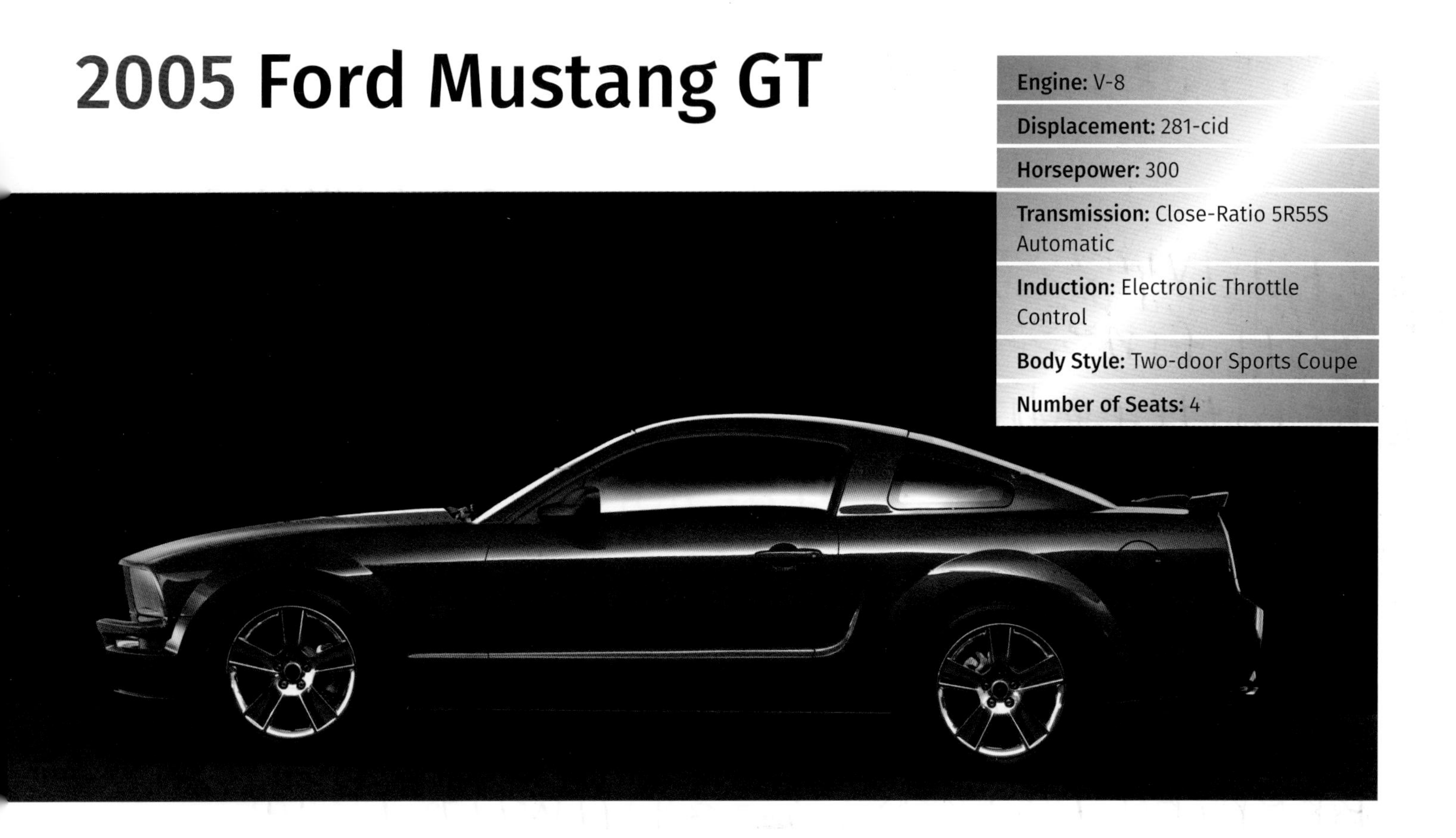

'Mustang attracts two kinds of drivers – those under 30 and those over 30. Really, that's its strength. America's most popular nameplate transcends demographics and socio-economic trends – because Mustang is really more than a car. It's an icon that's been woven into the fabric of America for 40 years and running.' (J. Mays, Group Vice President, Design, FoMoCo.)

The all-new, all-American 2005 Mustang is a bold, clean and contemporary version of possibly the most celebrated muscle car in the US. It is a purely American sports car. The new design is fresh and contemporary, but harks back to the venerable Mustangs of the past. Mustang was launched by Ford back in 1964 and just twelve months later it had broken auto manufacturing records by selling 419,000 cars in its first production year. By the end of its second year, 1,000,000 Mustangs were out on the roads of America. Somehow, Ford had managed to tap into a reservoir of un-catered for demand. Young, performance orientated, fun-loving, glamorous, the car reflected many buyer's view of themselves for the first time. It was the start of a love affair that would continue to the present. Through over 300 film appearances, including 'Bullit' and 'Gone in 60 Seconds' the image of the Mustang is entwined in American pop culture and continues to have resonance to this day – a bad-boy car, hip, smooth and definitely trouble.

The new car has undergone some fundamental development. The front wheel has been moved forward, significantly reducing the front overhang. The wheelbase has also been lengthened by six inches, pushing all the wheels further out to the corners. This has also increased the cabin size. Ford has tripled its investment in the car's interior, in order to offer the buyer an almost limitless customising opportunity.

So far as the exterior appearance is concerned, the shark-like nose reminds us of the 1967 model, while the jewelled round headlamps in trapezoidal housings look startlingly modern. The GT also has circular fog lamps in the black grille, and an upright lower fascia with an 'air dam' performance look. The body profile of the GT is low and aggressive looking, due to it body-color lower rocker panel. But power is at the heart of every Mustang. The new car will be equipped with an impressive 4.6-liter V-8 delivering 300 horsepower, which used to be the exclusive territory of the racing stock – the likes of Mach 1, Cobra and Boss Mustangs. This will make the car the most affordable 300 horsepower machine in the US. A SOHC V-6 will also be available, producing 202 horsepower.

2005 Chrysler 300-C

Engine: Hemi V-8

Displacement: 345 cid

Horsepower: 340

Transmission: 5-speed automatic

Wheelbase: 120 inches

Length: 196.8 inches

Base Price: $30,000 depending on options

Nothing is ever new under the sun and in 2005 Chrysler's original "beautiful brute" re-emerged once again fifty years after the first car in 1955.

The new car displays styling hints of the original car, like the deep square sectioned grille. As the company was owned by Mercedes-Benz at the time, the Chrysler 300 used components derived from the E-Class, which included the rear suspension design, front seat frames, wiring harnesses, steering column, and the 5-speed automatic transmission. The front suspension was a double wishbone system derived from the S-Class.

The Chrysler 300-C was first shown at the 2003 New York Auto Show as a concept car. Sales in the U.S. began in the spring of 2004 as an early 2005 model year car.

Designed by Ralph Gilles, the new 300-C was built in the same spirit as the original "letter car" as a high performance, full- size, sporty sedan. The car uses a 345 cu in Hemi V8 in common with its ancestor. Using the Multi displacement System (MDS), this engine can run on four-cylinders when less power is needed in order to reduce total fuel consumption. When all 8 cylinders are needed, the 300-C can make 340 hp and 390 ft/lbs of torque. It uses a 5-speed automatic transmission and comes standard with 18 inch chrome-clad alloy wheels. The Hemi engine features a pushrod induction tube, located on the side of the engine-block. This tube makes the 300-C more fuel efficient and quicker, because of the air being "pulled and pushed" into the engine's induction area. The engine uses a double rocker configuration, with a cam-in-block, overhead valve (OHV) pushrod design. There are two spark plugs per cylinder to effect a complete fuel/air mixture burn and to decrease emissions.

Chrysler also introduced the SRT-8 version of the car in 2004. This had a formidable 370-cubic-inch Hemi V8 producing 425 hp allowing a 0-60 mph time in the low 5-second range and quarter-mile in the high 13-second range. It went on sale in February 2005 at the not inconsiderable sum of US$43,695.

2008 Dodge Challenger SRT8

When Barry Newman (as Kowalski) heads into oblivion in his 1970 Dodge Challenger, a legend was born. A legend that would re-emerge in 2008.

The original concept of the model was as a "pony car competitor" for the Ford Mustang. The Mustang had its day of fame in the movie Bullitt (1968), the Challenger had its three years later in Vanishing Point. Both films achieved cult status. Vanishing Point was re-launched in 1998 as a not-so-successful TV film.

But the Challenger was to fare better with its metamorphosis in 2008. The new version of the car is a two-door coupe. It shares common design elements with the first generation Challenger, despite being significantly longer and taller. The chassis was a modified version of the LX platform (with a shortened wheelbase) that underpins the 2006 to the current Dodge Charger, the 2005-2008 Dodge Magnum, and the 2005-Current Chrysler 300. The LX platform was adapted from the Mercedes E Class, reflecting Mercedes-Benz's ownership of Chrysler at the time. All (7119) 2008 models were SRT8s, equipped with the 370 cu in Hemi and a 5-speed AutoStick automatic transmission. The entire 2008 US run of 6,400 cars was pre-sold.

The 2009 SRT8, while still equipped with the 370 cu in Hemi V8, was virtually identical to its 2008 counterpart. The main difference was the choice of either a 5-speed automatic or a 6-speed manual transmission. Standard features included Brembo brakes, a sport suspension, bi-xenon headlamps, heated leather sport seats, keyless ignition, Sirius satellite radio, and 20-inch forged aluminum wheels. The extras were in addition to the amenities already offered on the R/T and SE models, such as air conditioning and cruise control. Also, the 2009 had a proper "limited slip" differential. A "Spring Special" SRT8 Challenger was offered in B5 Blue, but due to rolling plant shutdowns, only 250 Spring Special Challengers were built before the end of the 2009 model year.

For 2010, SRT8 models added Detonator Yellow to the range of available colors (at extra cost), and only as an optional "Special Edition Group" car. Yellow Challengers would only be built for a limited time (October/December 2009) for the 2010 model year. Another retro color, Plum Crazy Purple, was also available during the Spring 2010 production, this was offered exclusively in the "Spring Special" package. Another retro color, Furious Fuchsia (similar to the Panther Pink of the 1970s) was limited to a single production day at Chrysler's Brampton, Ontario plant.

Engine: Hemi V-8

Displacement: 345 cid

Horsepower: 425

Transmission: 5-speed automatic/6-speed manual

Wheelbase: 116 inches

Length: 197.7 inches

Base Price: $40,095

2010 Chevrolet Camaro

Engine: (SS) GM LS3 V-8	
Displacement: 376 cid	
Horsepower: 426	
Wheelbase: 112.3 inches	
Length: 190.4 inches	
Weight: 3750 lbs	

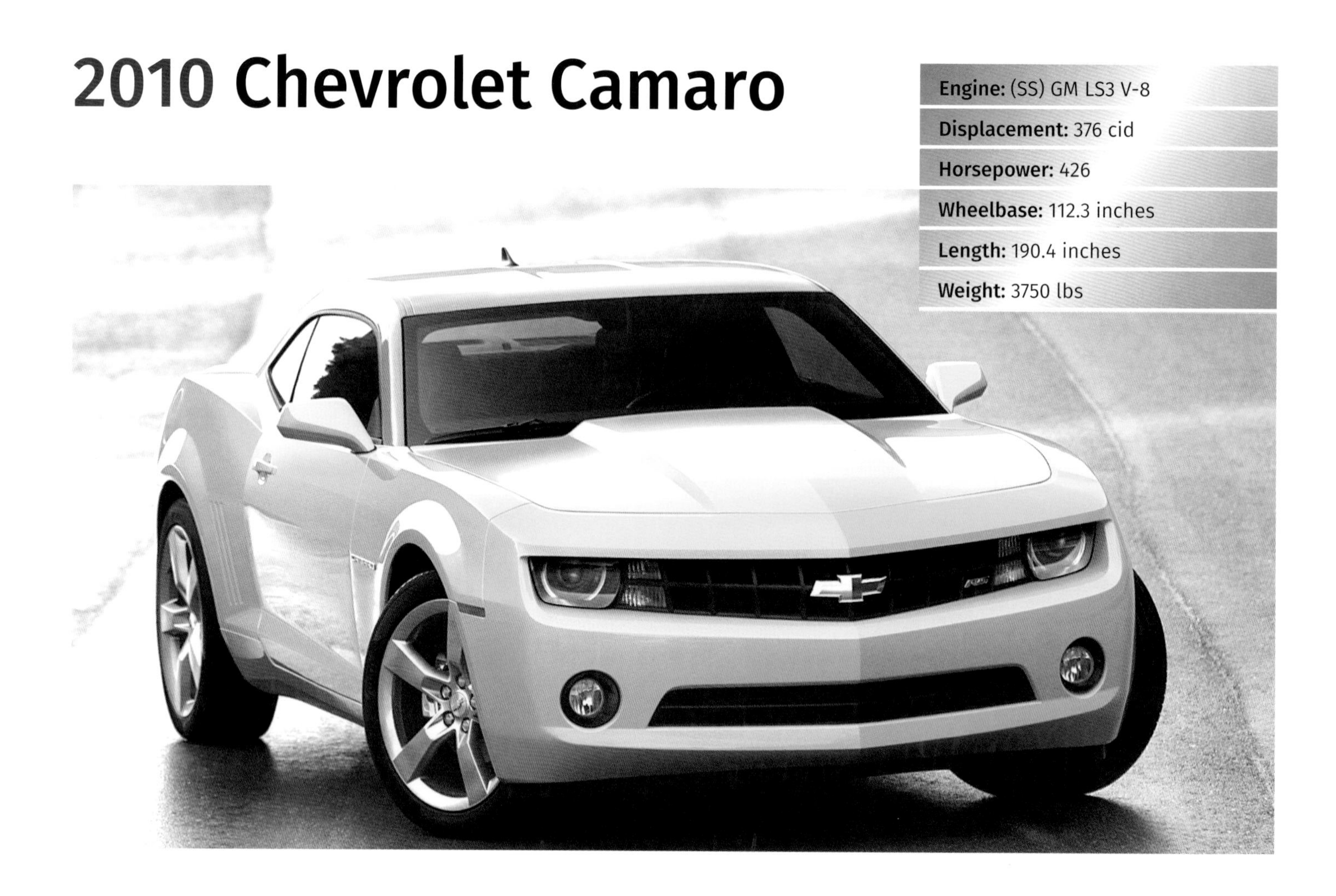

Chevrolet are keeping retro styling cues high on their agenda with this the fifth generation Camaro. The car happily just misses the complete retro look, combining retro styling cues with modern functionality by using the very latest technology and modern components. It seems that most informed reviewers are hard-pressed to discover anything disappointing with the exterior appearance of the car, and early customer response has been really good. However, while the exterior has been universally praised the interior has received some slightly critical reviews as it is felt to be "rather more vintage than retro" with its high-mounted gauges and center stack that look cool, but prove to be "ergonomically challenging" in operation. With the car's high belt line and low roof, coupled with its dark upholstery, there have been comments about the interior being gloomy.

This Coupe, based on the GM Zeta platform, comes in five different trims: the LS and LT versions are powered by a 217 cubic inch V-6 producing a very respectable 304 hp. The SS versions use two different V-8s. The SS, with 6-speed manual transmission, is powered by a GM LS3 unit of 376 cubic inches putting out 426 hp at 5900 rpm. Presumably, this version is aimed at all-out performance freaks with little concern for economy. By contrast, the auto version uses a GML99 V-8. This develops 400 horsepower, and features an Active Fuel Management system which allows the engine to run on just four cylinders during light-load driving conditions (such as highway cruising) to improve fuel economy. Other features of the model range include a fully independent four-wheel suspension system, variable-rate power steering, four-wheel disc brakes standard on all models (four-piston Brembo calipers on SS models), a StabiliTrak electronic stability/traction control system, Competitive/Sport modes for the stability system offered on SS models, launch control on SS models equipped with the six-speed manual transmission, and six standard air bags that include head curtain side-impact air bags and front seat-mounted thorax side air bags. An RS appearance package is available on LT and SS trim levels, which includes HID headlamps with integrated halo rings, a spoiler, and RS-specific tail lamps and wheels. Rally and hockey stripe packages are also available in several colors.

2011 Cadillac CTS-V Coupe

Introduced in summer 2011, the Cadillac CTS-V Coupe combines massive muscle performance disguised as a luxury sports coupe, complete with all the toys. Aimed at driving enthusiasts, the car oozes both finesse and power. As Bryan Nesbitt, Cadillac general manager said, the "The CTS-V Coupe marries our most potent and sophisticated technology with our most dramatic production design." The car is equipped with a whole of technological goodies including Magnetic Ride Control, racing-bred Brembo brakes, and an (optional) automatic transmission with paddle-shift control.

The V-Series Coupe includes a unique grille that doubles the car's air intake volume, an important enabler for its high-performance capabilities. It is also visually identified by a raised center section of the hood that provides clearance for the supercharger, as well as specific front and rear fascias. Like the CTS Coupe, its exhaust is centered inside the rear fascia, but with unique twin outlets. The car rolls on 19-inch cast aluminum wheels and Michelin Pilot Sport 2 performance tires.

The CTS-V Coupe's rear track is nearly an inch wider than the CTS-V Sport Sedan, adding to the car's aggressive stance and enabling an extra measure of handling capability. The current CTS-V Sport Sedan has quickly established itself among the world's highest performing luxury sedans, becoming the first production four-door on street tires to break the legendary 8-minute barrier at Germany's famed Nürburgring. The new V-Series Coupe's sleeker shape and slightly lower mass ensure a similarly high level of performance.

The CTS-V Coupe's design is a clear visual statement of the technical precision and performance intent of the car. Sharing the CTS-V Sport Sedan's wheel base, the Coupe is about two inches (51 mm) lower and two inches (51 mm) shorter. Key design features of the new Coupe include classic hardtop styling, with no conventional B-pillar; touch-pad operation for the doors, which removes the need for conventional door handles, a faster windshield angle (62.3 degrees) and a nearly horizontal backglass, which give the car a seriously aerodynamic profile.

The car's supercharged powertrain consists of a LSA 6.2L supercharged V-8, rated at 556 horsepower (415 kW) and 551 lb.-ft. of torque (747 Nm): at the time Cadillac's most powerful engine ever ,the unit has since been even further developed to produce 640 horsepower in the 2016 model year. But the car does not sacrifice comfort for power. The 2+2 cabin combines performance with refinement and luxury. Recaro performance driving seats are optional, providing excellent support for spirited driving, together with a microfiber-covered steering wheel.

Engine: Supercharged V-8

Displacement: 6.2 liter

Horsepower: 556

Acceleration: 0-60 mph in 3.9 seconds

2012 Dodge Charger

Engine: 6.4 liter 470 horsepower OHV 16-valve V-8
Transmission: Five-speed automatic
Brakes: Antilock Brembo brakes with vented and slotted front and rear discs
Performance: 4.5 seconds, quarter mile in 12.8 seconds at 115.5 mph
Wheelbase: 120.2 inches
Weight: 4,360 pounds
Base price: $47,425

Below: The SRT8 is equipped with a 6.4 liter hemi V-8 engine. With Fuel Saver Technology, the challenger powertrain can develop 470 horsepower and 470 pounds-per-feet of torque.

Below right: Split five-spoke twenty-inch forged aluminum wheels are equipped with Brembo four-piston Red calipers and vented and slotted rotors.

The 2005 SRT8 was equipped with a voracious 6.1 liter Hemi V-8, but the 2011 Charger weighed in with an even larger 6.4 liter hemi V-8. The newly styled 2012 car squeezed even more and far greater fuel efficiency from the same powertrain. The twenty-five percent fuel economy over that of the 2011 model was due to the engine's new cylinder-deactivation system. The hemi's aggressive engine power was channeled through a five-speed automatic.

Despite the car's full-sedan size and rear-wheel drive, the SRT8 was not only super quick in a straight line but, thanks to its fully hydraulic steering assembly, the car was highly maneuverable in bends and corners. Even at high speed, the car maintained complete stability, and thanks to its slippery aerodynamics and faster windshield angle, produced very little wind noise. The majority of the 2012 Challenger's suspension components had been modified by SRT and the car rode a half-inch lower when compared to the 2011 car. These modifications also made the ride far smoother than that of the car's predecessor. The new SRT8 also had improved stopping power due to its beefed up braking system, with Brembo calipers, and more precise steering. SRT CEO Ralph Gilles dubbed the 2012 SRT8 "the outspoken one." The car's revised muscle car styling was also comprehensively revised, inside and out. The car now sported side skirts, a rear spoiler, a diffuser-style rear bumper, fat chrome exhaust tips, and a hood equipped with functional vents at its leading edge. Inside, the car now featured a revised leather-covered dash, complete with an 8.4-inch touch screen display with a Performance Pages function.

The result of all of these changes was a blast from the muscle car past with a touch of modern refinement, a refined four-door sedan equipped with incredible performance. The car could be enjoyed as both a weekday commuter and weekend thrill machine that would surpass many German high-performance cars. The 2012 SRT8 brought the muscle car era up-to-date with a blend of heritage and innovation.

Above: The 2012 Dodge Charger SRT8 is a fully loaded muscle car sedan with a beautiful body and advanced technology

Left: The superfine interior of the 2012 Charger SRT8. The leather and perforated suede seats are both heated and ventilated, and have the SRT logo embossed in their seatbacks. The dash is also wrapped in leather, while the Harmon Kardon audio system is equipped with nineteen speakers.

2012 Tesla Model S

Engine: 310 KW electric motor
Horsepower: 416
Torque: 443lbs
Transmission: 1-speed fixed gear
Battery: 85kwh
Range: 265 miles
Wheelbase: 116.5 inches
Weight: 4647lbs

The Model S ranked as the world's second best selling plug-in electric vehicle, after the Nissan Leaf, in 2014. Global cumulative sales passed 90,000 units by October 2015.During the first nine months of 2015, the Model S ranked as the top selling plug-in electric car in the U.S. with about 17,700 units sold.

The Model S is produced at the "Tesla Factory" an assembly plant in Fremont, California. In June 2015, three years after the Model S introduction and with almost 75,000 Model S sedans delivered worldwide, Tesla announced that Model S owners have accumulated over 1 billion electric miles traveled, and saved more than half a million tons of CO2.

On October 9, 2014, Tesla announced the introduction of All Wheel Drive versions of the Model S' 60, 85, and P85 models, designated by a D at the end of the model number. The Model S 85D has a top speed of 155 mph and it accelerates from 0 to 60 miles per hour in 4.2 seconds, due to the improved traction of the all-wheel drive powertrain.

The Tesla Model S won awards such as the 2013 World Green Car of the Year, 2013 Motor Trend Car of the Year, Automobile Magazine's 2013 Car of the Year, Time Magazine Best 25 Inventions of the Year 2012 award and Consumer Reports' top-scoring car ever. In 2015, *Car and Driver* named the Model S the Car of the Century

2016 Dodge Charger

Engine: Hemi V8

Displacement: 6.2 liters

Horsepower: 707

Transmission: Six-speed manual

Brakes: Four-piston Brembo calipers and slotted rotors all round.

Base price: $65,945

The Dodge Charger in 2016 retains its traditional full-size four-door sedan status and can be factory equipped from mild to wild to suit all buyer's differing tastes and requirements. The SRT Hellcat is equipped with a 707-hp supercharged 6.2-liter V8.

The Hellcat engine is the most powerful regular-production V8 the Chrysler group has ever built, and it features upgrades like a forged-steel crankshaft, forged-alloy pistons and powder-forged connecting rods to provide reliability.

Like the naturally-aspirated V8, the supercharged Hellcat engine sends power to the rear wheels via a six-speed manual transmission or an eight-speed automatic unit.

Nought-to-60 mph is reached in 3.6 seconds and top speed is around of 200 mph.

Four-piston Brembo calipers and vented and slotted rotors measuring 14.2-inches up front and 13.8-inches in the rear bring the car to a halt.

Gas mileage checks in at 13 mpg in the city and a respectable 22 mpg on the highway.

Interestingly the Hellcat comes standard with a black key fob and a red key fob. The black fob limits the engine's performance, while the red fob allows the driver to unlock the V8's full potential. A valet mode allows the driver to reduce horsepower and torque, limit the engine to 4,000 rpm, turn off the launch control function, disable the shift paddles and lock out access to first gear in cars equipped with an automatic transmission.

2016 Chevrolet Malibu Hybrid

Engine: Four-cylinder gas plus two-motor electric	
Displacement: 1.8 liter	
Power: 182	
Transmission: Six-speed automatic	
Induction: Direct fuel injection	
Body Style: Four-door sedan	

Hopes of a world-beating hybrid in Chevrolet's previous Malibu model were largely unfulfilled but at last for 2016 the company achieved a gas-electric drivetrain that would challenge the best of the far-eastern competition. The new car could motor along at speeds of up to 55mph on the electric engine and fuel economy was estimated as up to 48mpg in town and 45mpg on the highway. Figures for the combined power output of gas and electric engines was 182 hp. The 80-cell, 1.5 kwh lithium battery located in the trunk received a top up charge from both a regenerative braking system and an exhaust gas heat exchanger. Significant weight savings were also made of around 300 pounds from the previous model coupled with a sleek aerodynamic body which was lowered slightly to achieve a lower drag factor all contributed to making the car a serious contender in the hybrid market.

Opposite: The clean lines of the Malibu are both easy on the eye and aerodynamically efficient.

Right: Ergonomically designed dash and steering wheel.

Below right: Tail end looks great too!

2016 Ford Fusion

The Fusion was first introduced in 2006 as Ford's midsize sedan. For 2016 Ford has given the popular car a makeover to maintain its reputation as a sporty looker with an infusion of styling tweaks inside and out, along with a new nine-speed automatic transmission.

One identifiable change is the deletion of the unique front quarter windows. Inside, a new dashboard design is complemented by a redesigned steering wheel and the automatic transmission upgrade boasts a rotary shift knob replacing the old shift lever.

The traditional array of engine choices also returns for the Fusion's upcoming trim offerings, with the base non-hybrid engine remaining a competent 2.5-liter inline 4-cylinder power plant driving the S and SE trims with 175 hp and 175 lb-ft of torque with the traditional 6-speed shiftable automatic to manage this entry-level car to the tune of 22 mpg city/34 highway.

Also available in the SE trim, but standard in the flagship Titanium trim, is the potent 2.0-liter turbocharged engine that develops 240 hp and 270 lb-ft of torque mated to the new nine speed auto box.

The Fusion Hybrid deploys a traditional 2.0-liter hybrid gasoline-fired engine, along with a single high-voltage electric motor. The continuously variable transmission again manages 188 total hp, while regenerative braking helps the engine recharge the 1.4 kilowatt/hour lithium-ion battery. Ford claims this hybrid will be able to scoot up to 85 mph on electric power alone for short distances, while maintaining market-leading mileage at 44/41.

Front-wheel drive remains standard across the lineup, but the non-hybrid SE and Titanium variants are also available in all-wheel drive. The Fusion Hybrids, meantime, remain available only with FWD.

2016 Ford Shelby Mustang GT350

The GT350 returned in 2016 after an absence of 47 years. The previous factory Ford Mustang to wear the vaunted badge appeared in 1969. Based on the Mustang GT, it was originally a street-legal track weapon produced in partnership with Carroll Shelby. The partnership between his name and Ford's pony car remain as strong as ever despite his death in 2012. The 2016 GT350, which picks up where the original GT350 left off must be rated one of the best-ever example of the pony car breed.

Ford's newest Shelby takes an entirely different approach from previous brute force models. It is based on the latest S550 Mustang chassis, which has a fully independent rear suspension. The GT350 builds on that with Ford's first suspension with magnetorheological dampers, and unique bodywork that is not only beautiful but aerodynamically efficient. The naturally aspirated 5.2-liter V-8 has a flat-plane crankshaft allowing the engine to reach maximum power via high revs rather than by turbo charging or supercharging. It's a thoroughly modern performance car built for racetracks with curves instead of the drag strip.

- **Engine:** V8 with flat plane crank
- **Displacement:** 5.2 liters
- **Horsepower:** 526@7,700 rpm
- **Torque:** 429ft lbs
- **Transmission:** 6-speed manual
- **Brakes:** Brembo with 15.5in front rotors
- **Fuel Consumption:** City 14 Highway 21mpg
- **Price:** $61,295

2017 Chevrolet Corvette Zora ZR-1

Engine: Small block V8 supercharged

Power: Est. 700 horsepower

Transmission: Dual clutch 7-speed automatic

Number of Seats/doors: 2

Base Price: Est $150,000

Named for Zora Arkus Duntov, the Father of the Corvette, the 2017 model C-8 Corvette finally gets to be mid-engined. Moving the engine to a mid point in the car improves the car's balance/weight distribution and thus significantly improves handling, allows more controlled acceleration and safer braking .The mid-engined package, together with fresh exterior proportions, is set to appeal to a wider (younger) market than the present car's 50 plus audience. Styling-wise it seemed sensible to adopt the marque's shark head heritage, dating back to the 1960s, in its cues. Nineteen-inch Star Chrome wheels and low suspension complete the sporty image.

The car's platform draws on Chevrolet's existing expertise with alumin space frame and composite coachwork construction but in reality only structural and chassis parts will graduate from the current C-7 Corvette. Lo production volumes will give GM an excellent excuse to show off advanced technology, such as door structures formed from sheets of magnesium.

The car competes in the power stakes with stiff opposition from Europe from the likes of Audi, Jaguar, Lamborghini, Mercedes-AMG, Porsche and at home Chrysler's Hellcat range (Charger and Challenger), plus the Ford Mustang GT350, and will need a fully developed Chevy small block to meet the challenge. Its design will also accommodate future alternate engine packages such as twin turbo V6 and electric hybrid options.